VOLUME FOUR

STRATIGRAPHY & TIMESCALES

Case Studies in Isotope Stratigraphy

VOLUME FOUR

STRATIGRAPHY & TIMESCALES

Case Studies in Isotope Stratigraphy

Edited by

MICHAEL MONTENARI

Department of Earth Sciences and Geography
Keele University
Newcastle
United Kingdom

ELSEVIER

ACADEMIC PRESS

An imprint of Elsevier

Academic Press is an imprint of Elsevier
50 Hampshire Street, 5th Floor, Cambridge, MA 02139, United States
525 B Street, Suite 1650, San Diego, CA 92101, United States
The Boulevard, Langford Lane, Kidlington, Oxford OX5 1GB, United Kingdom
125 London Wall, London, EC2Y 5AS, United Kingdom

First edition 2019

Notices

Knowledge and best practice in this field are constantly changing. As new research and experience broaden our understanding, changes in research methods, professional practices, or medical treatment may become necessary.

Practitioners and researchers must always rely on their own experience and knowledge in evaluating and using any information, methods, compounds, or experiments described herein. In using such information or methods they should be mindful of their own safety and the safety of others, including parties for whom they have a professional responsibility.

To the fullest extent of the law, neither the Publisher nor the authors, contributors, or editors, assume any liability for any injury and/or damage to persons or property as a matter of products liability, negligence or otherwise, or from any use or operation of any methods, products, instructions, or ideas contained in the material herein.

ISBN: 978-0-12-817552-1
ISSN: 2468-5178

For information on all Academic Press publications
visit our website at https://www.elsevier.com/books-and-journals

Publisher: Zoe Kruze
Acquisition Editor: Jason Mitchell
Editorial Project Manager: Joanna Collett
Production Project Manager: Denny Mansingh
Cover Designer: Alan Studholme

Typeset by SPi Global, India

Contents

Contributors

Carlos José Souza Alvarenga
Instituto de Geociências, Universidade de Brasília, Campus Darcy Ribeiro, Brasília, Brazil

Marly Babinski
Instituto de Geociências, Universidade de São Paulo, São Paulo, Brazil

Lucas Pinto Heckert Bastos
Basin Analysis and Mobile Belts Graduate Program, Universidade do Estado do Rio de Janeiro—UERJ, Rio de Janeiro, Brazil

Sérgio Bergamaschi
Department of Stratigraphy and Paleontology—DEPA, Universidade do Estado do Rio de Janeiro—UERJ, Rio de Janeiro, Brazil

Paulo César Boggiani
Instituto de Geociências, Universidade de São Paulo, São Paulo, Brazil

Jean R. Borgomano
UM 34 CEREGE CNRS (UMR 7630)—IRD (UMR 161), Aix-Marseille Université (Centre Saint-Charles), Marseille cedex 03, France

Fabrício de A. Caxito
Centro de Pesquisas Manoel Teixeira da Costa, Instituto de Geociências, Universidade Federal de Minas Gerais (CPMTC-IGC-UFMG), Campus Pampulha, Belo Horizonte, Brazil

Elton Dantas
Instituto de Geociências, Universidade de Brasília, Campus Darcy Ribeiro, Brasília, Brazil

Valderez P. Ferreira
NEG-LABISE, Departamento de Geologia, Universidade Federal de Pernambuco, Recife, Brazil

François Fournier
UM 34 CEREGE CNRS (UMR 7630)—IRD (UMR 161), Aix-Marseille Université (Centre Saint-Charles), Marseille cedex 03, France

Camille Frau
Groupement d'Intérêt Paléontologique, Science et Exposition, Toulon, France

João Pedro Hippertt
Centro de Pesquisas Manoel Teixeira da Costa, Instituto de Geociências, Universidade Federal de Minas Gerais (CPMTC-IGC-UFMG), Campus Pampulha, Belo Horizonte; Laboratório de Modelagem Tectônica (LabMod), Departamento de Geologia, Universidade Federal de Ouro Preto (UFOP), Campus Morro do Cruzeiro, Ouro Preto, Brazil

Robert A. Jackson
School of Chemical and Physical Sciences, Keele University, Keele, United Kingdom

Matheus Kuchenbecker
Centro de Pesquisas Manoel Teixeira da Costa, Instituto de Geociências, Universidade Federal de Minas Gerais (CPMTC-IGC-UFMG), Campus Pampulha, Belo Horizonte; Laboratório de Estudos Tectônicos/CeGeo/ICT/UFVJM, Diamantina, Brazil

Cyprien Lanteaume
UM 34 CEREGE CNRS (UMR 7630)—IRD (UMR 161), Aix-Marseille Université (Centre Saint-Charles), Marseille cedex 03, France

Philippe Léonide
UM 34 CEREGE CNRS (UMR 7630)—IRD (UMR 161), Aix-Marseille Université (Centre Saint-Charles), Marseille cedex 03, France

Wenjian Liu
Department of Mathematics and Computer Sciences, Queensborough Community College, CUNY, New York, NY, United States

Jean-Pierre Masse
UM 34 CEREGE CNRS (UMR 7630)—IRD (UMR 161), Aix-Marseille Université (Centre Saint-Charles), Marseille cedex 03, France

John M. McArthur
Department of Earth Sciences, University College London, London, United Kingdom

Michael Montenari
School of Geography, Geology and the Environment, Keele University, Keele, United Kingdom

Adina Paytan
Institute of Marine Science, University of California—Santa Cruz, Santa Cruz, CA, United States

Antônio Carlos Pedrosa-Soares
Centro de Pesquisas Manoel Teixeira da Costa, Instituto de Geociências, Universidade Federal de Minas Gerais (CPMTC-IGC-UFMG), Campus Pampulha, Belo Horizonte, Brazil

Egberto Pereira
Department of Stratigraphy and Paleontology—DEPA, Universidade do Estado do Rio de Janeiro—UERJ, Rio de Janeiro, Brazil

Humberto Reis
Laboratório de Modelagem Tectônica (LabMod), Departamento de Geologia, Universidade Federal de Ouro Preto (UFOP), Campus Morro do Cruzeiro, Ouro Preto, Brazil

René Rodrigues
Department of Stratigraphy and Paleontology—DEPA, Universidade do Estado do Rio de Janeiro—UERJ, Rio de Janeiro, Brazil

Janaína Rodrigues de Paula
Centro de Pesquisas Manoel Teixeira da Costa, Instituto de Geociências, Universidade Federal de Minas Gerais (CPMTC-IGC-UFMG), Campus Pampulha, Belo Horizonte; VALE S/A, Rio de Janeiro, Brazil

Jean-Paul Rolando
Total S.A., Centre Scientifique and Technique Jean Féger, Pau, France

Roberto Ventura Santos
Instituto de Geociências, Universidade de Brasília, Campus Darcy Ribeiro, Brasília, Brazil

Alcides N. Sial
NEG-LABISE, Departamento de Geologia, Universidade Federal de Pernambuco, Recife, Brazil

Marcus Paulo Sotero
Centro de Pesquisas Manoel Teixeira da Costa, Instituto de Geociências, Universidade Federal de Minas Gerais (CPMTC-IGC-UFMG), Campus Pampulha; CPRM—Serviço Geológico do Brasil (Brazilian Geological Survey), SUREG/BH, Belo Horizonte, Brazil

Anthony J.-B. Tendil
UM 34 CEREGE CNRS (UMR 7630)—IRD (UMR 161), Aix-Marseille Université (Centre Saint-Charles), Marseille cedex 03, France

Ricardo Trindade
Instituto de Astronomia, Geofísica e Ciências Atmosféricas, Universidade de São Paulo, São Paulo, Brazil

Alexandre Uhlein
Centro de Pesquisas Manoel Teixeira da Costa, Instituto de Geociências, Universidade Federal de Minas Gerais (CPMTC-IGC-UFMG), Campus Pampulha, Belo Horizonte, Brazil

Gabriel J. Uhlein
Centro de Pesquisas Manoel Teixeira da Costa, Instituto de Geociências, Universidade Federal de Minas Gerais (CPMTC-IGC-UFMG), Campus Pampulha, Belo Horizonte, Brazil

Lucieth Cruz Vieira
Instituto de Geociências, Universidade de Brasília, Campus Darcy Ribeiro, Brasília, Brazil

Ruoqi Wan
State Key Laboratory of Geological Process and Mineral Resources, China University of Geosciences, Beijing, China; Department of Earth and Atmospheric Sciences, CCNY, CUNY, New York, NY, United States

Zhengrong Wang
Department of Earth and Atmospheric Sciences, CCNY, CUNY, New York, NY, United States

Lucas Warren
Departamento de Geologia Aplicada, Instituto de Geociências e Ciências Exatas, Universidade Estadual Paulista, Rio Claro, Brazil

Ulrich G. Wortmann
Department of Earth Sciences, University of Toronto, Toronto, ON, Canada

Weiqi Yao
Department of Earth Sciences, University of Toronto, Toronto, ON, Canada

Preface

The current book on *Case Studies in Isotope Stratigraphy* is aiming to present an overview on the progress achieved over the past years within various fields of isotope stratigraphy. In the following, the chapters and their abstracts will be briefly introduced.

In the first chapter, Yao et al. point out that "the sulfur isotopic composition of seawater sulfate oscillates over geological time, responding to changes in the redox state and biogeochemical processes on the Earth's surface. During times of rapid perturbations in sulfur isotopes where biostratigraphic controls do not provide good temporal resolution, sulfur isotopes of seawater sulfate can be a useful tool for stratigraphic correlations. Secular data of seawater sulfur isotopes archived in evaporite, carbonate-associated sulfate, and marine barite from worldwide locations allow reconstruction of a continuous seawater sulfate S-isotope curve for the Phanerozoic and Protozoic. For some time intervals (e.g., the Cenozoic), excursions in high resolution records are identified and these short-term fluctuations can be applied for stratigraphy and correlation. High resolution S-isotope records can also substantially enhance our understanding of temporal variations in the global sulfur cycle and earth system evolution." The second chapter is followed by the contribution of Wan et al., who establish that "the ages of carbonate rocks can be determined by comparing their $^{87}Sr/^{86}Sr$ ratios with well-established seawater $^{87}Sr/^{86}Sr$ curve. Current models and databases provide no explicit solution to the problem when the $^{87}Sr/^{86}Sr$ ratios of carbonate rocks contain analytical errors. This is particularly concerning as significant amount of $^{87}Sr/^{86}Sr$ ratios are being measured by multicollector inductively coupled-plasma mass-spectrometry (MC-ICP-MS) which typically has a lower precision (1σ is typically several tens of ppm) compared with thermoionization mass spectrometry (TIMs, 1σ can be as low as several ppm) used for calibrating the seawater $^{87}Sr/^{86}Sr$ curve. Deriving ages from carbonate $^{87}Sr/^{86}Sr$ ratios with large analytical errors could introduce significant, sometimes asymmetric errors following non-Gaussian distribution. More importantly, errors around the local maximum/minimum and saddle points on the Phanerozoic seawater $^{87}Sr/^{86}Sr$ curve are difficult to examine using conventional methods. In this study, we developed a rigorous mathematical framework (including perturbation and probability density methods) to estimate 95% confidence interval (or any confidence interval of interest) of the derived age with any given analytical uncertainty. We applied this method to estimating errors for one version of the Phanerozoic seawater $^{87}Sr/^{86}Sr$ curve, assessing the sources and amplitudes of the uncertainties. Our results indicate this method has the potential to be widely applied to error estimates involving inversions using an empirical relationship. A MATLAB program is developed to calculate the age of carbonate rocks (based on their $^{87}Sr/^{86}Sr$ ratios and analytical uncertainties), the age gaps, and their confidence interval of choice in the last 509 Ma." The third chapter by Caxito et al. highlights that "Brazil is a natural laboratory for the application of isotope stratigraphy tools on the study of Precambrian sedimentary successions, with excellent registers of Archean, Paleo, Meso, and Neoproterozoic age. Those successions witnessed the most extreme and unidirectional events in Earth's history, such as the Paleoproterozoic and Neoproterozoic oxygenation events (GOE and NOE), the main stages of iron formation deposition, the global Cryogenian glaciations and the explosive diversification of metazoans near the Precambrian/Cambrian border. Chemical sedimentary rocks such as carbonates, iron formations, and evaporites that can act as natural recorders of seawater composition are present in all of those successions and furnish important pieces of evidence in order to reconstruct the chemostratigraphic record. The field of isotope stratigraphy keeps expanding in Brazil and novel information from developing isotopic systems will surely add to the growing database and provide important information on Earth's ancient biogeochemical cycles in the years to come." In the fourth chapter Rodrigues et al. highlight "the application of stable isotopes in stratigraphic studies is here demonstrated with selected sedimentary Brazilian sections from the Paleozoic, Mesozoic, and Cenozoic. In those different age case, we present the application of $\delta^{13}C_{org}$ and $\delta^{13}C_{carb}$, $\delta^{15}N_{total}$ and $\delta^{18}O_{carb}$ stable isotopes showing how past global events and environmental conditions variation can imprint characteristic isotopic signatures in sedimentary rocks profiles and how they can be useful in stratigraphic studies. Also, key points for the interpretation of stable isotope data derived from sedimentary organic matter and carbonates are discussed demonstrating how bulk-rocks analyses can be really useful for stratigraphic correlations and paleoenvironmental interpretations when a high resolution interpretation is conducted." The following contribution by Tendil et al. point out that "in the recent years, $\delta^{13}C$ and $\delta^{18}O$ chemostratigraphy has become a powerful tool used to solve controversial results using different biological time markers from shallow-marine carbonate environments and establish long-distance correlations across proximal and distal settings. However, its reliability was recently questioned due to the impact of diagenesis on stable isotope records. The ammonite-calibrated, platform-to-basin transect of the late Barremian–early Aptian Urgonian Provence platform is made particularly relevant to challenge chemostratigratigraphy-based correlation methods in shallow carbonate series at different scales. The origin, preservation and stratigraphic potential of isotopic signatures are herein discussed along 11 shallow-marine and basinal sections. In the proximal settings of the Provence domain, tight carbonates—either resulting from muddy depositional textures or early cementation—likely preserve original marine signatures. Contrasting observations arise from porous carbonates whose stable isotope signal is heavily affected by micrite recrystallization (i.e., microporosity genesis) during a mid-Cretaceous tectonic uplift event. Locally, the C-isotope signal of microporous Urgonian limestones can be controlled by the vertical distribution of porosity. Although basinal series are more prone to preserve primary geochemical signatures, extreme values for carbon and oxygen isotopes are observed in correspondence with fine-grained bioclastic material input originating from the adjacent platform, following the latest Barremian demise of the rudist-dominated platform. This study suggests that stable isotope chemostratigraphy may not constitute a suitable tool for calibrating and correlating Urgonian shallow-marine series." This chapter is followed by the more theoretical contribution of Jackson and Montenari who explain "the structure of zircon, $ZrSiO_4$ is modeled using interatomic potentials. The uranium end-member, coffinite ($USiO_4$) and intermediate solid solutions of zircon and coffinite ($U_xZr_1-xSiO_4$) are then modeled, allowing the prediction of lattice parameters as a function of uranium concentration. Finally, possible structures resulting from the radioactive decay of uranium to lead in coffinite are considered."

Michael Montenari
Keele University
United Kingdom

CHAPTER ONE

Sulfur isotopes—Use for stratigraphy during times of rapid perturbations

Weiqi Yao[a], Ulrich G. Wortmann[a], Adina Paytan[b,*]
[a]Department of Earth Sciences, University of Toronto, Toronto, ON, Canada
[b]Institute of Marine Science, University of California—Santa Cruz, Santa Cruz, CA, United States
*Corresponding author: e-mail address: apaytan@ucsc.edu

Contents

Abstract

The sulfur isotopic composition of seawater sulfate oscillates over geological time, responding to changes in the redox state and biogeochemical processes on the Earth's surface. During times of rapid perturbations in sulfur isotopes where biostratigraphic controls do not provide good temporal resolution, sulfur isotopes of seawater sulfate

Stratigraphy & Timescales, Volume 4
ISSN 2468-5178
https://doi.org/10.1016/bs.sats.2019.08.004

can be a useful tool for stratigraphic correlations. Secular data of seawater sulfur isotopes archived in evaporite, carbonate associated sulfate, and marine barite from worldwide locations allow reconstruction of a continuous seawater sulfate S-isotope curve for the Phanerozoic and Protozoic. For some time-intervals (e.g., the Cenozoic), excursions in high resolution records are identified and these short-term fluctuations can be applied for stratigraphy and correlation. High resolution S-isotope records can also substantially enhance our understanding of temporal variations in the global sulfur cycle and earth system evolution.

1. Introduction

Sulfur is the fourth most abundant element dissolved in seawater and the ninth most abundant element in the entire Earth (McDonough, 2000). As a multivalent element, sulfur has a broad range of oxidation states ranging between the fully reduced state (−2) and the fully oxidized state (+6). Due to the variety of redox states and reactivity toward both metals and non-metal elements, sulfur is ubiquitous in the Earth's lithosphere, biosphere, hydrosphere, and atmosphere. The fully reduced form of sulfur is present as metal sulfides (e.g., pyrite) and as dissolved sulfide species which are toxic to most of the life forms (H_2S and HS^-; Smith et al., 1976), whereas the fully oxidized form of sulfur is predominantly present as dissolved sulfate (SO_4^{2-}) in natural waters and as sulfate minerals (e.g., gypsum, barite).

The partitioning of sulfur between the reduced and oxidized reservoirs through microbial sulfate reduction (MSR: $2CH_2O + SO_4^{2-} \rightarrow 2HCO_3^- + H_2S$) is one of the most important processes for organic matter remineralization in the ocean and the dominant respiration process in organic-rich anoxic sediments. When organic matter respiration has depleted all oxygen (as well as dissolved nitrate, iron, and manganese) in the interstitial water, sulfate serves as the terminal electron acceptor to respire organic matter back to carbon dioxide (CO_2) with sulfide as the main metabolite end-product (Jørgensen, 1983; Reeburgh, 1983). In cold-seep environments and in anaerobic sediments sulfate is also the electron acceptor driving anaerobic oxidation of methane (AOM: $CH_4 + SO_4^{2-} \rightarrow HCO_3^- + HS^- + H_2O$; Valentine, 2002; Jørgensen and Kasten, 2006).

Some fractions of the sulfide produced by microbial sulfate reduction can form iron polysulfides and in the presence of reactive iron pyrite (Berner, 1982; Jørgensen, 1982), while the majority (80%–95%) of the sulfide is subsequently reoxidized back to sulfate when exposed to more oxic conditions either through mechanical mixing (bioturbation) or through diffusion into oxic sections (Jørgensen, 1982). This reoxidation process is also known as the oxic sulfur cycle. In the modern ocean, active oxic sulfur cycling is limited to shelf areas where the anoxic part of the sediment column is shallow enough to overlap with the bioturbation zone and allows oxygen penetration to the MSR zone.

As such, biogeochemical cycles of sulfur and carbon constitute the dominant control of oxygen in the ocean and atmosphere (Berner, 1989; Berner and Raiswell, 1983; Canfield, 2005; Garrels and Lerman, 1984; Hurtgen, 2012; Walker, 1986), and are intricately linked to cycling of other elements like oxygen, nitrogen, and phosphorus (Berner and Raiswell, 1983; Canfield et al., 2010b; Petsch and Berner, 1998; Wortmann and Paytan, 2012). Reconstructing the evolution of the sulfur cycle can in turn reflect the history of oceanic and atmospheric chemistry.

Since the residence time of sulfate in the ocean exceeds the oceanic mixing time (Claypool et al., 1980; Jørgensen and Kasten, 2006; see below for details), changes in the marine sulfate S-isotope signal are synchronous across ocean basins. Secular variations in the sulfur isotope ratio of seawater sulfate stemming from changes in earth processes over time are usually too slow to serve as a high temporal resolution stratigraphic marker. However, recent work showed that during times of rapid perturbations in sulfur cycling, short-term variations in seawater sulfur isotopes provide signals on a timescale as short as hundreds of thousand years, which can be a powerful tool for stratigraphic correlations (e.g., Luo et al., 2010; Newton et al., 2004; Paytan et al., 1998, 2004; Sim et al., 2015; Wortmann and Paytan, 2012; Yao et al., 2018).

1.1 Stable sulfur isotopes of seawater sulfate

Sulfur has four stable isotopes (^{32}S, ^{33}S, ^{34}S, and ^{36}S) with an average atomic weight of 32.06 (Table 1; Meija et al., 2016). The sulfur isotope ratio of

Table 1 The abundance of stable sulfur isotopes for Vienna Canyon Diablo Troilite.

Isotope	Molar fraction [%]
^{32}S	95.02
^{33}S	0.760
^{34}S	4.22
^{36}S	0.0136

$^{34}S/^{32}S$ is expressed as a per mil (‰) difference relative to the international reference standard—Vienna Canyon Diablo Troilite (VCDT) using the standard delta notation ($\delta^{34}S$):

$$\delta^{34}S = \frac{\left(\frac{^{34}S}{^{32}S}\right)\text{sample} - \left(\frac{^{34}S}{^{32}S}\right)\text{standard}}{\left(\frac{^{34}S}{^{32}S}\right)\text{standard}} \times 1000‰$$

where $^{34}S/^{32}S$ denotes the mass ratio of ^{34}S over ^{32}S.

The relative abundances of sulfur isotopes vary between coexisting sulfur phases as a result of mass-independent isotope fractionation (e.g., Farquhar et al., 2000; Pavlov and Kasting, 2002) and mass-dependent isotope fractionation (including kinetic and equilibrium fractionation). Kinetic fractionation is frequently associated with biogeochemical redox reactions and commonly observed in biological systems. Typically, lighter isotopes are more enriched in the products of the reaction because bonds of lighter masses have a higher vibrational energy and are more easily broken. Equilibrium fractionation is also governed by thermodynamics.

The microbially mediated sulfate reduction proceeds through a chain of reversible enzymatic reactions (Brunner and Bernasconi, 2005; Eckert et al., 2011; Holler et al., 2011; Rees, 1973) which preferentially break the ^{32}S–O bond, leaving the residual sulfate pool enriched in ^{34}S (Chambers and Trudinger, 1979). The isotopic enrichment during this process is expressed as the fractionation factor alpha, α (or the enrichment factor epsilon, ε). The maximum attainable fractionation factor was controversial, but it is now understood to be similar to the equilibrium fractionation factor (Brunner and Bernasconi, 2005; Canfield et al., 2010a; Kaplan et al., 1963; Rudnicki et al., 2001; Sim et al., 2011; Wortmann et al., 2001, 2007). How much of the equilibrium fractionation factor will be expressed during MSR, depends on the ratio of the backward to forward fluxes during MSR (Brunner and Bernasconi, 2005). As a rule of thumb, the ratio of backward to forward fluxes will be small as long as the net reduction flux is fast, and will approach unity once microbial reduction comes almost to a standstill. In natural environments, a complete expression of the S-isotope fractionation between the oxidized (sulfate) and reduced (sulfide) reservoir reaches up to 72‰ (e.g., Kaplan et al., 1963; Rudnicki et al., 2001; Sim et al., 2011; Wortmann et al., 2001).

1.2 The global sulfur cycle

At the present day and likely throughout the Phanerozoic, sulfate is the main sulfur species in the hydrosphere and the second most abundant anion in the ocean. Sulfate in the modern open ocean has an average concentration of ~28 mM and a $\delta^{34}S$ value of +21‰ (Horita et al., 2002; Lowenstein et al., 2003; Paytan et al., 1998; Rees, 1973). Fluctuations in concentrations and stable sulfur isotope ratios of seawater sulfate have occurred over geological time in response to changes in geological, geochemical and biological processes (e.g., Berner and Raiswell, 1983; Paytan et al., 1998, 2004; Strauss, 1997; Wortmann and Chernyavsky, 2007; Wortmann and Paytan, 2012; Yao et al., 2018). Sulfate concentrations in the Proterozoic ocean were much lower than during the Phanerozoic (e.g., Canfield and Farquhar, 2009; Habicht et al., 2002; Kah et al., 2004), hence it is likely that the oceanic water column was not homogenous with respect to sulfur isotopes limiting the applicability of S-isotopes for stratigraphy and correlation.

The concentration and sulfur isotopic ratios of seawater sulfate are controlled by sulfur delivered to the oceans via weathering and volcanic degassing and sulfur removed from the oceans via burial of sulfur-bearing minerals, and their respective isotopic compositions. Sedimentary sulfur is present in its oxidized form as sulfate-bearing minerals (e.g., gypsum) and in its reduced form as metal sulfides (e.g., pyrite). Today, continental weathering of evaporite and pyrite constitutes the dominant source of sulfur to the ocean with an estimated annual input of 1.5×10^{12} to 3.5×10^{12} mol via river runoffs (Garrels and Lerman, 1984; Kump and Garrels, 1986; Markovic et al., 2015; Wortmann and Paytan, 2012). Additional sulfur input is through volcanic or hydrothermal activities, which transfers 3.4×10^{11} and 1×10^{12} mol sulfur per year from the mantle back to the Earth's surface, respectively (Edmond et al., 1979; Hansen and Wallman, 2003; Walker, 1986). Precipitation of pyrite and evaporite constitutes the main sink, removing 0.5×10^{12}–1.6×10^{12} mol and 1×10^{12}–2.7×10^{12} mol sulfur per year from seawater, respectively (Garrels and Lerman, 1984; Kump and Garrels, 1986; Markovic et al., 2015; Walker, 1986).

The $\delta^{34}S$ signature of seawater sulfate responds to changes in the relative proportion of sulfate and sulfide input and output to and from the ocean through time. Specifically, volcanic and hydrothermal inputs contribute sulfate with $\delta^{34}S$ values of 0–2‰ to the ocean whereas input from evaporites typically has higher $\delta^{34}S$ values (Hansen and Wallman, 2003; Sakai et al., 1982). On the other hand, sulfate-bearing evaporites usually precipitate with little or no fractionation relative to contemporaneous seawater sulfate, and thus have only minor effects on the seawater $\delta^{34}S$ value (Claypool et al., 1980). In contrast, sulfur in pyrite is isotopically light with a $\delta^{34}S$ value ranging between −60‰ and −25‰ (Canfield et al., 2010a; Kaplan et al., 1963; Strauss, 1997). Due to the large isotopic fractionation associated with MSR and hence pyrite formation, weathering and precipitation of sedimentary pyrite are the dominant controls of the seawater $\delta^{34}S$ (e.g., Berner and Raiswell, 1983; Kurtz et al., 2003; Paytan et al., 2004; Wortmann and Paytan, 2012; Wu et al., 2010). In addition, the sulfur isotope ratio of seawater sulfate is sensitive to rapid changes in oceanic redox conditions. The development of ocean internal reservoirs of reduced sulfur can exert temporary but significant influence on the seawater sulfate $\delta^{34}S$ (e.g., during the Paleocene–Eocene (Yao et al., 2018) and the Permian–Triassic boundaries (Algeo et al., 2007; Bernasconi et al., 2017; Luo et al., 2010; Newton et al., 2004)).

The size of the current oceanic sulfate reservoir is large (4×10^{19} mol) compared to the sulfur input and output fluxes (Kump and Garrels, 1986; Kurtz et al., 2003). As such, the residence time of sulfate in the current ocean is approximately 20 million years, about four orders of magnitude longer than the mixing time of the ocean water (~1700 years; Claypool et al., 1980; Jørgensen and Kasten, 2006). Therefore, the sulfur isotope ratio of seawater sulfate is homogenous (i.e., conservative) throughout the global ocean. The seawater $\delta^{34}S$ fingerprints stored in sediments and sedimentary rocks are useful proxies for stratigraphic correlations. However, since the sulfate concentration changed throughout Earth's history (e.g., Horita et al., 2002; Lowenstein et al., 2003), care must be taken to evaluate the actual residence time of sulfate in the ocean at any given time.

2. Archives for seawater sulfur isotopes

Sulfur isotope records of ancient marine sulfate are available through the analyses of marine evaporites, carbonate associated sulfate (CAS), and marine biogenic barite from worldwide locations (e.g., Claypool et al., 1980; Kampschulte and Strauss, 2004; Paytan et al., 1998, 2004; Rennie et al., 2018; Strauss, 1997). Compared to the data based primarily on marine evaporite sulfate minerals (e.g., gypsum, anhydrite, halite), CAS and marine barite in stratigraphically well-constrained samples provide more continuous and detailed data especially for the Cenozoic and Mesozoic.

2.1 Marine evaporite sulfate records

Marine evaporite sulfate incorporates the sulfur isotopic signature from the seawater in which the mineral precipitated. Since marine evaporite sulfate minerals have high concentrations of sulfate, it is straightforward to use them for isotope analyses. Indeed, fluctuations of the seawater $\delta^{34}S$ through time have been initially revealed by evaporite-derived sulfate $\delta^{34}S$ data (Claypool et al., 1980; Strauss, 1997; Strauss et al., 2001). Claypool et al. (1980) first presented a compilation of marine evaporite sulfate $\delta^{34}S$ data for the terminal Precambrian and the Phanerozoic. The age curve of the evaporite-derived $\delta^{34}S$ is characterized by a discernible increase from approximately 18‰ to an average value up to 35‰ during the late Neoproterozoic, followed by a pronounced 25‰ decline throughout the Paleozoic to a Permian minimum at ~10‰, and a gradual rise toward the modern oceanic value of 21‰ (Fig. 1; Claypool et al., 1980). Although large temporal gaps exist in the evaporite record, their work demonstrated that the sulfur isotope ratio of seawater sulfate had undergone substantial changes over geological time and laid the groundwork for our current understanding of the Phanerozoic sulfur cycle and S-isotope stratigraphy. Further studies presented more comprehensive evaporite sulfate $\delta^{34}S$ records with a thorough compilation of the evaporite data spanning the Proterozoic and the Phanerozoic (Crockford et al., 2019; Strauss, 1997, 1999; Strauss et al., 2001), which considerably extend the applications for stratigraphic correlations.

However, there are some fundamental limitations when considering the use of the evaporite records for stratigraphy. First, the evaporite-derived $\delta^{34}S$ records are not continuous for some geological intervals. Because of the sporadic occurrence of evaporite deposits in sedimentary records, the temporal resolution of this record is relatively low (but also see Bernasconi et al., 2017). This disadvantage was partially overcome by extrapolating between the discrete data sets to approximate a best-estimated age curve (Claypool et al., 1980). However, this smoothing results in masking of finer changes in the $\delta^{34}S$ signatures during times of rapid perturbations. Second, the age control on this record is imperfect. Due to the scarcity of fossils in evaporite settings, it is difficult to correlate to biostratigraphy which constrains the stratigraphic position of the boundaries, and thus evaporite records are typically accompanied with large age uncertainties. Depending on the specific time intervals, the age uncertainties presented in the evaporite record are on the order of tens of million years or higher (Strauss, 1997). Third, the spread in the $\delta^{34}S$ data at each time point interval is very large, making it difficult for stratigraphic correlations. The range of

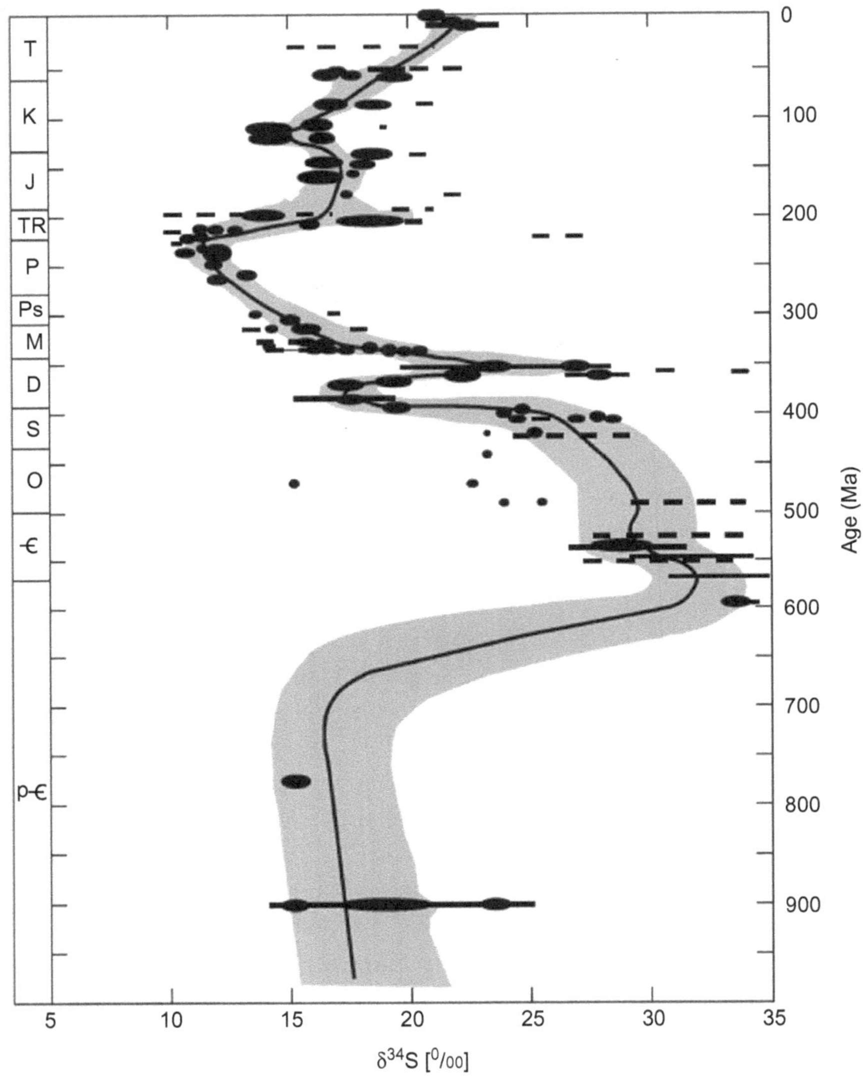

Fig. 1 The age curve for the evaporite-derived $\delta^{34}S$ data of Claypool et al. (1980). The solid line denotes the best-estimated $\delta^{34}S$ plotted against their most probable ages. Dash lines denote the range of measurements. Shaded areas denote the uncertainty related to the curve.

the $\delta^{34}S$ values is approximately 5‰ for most of the time intervals and can be as high as 25‰ for some older sections (e.g., the Devonian; Claypool et al., 1980; Strauss, 1997). This is despite minor sulfur fractionation (0–2.5‰) during evaporite crystallization (Raab and Spiro, 1991) indicating other controls on the measured $\delta^{34}S$.

2.2 Carbonate associated sulfate records

Since the 1980s, carbonate associated sulfate has been used as an archive for reconstructing the sulfur isotope composition of seawater sulfate (Burdett et al., 1989; Kampschulte and Strauss, 2004; Kampschulte et al., 2001; Rennie and Turchyn, 2014; Rennie et al., 2018). Since sulfate can replace carbonate ions in marine carbonates in trace amounts, CAS is expected to record the contemporaneous seawater sulfate. In contrast to the intermittent distribution of marine evaporite deposits, carbonate sequences are much more widespread and continuously available in geological records. Relatively rapid rates of carbonate precipitation enable the development of higher-resolution continuous CAS records. In addition, the abundance of microfossils, primarily calcareous and siliceous microfossils, in carbonate sequences, provides much better biostratigraphic constraints, improving the temporal resolution of the secular $\delta^{34}S$ reconstructions (Fig. 2).

CAS is a reliable tracer of seawater sulfate $\delta^{34}S$ at least as far back as 65 million years ago (Rennie et al., 2018). However, the application of CAS in older sediments can be hampered by post-depositional alteration (e.g., Bernasconi et al., 2017; Bottrell and Newton, 2006). This is evident by a wider range of $\delta^{34}S$ for older samples of the same age and more scatter in the data (e.g., >20‰ for the Ordovician and older) compared to younger samples (e.g., <10‰ for the Mesozoic and the Cenozoic) (Fig. 2). The observed range in values can be attributed to large variations of ancient seawater $\delta^{34}S$ values (e.g., Canfield and Farquhar, 2009; Horita et al., 2002; Lowenstein et al., 2003), poorly defined ages, and/or more likely local effects of diagenesis (e.g., Goldberg et al., 2005; Mazumdar and Strauss, 2006). Indeed, carbonate minerals undergo extensive post-depositional alteration in many locations (both through dissolution and reprecipitation and overgrowth of carbonate in pore waters). Although the mechanisms of early marine diagenesis have not been fully understood to date, previous studies demonstrated certain extents of pore water sulfate incorporation into the carbonate mineral lattice during carbonate recrystallization (Bottrell and Newton, 2006; Rennie and Turchyn, 2014).

Another factor that may contribute to the scatter is species-specific fractionation of sulfur isotopes between CAS and seawater sulfate during biomineralization. Depending on the species of *Foraminifera*, there are up to 2‰ offsets of the CAS $\delta^{34}S$ within different foraminiferal species of the same age (Paris et al., 2014). Recent work largely overcame these disadvantages using CAS from single shells of different species of *Foraminifera* and correcting the data for offsets between species (Rennie et al., 2018).

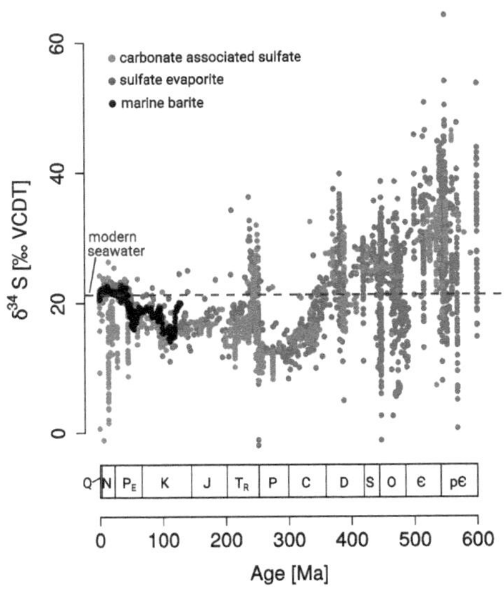

Fig. 2 The Phanerozoic seawater sulfate δ^{34}S record. Green circles, CAS data (Gill et al., 2007; Goldberg et al., 2005; Hurtgen et al., 2009; Kah et al., 2016; Kampschulte and Strauss, 2004; Mazumdar and Strauss, 2006; Present et al., 2015; Rennie et al., 2018; Schobben et al., 2017; Sim et al., 2015; Strauss, 1999; Thompson and Kah, 2012; Turchyn et al., 2009; Ueda et al., 1987; Wotte et al., 2012; Wu et al., 2010, 2014); gray circles, evaporites data (Bernasconi et al., 2017; Claypool et al., 1980; Cortecci et al., 1981; Crockford et al., 2019; Das et al., 1990; Fox and Videtich, 1997; Holser and Kaplan, 1966; Kampschulte et al., 1998; Longinelli and Flora, 2007; Orti et al., 2010; Peryt et al., 2005; Pierre and Rouchy, 1986; Rick, 1990; Sakai, 1972; Strauss, 1997, 1999; Strauss et al., 2001; Surakotra et al., 2018; Utrilla et al., 1992; Worden et al., 1997); black circles = barite data (Paytan et al., 1998, 2004; Turchyn et al., 2009; Markovic et al., 2015, 2016; Masterson et al., 2016; Yao et al., 2018; Yao et al., 2019, under review). Blue dash line, the modern seawater sulfate δ^{34}S value of ~21‰ (Markovic et al., 2015; Rennie et al., 2018; Tostevin et al., 2014).

In addition, contamination from pyrite in the samples which is incorporated into the analyzed samples in the CAS extraction processes may also limit the reliability of the CAS records (Marenco et al., 2008). CAS is quantitatively recovered from a bulk sample through carbonates dissolution (Burdett et al., 1989; Marenco et al., 2008). However, a systematically controlled investigation of pyrite oxidation during CAS processing showed that pyrite is oxidized in both strong and weak acids during the extraction processes (Marenco et al., 2008). Oxidation of pyrite with isotopically light sulfur can impart a δ^{34}S offset of a few per mil even in sediments with low pyrite content, drawing the CAS δ^{34}S toward lower values (Marenco et al., 2008). Therefore, careful sample selection and inspection, as well as cautious in the extraction process of CAS from bulk sediment samples, are required to ensure that the recovered sulfate accurately represents only the original CAS signature.

2.3 Marine barite records

Similar to carbonate, marine barite continuously forms in the water column directly from seawater with little $\delta^{34}S$ fractionation between barite and coeval seawater sulfate (Bishop, 1988; Dehairs et al., 1990; Ganeshram et al., 2003; Kusakabe and Robinson, 1977). Due to its low solubility at normal marine temperature and pH conditions, barite is stable during diagenesis as long as the interstitial water is saturated with respect to barium sulfate and has not undergone sulfate reduction (Dean and Schreiber, 1977; Dymond et al., 1992; Paytan et al., 1993, 2002). Continuous precipitation and resistance to post-depositional alteration allow barite to be readily available in pelagic sediments. Therefore, marine barite records the $\delta^{34}S$ signature of seawater sulfate at the time of its formation and thus can be used for reconstructing variations in isotope compositions of seawater sulfate (Markovic et al., 2015, 2016; Masterson et al., 2016; Paytan et al., 1998, 2004; Turchyn et al., 2009; Yao et al., 2018).

The $\delta^{34}S$ records derived from marine barite are often more robust with good constraints on age compared to the evaporites records (Markovic et al., 2015; Paytan et al., 1998, 2004; Yao et al., 2018). However, since so far barite has been recovered predominantly from pelagic sediments, the application of marine barite as a recorder of the seawater $\delta^{34}S$ and its use for correlation and stratigraphy are limited to ages of ~130 Ma, represented by deep sea sediments (Paytan et al., 2011). The barite records compiled by Paytan et al. (1998, 2004) with further work (Markovic et al., 2015, 2016; Masterson et al., 2016; Turchyn et al., 2009; Yao et al., 2018) provide detailed information on the variations in the seawater $\delta^{34}S$ for the past 130 Ma (Fig. 3). The $\delta^{34}S$ values recorded within each time interval scatter at approximately 0.5‰ and the current barite-derived curve is continuous with a temporal resolution of <1 million years, filling the gaps that are missed in the previous records.

2.4 Additional notes

During times when sulfate concentrations are very low and/or substantial areas of the ocean become stratified, the development of a transient reservoir of reduced sulfur (e.g., within oxygen minimum zones) and large ocean

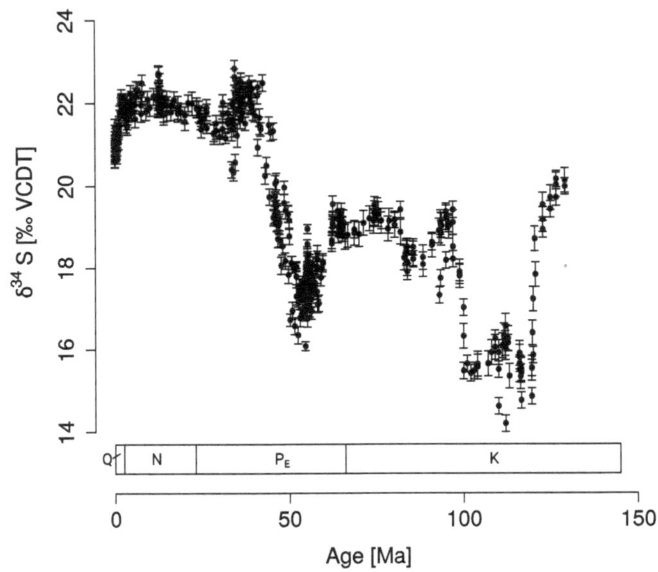

Fig. 3 The marine barite-derived δ^{34}S record for the past 130 Ma. Error bars are 1σ. *After Paytan, A., Kastner, M., Campbell, D., Thiemens, M.H., 1998. Sulfur isotopic composition of Cenozoic seawater sulfate. Science 282, 1459–1462; Paytan, A., Kastner, M., Campbell, D., Thiemens, M.H., 2004. Seawater sulfur isotope fluctuations in the cretaceous. Science 304, 1663–1665; Turchyn, A.V., Schrag, D.P., Coccioni, R., Montanari, A., 2009. Stable isotope analysis of the Cretaceous sulfur cycle. Earth Planet. Sci. Lett. 285, 115–123; Markovic, S., Paytan, A., Wortmann, U.G., 2015. Pleistocene sediment offloading and the global sulfur cycle. Biogeosciences 12, 3043–3060.; Markovic, S., Paytan, A., Li, H., Wortmann, U.G., 2016. A revised seawater sulfate oxygen isotope record for the last 4 Myr. Geochim. Cosmochim. Acta 175, 239–251; Masterson, A.L., Wing, B.A., Paytan, A., Farquhar, J., Johnston, D.T., 2016. The minor sulfur isotope composition of Cretaceous and Cenozoic seawater sulfate. Paleoceanography 31, 779–788; Yao, W., Paytan, A., Wortmann, U.G., 2018. Large-scale ocean deoxygenation during the Paleocene–Eocene thermal maximum. Science 361, 804–806; Yao, W., Paytan, A., Griffith, E.M., Martínez-Ruiz, F., Markovic, S., Wortmann, U.G., 2019, under review. A revised seawater sulfate S-isotope curve for the Eocene. Chem. Geol.*

internal fluxes can render the residence time of sulfur species in seawater much shorter than millions of years (e.g., the Paleocene–Eocene and Permian–Triassic boundaries; Newton et al., 2004; Yao et al., 2018). The ocean during these time intervals may no longer be well mixed with respect to sulfate. Therefore, it is important to consider the environment where the minerals precipitate and form during such events when samples are selected for analyses. Correlations between multiple sites at each time point also help to reduce uncertainties caused by local variations and to identify the global nature of seawater sulfate δ^{34}S.

3. Stratigraphic correlations of specific age intervals

Stratigraphic boundaries are characterized by profound perturbations in geological, geochemical, and biological processes. Specifically, several major extinction events have been observed across the Precambrian–Cambrian, the Ordovician–Silurian, the late Devonian, the Permian–Triassic, and the Cretaceous–Tertiary boundaries. Widespread and rapid decreases in biomass and biodiversity exert devastating influence on the biosphere, but also affect microbial sulfate reduction and often biogeochemical sulfur cycling (Canfield and Farquhar, 2009). Limited by the availability of organic matter, the MSR rate becomes slower with a decrease in organic matter supply, resulting in a reduction of the subsequent formation and burial of pyrite (i.e., in the modern ocean, but also see Wortmann and Chernyavsky, 2007), and vice versa. As a result, substantial fluctuations in $\delta^{34}S$ values of sulfate in the open oceans could be identified for these distinct age intervals and linked to catastrophic extinction events. Here we provide a compilation of the S-isotope data from marine evaporite sulfate, CAS and marine barite to show variations in seawater $\delta^{34}S$ signatures from the Protozoic to the Quaternary, and discuss the potential use of the ranges and general trends of the $\delta^{34}S$ records for broad age assignments and correlations, particularly during geological times of rapid perturbations.

3.1 Proterozoic (2500–541 Ma)

The $\delta^{34}S$ values of the Proterozoic seawater sulfate cluster between +4‰ and +42‰ with an average of about 16‰ (Fig. 4; Crockford et al., 2019 and references therein), which is slightly lower than the modern seawater $\delta^{34}S$ value of ~21‰ (Markovic et al., 2015; Rennie et al., 2018; Tostevin et al., 2014). There is a slight increase in the $\delta^{34}S$ value around the Great Oxygenation Event (GOE, 2450–2000 Ma). Following that, the $\delta^{34}S$ values span a wider range between −10‰ and +40‰ during the middle-Proterozoic and show positive excursions in the Neoproterozoic (Fig. 4, purple box). These increases coincide with at least two widespread and enduring glaciation events, i.e., the Sturtian (~700–690 Ma) and Marinoan (600–595 Ma) glacial periods (e.g., Hurtgen et al., 2002).

The overall variations in the $\delta^{34}S$ value in the middle Proterozoic are significantly larger compared to the GOE and lower Neoproterozoic. This is consistent with the anoxic and sulfidic bottom ocean (the so-called "Canfield Ocean") in the middle Proterozoic, which persisted until the deposition of banded iron formations (BIF) ceased (Canfield, 1998). The

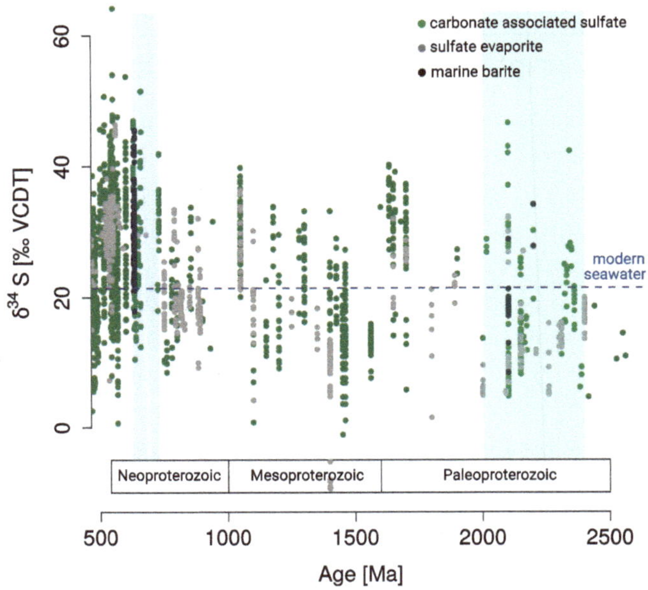

Fig. 4 The Proterozoic seawater sulfate δ^{34}S record. Green circles, CAS data; Gray circles, evaporites data; Black circles, barite data; Blue dash line, the modern seawater sulfate δ^{34}S value of ~21‰ (Markovic et al., 2015; Rennie et al., 2018; Tostevin et al., 2014). The blue and purple boxes denote the periods of the Great Oxygenation Event (2450–2000 Ma) and Cryogenian (635–717 Ma), respectively. *After Crockford, P.W., et al., 2019. Claypool continued: Extending the isotopic record of sedimentary sulfate. Chem. Geol. 513, 200–225 and references therein.*

narrower range of the δ^{34}S value at the GOE and the late Proterozoic supports the growth in the size of the seawater sulfate reservoir as a consequence of the significant oxidation events of the Earth's surface (e.g., Blättler et al., 2018; Bottrell and Newton, 2006; Crockford et al., 2019; Kah et al., 2004; Schröder et al., 2008; Scott et al., 2014). With an increased seawater sulfate level (from 0 to over 10 mM; e.g., Bottrell and Newton, 2006 and references therein), the residence time of sulfate in the ocean would have become longer, potentially resulting in a more isotopically homogenous system. In such a system, seawater δ^{34}S value could have been less prone to transient shifts in response to changes in input and output fluxes (Kah et al., 2004).

In general, the use of the seawater δ^{34}S record of the Proterozoic for stratigraphy and correlation is limited due to the scarcity of sulfate data and the substantial variations over very small stratigraphic intervals (e.g., >39‰ over 55 m; Kah et al., 2004). However, examination of δ^{34}S values in conjunction with estimates of sedimentation rates can provide a potential for stratigraphic applications, albeit at a coarse resolution. Additionally, combining the $\delta^{34}S$ value derived from coeval sedimentary sulfate and sulfide minerals help to constrain the S-isotope fractionation, which can in turn be used to estimate the evolution of marine sulfate concentrations and to link to the oxygenation of the Earth's biosphere (e.g., Canfield, 1998; Hurtgen et al., 2005; Kah et al., 2004; Planavsky et al., 2012). Recent studies also showed that multiple sulfur isotopes (^{33}S and ^{36}S) of sulfate in the Proterozoic could be powerful tools for stratigraphy (e.g., Crockford et al., 2019; Farquhar and Wing, 2003; Johnston, 2011 and references therein). However, the use of ^{33}S and ^{36}S has so far been limited and will not be further discussed here.

3.2 Cambrian (541–485 Ma)

The seawater $\delta^{34}S$ records for the Cambrian are derived from carbonate and evaporite rocks (and a few from barite) in Australia, Canada, China, India, Russia, Spain, and France (Goldberg et al., 2005; Hough et al., 2006; Hurtgen et al., 2009; Mazumdar and Strauss, 2006; Peryt et al., 2005; Wotte et al., 2012). The seawater $\delta^{34}S$ value is characterized by the wide range and globally recognized ^{34}S-enrichment (average >30‰) for the entire Cambrian, which may partially result from the intrabasinal microbial sulfate reduction under sulfate limitation or diagenetic processes (Goldberg et al., 2005; Mazumdar and Strauss, 2006). Data obtained worldwide suggested a slight rise in seawater $\delta^{34}S$ to over 40‰ in the lower Cambrian, followed by a systematic >15‰ decrease across the middle-upper Cambrian. The $\delta^{34}S$ increase has been attributed to a more pronounced basinal euxinia and an enhanced draw-down of ^{32}S via intense MSR in the water column, as a result of increased primary productivity and subsequent utilization of O_2 (Goldberg et al., 2005; Hough et al., 2006; Mazumdar and Strauss, 2006; Peryt et al., 2005).

Due to the absence of an internationally accepted biostratigraphic correlations for the Cambrian, some studies attempted to correlate carbonate $\delta^{13}C$, $\delta^{18}O$, $\delta^{87}Sr$, and other methods to the $\delta^{34}S$ records for inter-basinal chemostratigraphic correlations (e.g., Hurtgen et al., 2009; Mazumdar and Strauss, 2006; Wotte et al., 2012). However, in general, identifying sample ages and the global nature of seawater sulfate $\delta^{34}S$ remains challenging as sulfate was most likely a non-conservative anion in the Cambrian ocean (Wotte et al., 2012).

3.3 Ordovician (485–444 Ma) and Silurian (444–419 Ma)

The seawater $\delta^{34}S$ value appears to be more stable during the Ordovician and Silurian. The CAS based $\delta^{34}S$ data show a downward trend from approximately 30‰ in the lower Ordovician to about 24‰ in the uppermost Ordovician, and then remain between 21‰ and 36‰ in the Silurian (Kampschulte and Strauss, 2004). The overall scatter in the $\delta^{34}S$ value is large and the temporal changes are relatively small in comparison, thus it is difficult to utilize the $\delta^{34}S$ records to distinguishing older and younger samples within the Ordovician and Silurian.

3.4 Devonian (419–359 Ma)

Following the Ordovician/Silurian plateau with $\delta^{34}S$ values of around 26‰, the seawater $\delta^{34}S$ value shows a considerable drop to approximately 18‰ in the lower-middle Devonian and a subsequent increase to another plateau at about 23‰ in the upper Devonian. The age resolution of the Devonian data is poorly constrained. High stratigraphic $\delta^{34}S$ oscillations (between 10‰ and 40‰), coupled with strong reservoir effects, suggests a relatively small oceanic sulfate reservoir during the Devonian. It is noteworthy that Sim et al. (2015) correlated the S-isotope record among sections throughout the world representing the Fransnian–Famennian boundary, despite relatively low-resolution data available at that time. The generally similar ~5‰ decline in seawater $\delta^{34}S$ has been reported for sections in the United States, Belgium, and Poland, which has the potential for correlation applications as seen in Fig. 5 (Sim et al., 2015 and references therein). Moreover, the $\delta^{34}S$ and $\delta^{13}C$ excursions may be linked to the late Devonian mass extinction (Sim et al., 2015). It is, however, important to obtain more data with better-defined ages from diverse sites to verify a global trend.

3.5 Carboniferous (359–299 Ma)

The Carboniferous is marked by a substantial decline in the seawater $\delta^{34}S$ value from around 23‰ at the Devonian–Carboniferous boundary to around 15‰ in the middle Carboniferous (~330 Ma) which remains stable until decreasing to a minimum value of ~12‰ at the Carboniferous–Permian boundary (Kampschulte and Strauss, 2004; Kampschulte et al., 2001; Surakotra et al., 2018). The temporal evolution of seawater $\delta^{34}S$ is well defined for the Carboniferous in both CAS and evaporite based $\delta^{34}S$ records. Assuming

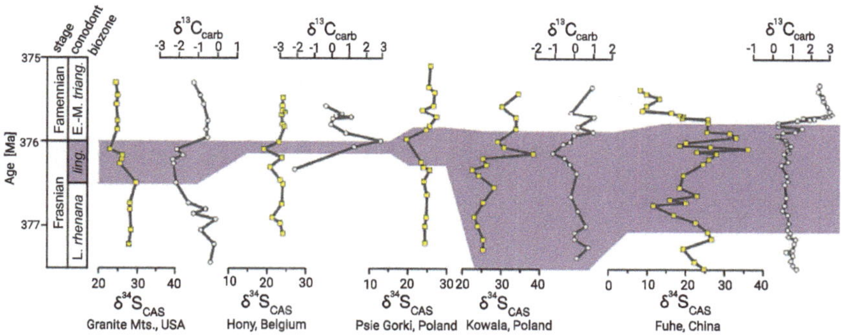

Fig. 5 Sulfur and carbon isotope records across the Fransnian–Famennian boundary. There is a brief $\delta^{34}S$ drop throughout the linguiformis biozone and a positive $\delta^{13}C$ excursion starting in the uppermost part of this biozone. The shaded area denotes the linguiformis conodont biozone. Abbreviations: L. rhenana, Late rhenana; ling., linguiformis; E.-M. triang., Early to middle triangularis. *Figure after Sim, M.S., Ono, S., Hurtgen, M.T., 2015. Sulfur isotope evidence for low and fluctuating sulfate levels in the Late Devonian Ocean and the potential link with the mass extinction event. Earth Planet. Sci. Lett. 419, 52–62.*

the available data are representative of the global ocean signature, the Carboniferous $\delta^{34}S$ data can be used for constraining the stratigraphic position of the boundaries.

3.6 Permian (299–252 Ma)

The sulfur isotope ratio of seawater sulfate persists at a relatively low value throughout the Permian with an average value of around 12‰ (Cortecci et al., 1981; Kampschulte et al., 1998; Longinelli and Flora, 2007). This plateau in the record is distinctive for the entire Permian, which is useful for dating the period as a whole but does not allow for more precise correlation within the Permian.

3.7 Triassic (252–201 Ma)

The transition from the Paleozoic to the Mesozoic is characterized by an abrupt increase in the seawater $\delta^{34}S$ value from 12‰ in the upper Permian to a maximum value of ~30‰ across the Permian–Triassic boundary (PTB; Algeo et al., 2007; Bernasconi et al., 2017; Cortecci et al., 1981; Kampschulte and Strauss, 2004; Longinelli and Flora, 2007; Luo et al., 2010; Newton et al., 2004; Schobben et al., 2017; Song et al., 2014; Worden et al., 1997). This peak value occurs at the top of the Permian–Triassic extinction (P-Tr extinction)

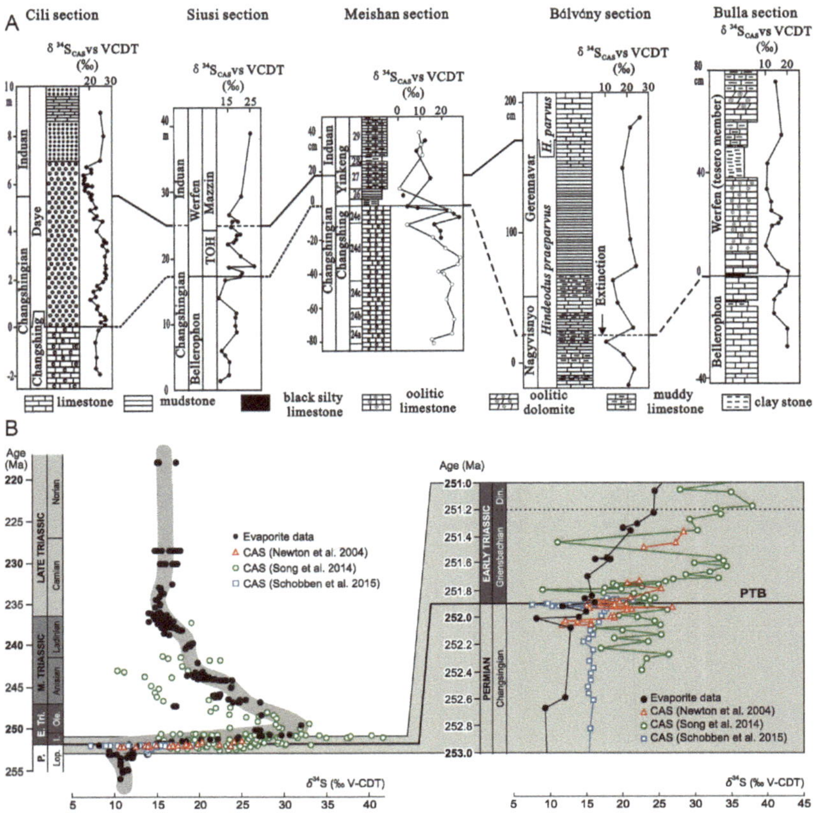

Fig. 6 (A) The CAS-based sulfur isotope records across the Permian-Triassic boundary at different sections from worldwide locations. (B) Comparison of the evaporite-based and CAS-based sulfur isotope records across the Permian–Triassic boundary. *Panel (A): After Luo, G., Kump, L.R., Wang, Y., Tong, J., Arthur, M.A., Yang, H., Huang, J., Yin, H., Xie, S., 2010. Isotopic evidence for an anomalously low oceanic sulfate concentration following end-Permian mass extinction. Earth Planet. Sci. Lett. 300, 101–111. Panel (B): After Bernasconi, S.M., Meier, I., Wohlwend, S., Brack, P., Hochuli, P.A., Bläsi, H., Wortmann, U.G., Ramseyer, K., 2017. An evaporite-based high-resolution sulfur isotope record of Late Permian and Triassic seawater sulfate. Geochim. Cosmochim. Acta 204, 331–349.*

interval followed by a sharp drop to around a mean of 17‰ in the lower and middle Triassic. These data have been sampled from worldwide locations at a temporal resolution of <1 million years (Fig. 6), indicating that the striking fluctuation is a predominant and global signal.

Previous studies interpreted such extreme changes as evidence for the development of a sizable, relatively short-lived reservoir of reduced sulfur in the deep oceanic water column followed by oceanic overturning and sulfide reoxidation (Algeo et al., 2007; Bernasconi et al., 2017; Luo et al., 2010; Newton et al., 2004). The estimated seawater sulfate concentrations were relatively low for the end Permian and the early Triassic, varying between 2 and 6 mM (Bernasconi et al., 2017). More importantly, the positive excursion of more than 10‰ over a time scale of a few million years or even less allows for robust stratigraphic correlations (e.g., Luo et al., 2010).

For the remainder of the Triassic, the seawater $\delta^{34}S$ value remains relatively constant at approximately 16‰, followed by short-term fluctuations between 11‰ and 25‰ in the uppermost Triassic. The period of distinct variations is potentially suitable for correlations (Kampschulte and Strauss, 2004).

3.8 Jurassic (201–145 Ma)

The $\delta^{34}S$ data for the Jurassic seawater sulfate cluster between 14‰ and 18.0‰ with two maxima of 23.4‰ in the lower middle Jurassic (Toarcian) and 20.7‰ in the upper middle Jurassic (Bathonian) (Claypool et al., 1980; Gill et al., 2011; Kampschulte and Strauss, 2004; Newton et al., 2011; Williford et al., 2009). Studies attributed the positive excursion in the Toarcian to the early Toarcian Oceanic Anoxic Event (T-OAE, ~183 Ma) with the spread of euxinic (i.e., anoxic and sulfidic) bottom waters and thus increases in pyrite burial (Gill et al., 2011; Jenkyns, 1988, 2010; Newton et al., 2011; Williford et al., 2009). This drastic change coincides with the widespread extinction of benthic organisms in the Northern Europe (Jenkyns, 1988). The temporal resolution of the evaporite and CAS data for the Toarcian and Pliensbachian is constrained on the sub-million year scale providing more precise information of seawater $\delta^{34}S$ variations, which could be used for stratigraphy. However, for the rest of the Jurassic the overall age uncertainty is relatively large, and more data are required to show finer $\delta^{34}S$ changes.

3.9 Cretaceous (145–65 Ma)

The barite-based data for the Cretaceous are more robust compared to the evaporite and CAS records, illuminating finer features of $\delta^{34}S$ variations (DeBond et al., 2012; Paytan et al., 2004). The lower Cretaceous is characterized by very pronounced fluctuations with the seawater $\delta^{34}S$ value falling from 20‰ to 15‰ between 130 and 120 Ma (Berremian and lower Aptian) and remaining at around 16‰ until 104 Ma (middle-upper Albian). This low value is followed by an abrupt rise to 19‰ over the next 10 million years, defining the boundary of the lower and upper Cretaceous (Albian and Cenomanian). There is also a subsequent decline to a minimum value of 18‰ at 87 Ma (Cenomanian, Turonian, and Coniacian) before returning to the pre-excursion value of 19‰ at about 80 Ma (Santonian and lower Campanian), which remains relatively constant for the remainder of the Cretaceous (upper Campanian and Maastrichtian) (Fig. 3). Both negative excursions (130–120 Ma and 80–87 Ma) occur on relatively short time scales, which are in line with the expected shorter-than-present residence time of sulfate in the ocean (5–10 million years) due to the lower seawater sulfate concentration in the Cretaceous (Horita et al., 2002).

The Cretaceous barite data are sampled at a resolution of ~1 million years, providing tight constraints on the timing and magnitude of the $\delta^{34}S$ excursions that could be used for refined stratigraphic correlations. The Cretaceous ocean is characterized by numerous oceanic anoxic events, among which the three major ones occurred in the early Aptian (Selli event, OAE 1a, ~120 Ma), the early Albian (Paquier event, OAE 1b, ~111 Ma), the Cenomanian-Turonian boundary (C/T OAE, or OAE 2, ~93 Ma) (Jenkyns, 2010; Owens et al., 2013). These OAEs are accompanied with widespread marine organic-carbon burial and the existence of an equable global climate and correlate closely with transgressions which can be seen throughout the stratigraphical column (e.g., Jenkyns, 1980; Jenkyns, 2010; Owens et al., 2013; Schlanger and Jenkyns, 1976; Tedeschi et al., 2017).

Indeed, many studies attempted to link the sulfur and carbon isotope records to investigate the short-term perturbations of the biogeochemical cycles and OAEs in the Cretaceous (e.g., DeBond et al., 2012; Hansen and Wallman, 2003; Jenkyns, 2018; Owens et al., 2013; Wortmann and Chernyavsky, 2007; Wortmann and Paytan, 2012). For instance, Paytan et al. (2004) attributed the drastic decrease in $\delta^{34}S$ between 130 and 120 Ma (pre-OAE 1a) to lower rates of pyrite burial resulted from a shift in the location of organic carbon burial to terrestrial or open ocean settings where reactive iron is limited. Wortmann and Chernyavsky (2007) proposed that the reduction of microbially mediated pyrite burial and organic matter remineralization could also be a result of the temporary but drastic changes in the seawater sulfate concentration. On the other hand, the positive S-isotope excursion (roughly parallel the C-isotope excursion) across the Cenomanian–Turonian boundary (OAE 2) is likely due to the transiently increased pyrite burial under euxinic conditions (Owens et al., 2013). Compared to the modern ocean, the Cretaceous ocean is generally enriched in isotopically light sulfur, which could be a result of increased volcanic and hydrothermal activities and slightly higher weathering rates (Hansen and Wallman, 2003; Jenkyns, 2010; Paytan et al., 2004) or lower organic matter and pyrite burial rates (Wortmann and Paytan, 2012).

3.10 Cenozoic (65–0 Ma)

The age resolution for the sulfur isotope curve for the Cenozoic record is much higher compared to that of the other geological periods. The most prominent features of the Cenozoic sulfur isotope record are a ~2‰ decrease during the Paleocene and a pronounced 5‰ increase in the early to middle Eocene (Markovic et al., 2015; Paytan et al., 1998; Rennie et al., 2018; Yao et al., 2019, under review). In the previous barite record, the Eocene rise of seawater $\delta^{34}S$ is defined by only a few samples from Deep Sea Drilling Project (DSDP) Site 366 (Paytan et al., 1998), where the biostratigraphy is not well constrained (Lancelot et al., 1977; Langton et al., 2016). In addition, the decreasing porewater sulfate concentrations with depth, generally higher sedimentation rates (29–41.5 m/Myr), and observable pyrite occurrences at Site 366 throughout the middle to lower Eocene sections (38–56 Ma) imply an organic-rich and reducing environment during this time (Boersma and Shackleton, 1977; Couture et al., 1977; Lancelot et al., 1977), which suggest that the barite in that section could have been diagenetically altered.

Taking advantage of more recently retrieved cores and a much improved biostratigraphic framework, Yao et al. (2019, under review) recently evaluated and refined the Eocene $\delta^{34}S$ data with a new high-resolution barite-based $\delta^{34}S$ record between 60 and 30 Ma. They showed anomalously high $^{87}Sr/^{86}Sr$ ratios of Site 366 barites older than 38 Ma, indicating that the local conditions at Site 366 during the Eocene allowed for sulfate reduction and the formation of diagenetic barite. Hence, these samples should be excluded from the barite record. With a temporal resolution better than 0.5 million years, their new data from Ocean Drilling Program (ODP) Leg 198 Site 1211, ODP Leg 199 Site 1219, and Integrated Ocean Drilling Program (IODP) Expedition 320 Site U1333 now suggest a slower rise in seawater $\delta^{34}S$ (Fig. 7). Following the Cretaceous/Paleogene boundary,

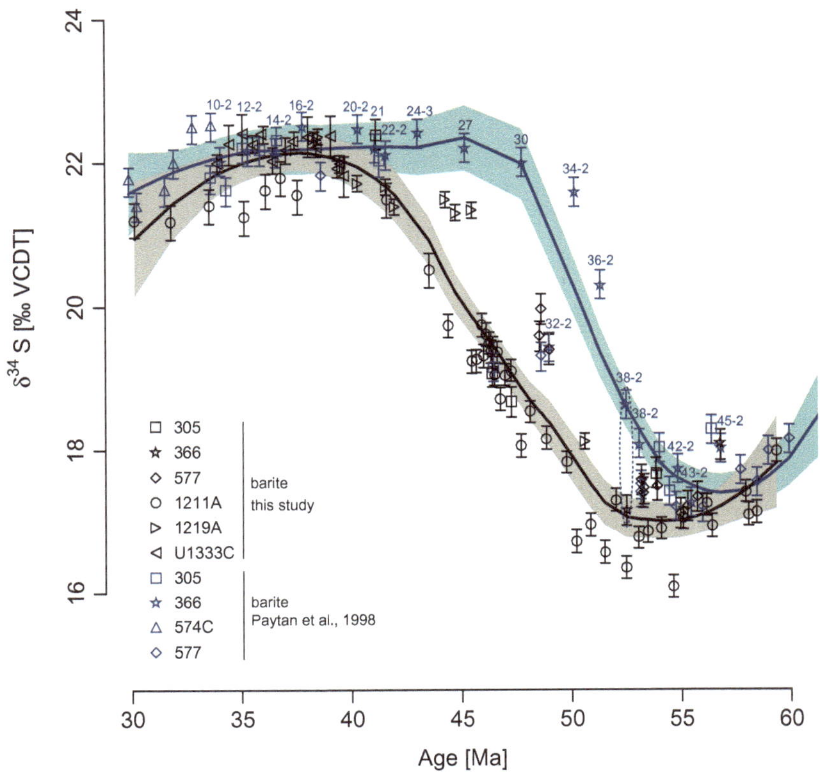

Fig. 7 The refined barite record (gray) versus the previous barite record (blue) between 60 and 30 Ma. Black and blue symbols denote measurements from this study and from Paytan et al. (1998), respectively: DSDP Site 305 (squares), Site 366 (pentagrams), Site 574 Hole C (upward-pointing triangles), Site 577 (diamonds), ODP Site 1211 Hole A (circles), Site 1219 Hole A (right-pointing triangles), and IODP Site U1333 Hole C (left-pointing triangles). Barite from DSDP Site 366 is labeled as core and section number (e.g., 32-2). Error bars are 1σ. The gray and the blue envelops denote the 95% confidence intervals of the LOESS regressions for the respective data set. Data for all the sites are mapped onto the GTS2012 timescale (Gradstein et al., 2012; Lazarus, 1994; Spencer-Cervato, 1999). *After Paytan, A., Kastner, M., Campbell, D., Thiemens, M.H., 1998. Sulfur isotopic composition of Cenozoic seawater sulfate. Science 282, 1459–1462.; Yao, W., Paytan, A., Griffith, E.M., Martínez-Ruiz, F., Markovic, S., Wortmann, U.G., 2019, under review. A revised seawater sulfate S-isotope curve for the Eocene. Chem. Geol.*

the seawater δ^{34}S drops precipitously from 19‰ to ~17‰ between 65 and 53 Ma and then increases to a peak value of 22.4‰ at ~40 Ma. This peak value remains relatively constant over the next 30 million years until the 1‰ decline in the Quaternary (Markovic et al., 2015; Paytan et al., 1998).

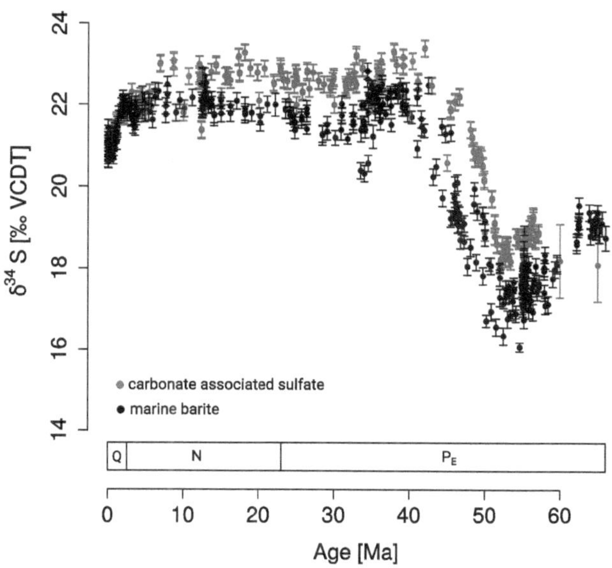

Fig. 8 The refined barite record (black) versus the carbonate-associated sulfate record (green) for the Cenozoic. Error bars are 1σ. All the data are mapped onto the GTS2012 timescale (Gradstein et al., 2012; Lazarus, 1994; Spencer-Cervato, 1999). *After Paytan, A., Kastner, M., Campbell, D., Thiemens, M.H., 1998. Sulfur isotopic composition of Cenozoic seawater sulfate. Science 282, 1459–1462; Turchyn, A.V., Schrag, D.P., Coccioni, R., Montanari, A., 2009. Stable isotope analysis of the Cretaceous sulfur cycle. Earth Planet. Sci. Lett. 285, 115–123; Markovic, S., Paytan, A., Wortmann, U.G., 2015. Pleistocene sediment offloading and the global sulfur cycle. Biogeosciences 12, 3043–3060; Markovic, S., Paytan, A., Li, H., Wortmann, U.G., 2016. A revised seawater sulfate oxygen isotope record for the last 4 Myr. Geochim. Cosmochim. Acta 175, 239–251; Masterson, A.L., Wing, B.A., Paytan, A., Farquhar, J., Johnston, D.T., 2016. The minor sulfur isotope composition of Cretaceous and Cenozoic seawater sulfate. Paleoceanography 31, 779–788; Yao, W., Paytan, A., Wortmann, U.G., 2018. Large-scale ocean deoxygenation during the Paleocene–Eocene thermal maximum. Science 361, 804–806; Rennie, V.C.F., Paris, G., Sessions, A.L., Abramovich, A., Turchyn, A.V., Adkins, J.F., 2018. Cenozoic record of δ^{34}S in foraminiferal calcite implies an early Eocene shift to deep-ocean sulfide burial. Nat. Geosci. 11, 761–765; Yao, W., Paytan, A., Griffith, E.M., Martínez-Ruiz, F., Markovic, S., Wortmann, U.G., 2019, under review. A revised seawater sulfate S-isotope curve for the Eocene. Chem. Geol.*

The refined barite data (Fig. 8, black symbols) faithfully track the recently published species-adjusted CAS record (Fig. 8, green symbols; Rennie et al., 2018) with respect to the magnitude and duration of the excursion. More importantly, the slope of the refined S isotope curve (~0.4‰/Myr) is now consistent for barite and CAS. On the other hand, the previously observed 1‰ offset between CAS and barite data (Rennie et al., 2018), now extends past the Miocene well into the Paleocene (Fig. 8). This observation is intriguing as it strengthens the case for a systematic difference between the CAS and barite archives. One possibility for the offset is that the formation of CAS may involve species-specific S-isotope fractionation (e.g., Rennie et al., 2018; Rennie and Turchyn, 2014). The fractionation from seawater for both phases (if exists) needs to be further examined via laboratory precipitation experiments using isotope spike solutions of known composition. Nevertheless, the consistent and finer trends reported in both data sets from worldwide sites indicate that such S-isotope variations are primary and, most importantly, global, which is particularly useful for stratigraphic correlations.

It is noteworthy that recent work on marine barite has also generated high-resolution $\delta^{34}S$ data for the Paleocene–Eocene Thermal Maximum (PETM; ~55 Ma) from Ocean Drilling Program (ODP) Leg 199 Hole 1221C (Core 11X-3) in the equatorial Pacific, and ODP Leg 208 Hole 1263C (Core 14H-2 and 14H-CC) and Hole 1265A (Core 29H-6 and 29H-7) in the South Atlantic (Yao et al., 2018). The PETM is characterized by a transient positive 1‰ excursion in seawater $\delta^{34}S$ within 100 thousand years (kyrs) at the onset of the Paleocene Eocene boundary. The seawater $\delta^{34}S$ value increased from 17.5‰ to a peak value of ~18.3‰ (except for one extremely high value of 18.9‰) during the main-PETM interval (~50 kyrs) and returned over the next ~50 kyrs to its pre-excursion value. The $\delta^{34}S$ peak precedes the most negative $\delta^{13}C$ excursion by about 10 kyrs (Fig. 9).

Although the excursion appears small, it takes place within a stratigraphic length of 90 cm at ODP Hole 1221A and 200 cm at Hole 1265A, equivalent to the total PETM interval of 230 kyrs. Such rapid variations in seawater $\delta^{34}S$ indicate a much more dynamic biogeochemical sulfur cycle, where a transient sulfide reservoir such as sulfur cycling within the oxygen minimum zone was sufficiently large to have released substantial amounts of reduced sulfur to the rest of the ocean (Yao et al., 2018). The transient positive $\delta^{34}S$ excursion, along with the negative $\delta^{13}C$ excursion, could be indicative of the onset, main, and recovery stages of the PETM. Furthermore, widespread oxygen loss and the toxicity of hydrogen sulfide could have rendered the marine ecosystem uninhabitable by many marine species (Yao et al., 2018), which may be associated with the Paleocene–Eocene benthic foraminiferal extinction.

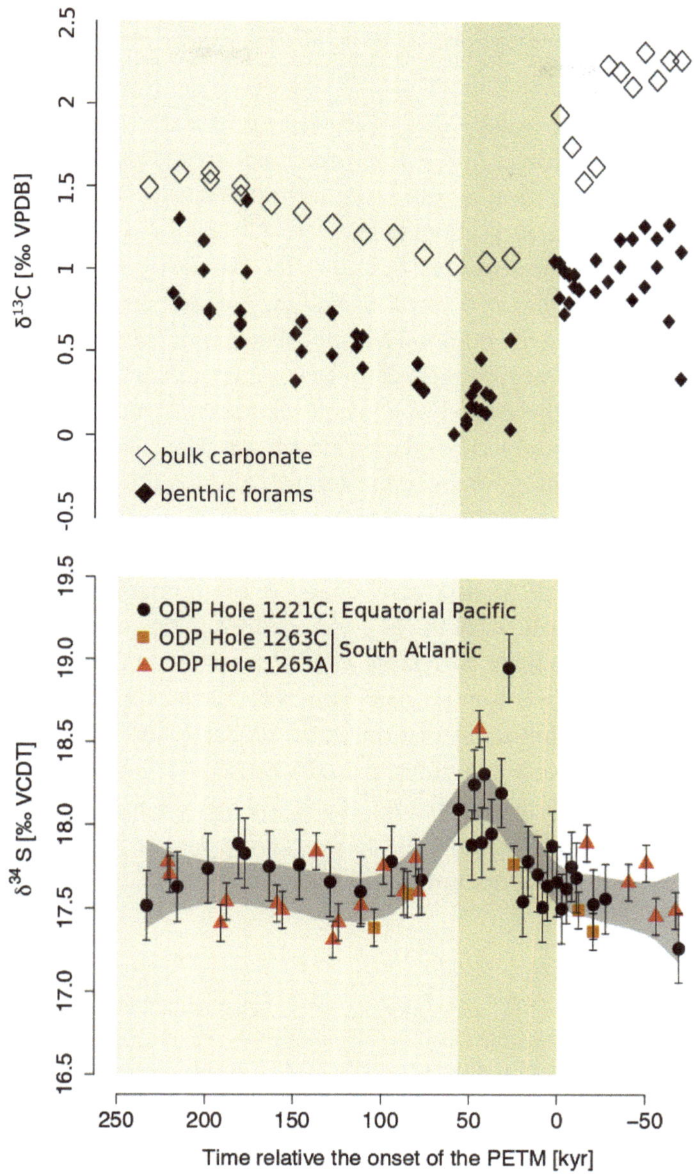

Fig. 9 The sulfur and carbon isotope records across the PETM. Open and solid diamonds denote the $\delta^{13}C$ data derived from bulk carbonate and benthic foraminifera from ODP Hole 1221A (Nunes and Norris, 2005). Black circles, yellow squares and red triangles denote the barite-based seawater $\delta^{34}S$ data (1σ) from ODP Hole 1221A, 1263C, and 1265A (Yao et al., 2018). The gray envelope denotes the 95% confidence interval of the LOESS regression for the total $\delta^{34}S$ data. Ages and the PETM stages (shaded boxes) as defined by Nunes and Norris (2005).

4. Summary

We compile S-isotope data of seawater sulfate derived from different archives for the Protozoic and Phanerozoic and assess possible alterations of the signals to constrain the secular seawater sulfur isotope curve. Long-term trends and short-term fluctuations of seawater $\delta^{34}S$ values have been observed. Correlations using the seawater S-isotope records for strata older than the Carboniferous are difficult due to uncertainties of samples ages (>5 million years), diagenesis, and the questionable global nature of seawater sulfate sulfur isotope value. The S-isotope records for ages younger than Carboniferous can be used for dating the period as a whole or for constraining positions for some specific stratigraphic boundaries, which are particularly useful for the sections with deficient biostratigraphic controls. During times of rapid perturbations (e.g., the Eocene, PETM, Permian–Triassic boundary), much higher temporal resolution records ($\geq$0.5 million years) are identified primarily based on marine barite and CAS data, providing precise information of the magnitude, duration and gradient of the $\delta^{34}S$ excursions and extending the utility of this system to stratigraphic correlations. While expanding the S-isotope data available to seawater S-isotope reconstruction allows unprecedented new insights into high-resolution variations in the global sulfur cycle, it also draws attention to newly raised questions (e.g., the constant 1‰ offset between the Cenozoic CAS and marine barite). Experiments are further required to explore the effects of changing microbial ecology and redox conditions on biomineralization of sulfur-bearing minerals.

Acknowledgments

This work was supported by the National Science Foundation (NSF) CAREER grant OCE-0449732 to A. P. and the Discovery Grant of the Natural Sciences and Engineering Research Council of Canada (NSERC) to U. G. W.

References

Algeo, J., Ellwood, B.B., Nguyen, T.K.T., Rowe, H., Maynard, J.B., 2007. The Permian–Triassic boundary at Nhi Tao, Vietnam: evidence for recurrent influx of sulfidic watermasses to a shallow-marine carbonate platform. Palaeogeogr. Palaeoclimatol. Palaeoecol. 252, 304–327.

Bernasconi, S.M., Meier, I., Wohlwend, S., Brack, P., Hochuli, P.A., Bläsi, H., Wortmann, U.G., Ramseyer, K., 2017. An evaporite-based high-resolution sulfur isotope record of Late Permian and Triassic seawater sulfate. Geochim. Cosmochim. Acta 204, 331–349.

Berner, R.A., 1982. Burial of organic-carbon and pyrite sulfur in the modern ocean—its geochemical and environmental significance. Am. J. Sci. 282, 451–473.

Berner, R.A., 1989. Biogeochemical cycles of carbon and sulfur and their effect on atmospheric oxygen over phanerozoic time. Palaeogeogr. Palaeoclimatol. Palaeoecol. 75, 97–112.

Berner, R.A., Raiswell, R., 1983. Burial of organic carbon and pyrite sulfur in sediments over Phanerozoic time: a new theory. Geochim. Cosmochim. Acta 47, 855–862.

Bishop, J.K.B., 1988. The barite-opal-organic carbon association in oceanic particulate matter. Nature 331, 341–343.

Blättler, C.L., et al., 2018. Two-billion-year-old evaporites capture Earth's great oxidation. Science 360, 320–323.

Boersma, A., Shackleton, N., 1977. 36. Oxygen and carbon isotope record through the Oligocene, DSDP Site 366, equatorial Atlantic. In: Initial Reports of the Deep Sea Drilling Project. vol. 41. Texas A & M University, Ocean Drilling Program, College Station, TX, pp. 957–962.

Bottrell, S.H., Newton, R.J., 2006. Reconstruction of changes in global sulfur cycling from marine sulfate isotopes. Earth Sci. Rev. 75, 59–83.

Brunner, B., Bernasconi, S.M., 2005. A revised isotope fractionation model for dissimilatory sulfate reduction in sulfate reducing bacteria. Geochim. Cosmochim. Acta 69, 4759–4771.

Burdett, J.W., Arthur, M.A., Richardson, M., 1989. A Neogene seawater sulfur isotope age curve from calcareous pelagic microfossils. Earth Planet. Sci. Lett. 94, 189–198.

Canfield, D.E., 1998. A new model for Proterozoic ocean chemistry. Nature 396, 450–453.

Canfield, D.E., 2005. The early history of atmospheric oxygen: homage to Robert M. Garrels. Annu. Rev. Earth Planet. Sci. 33, 1–36.

Canfield, D.E., Farquhar, J., 2009. Animal evolution, bioturbation, and the sulfate concentration of the oceans. Proc. Natl. Acad. Sci. U. S. A. 106, 8123–8127.

Canfield, D.E., Farquhar, J., Zerkle, A.L., 2010a. High isotope fractionations during sulfate reduction in a low-sulfate euxinic ocean analog. Geology 38, 415–418.

Canfield, D.E., Glazer, A.N., Falkowski, P.G., 2010b. The evolution and future of Earth's nitrogen cycle. Science 330, 192–196.

Chambers, L.A., Trudinger, P.A., 1979. Microbiological fractionation of stable sulfur isotopes: a review and critique. Geomicrobiol J. 1, 249–293.

Claypool, G.E., Holser, W.T., Kaplan, I.R., Sakai, H., Zak, I., 1980. The age curves of sulfur and oxygen isotopes in marine sulfate and their mutual interpretation. Chem. Geol. 28, 199–260.

Cortecci, G., Reyes, E., Berti, G., Casati, P., 1981. Sulfur and oxygen isotopes in Italian marine sulfates of Permian and Triassic ages. Chem. Geol. 34, 65–79.

Couture, R., Miller, R.S., Gieskes, J.M., 1977. 32. Interstitial water and mineralogical studies, Leg 41. In: Gardner, J., Herring, J. (Eds.), Initial Reports of the Deep Sea Drilling Project. vol. 41, pp. 907–914. United States Government Printing Office, Washington, D.C.

Crockford, P.W., et al., 2019. Claypool continued: extending the isotopic record of sedimentary sulfate. Chem. Geol. 513, 200–225.

Das, N., Horita, J., Holland, H.D., 1990. Chemistry of fluid inclusions in halite from the Salina Group of the Michigan Basin: implications for Late Silurian seawater and the origin of sedimentary brines: Geochim. Cosmochim. Acta 54, 319–327.

Dean, W.E., Schreiber, B.C., 1977. Authigenic barite. In: Gardner, J., Herring, J. (Eds.), Initial Reports of Deep Sea Drilling Project. vol. 41. Government Printing Office, Washington, D.C., pp. 915–931.

DeBond, N., Oakes, R.L., Paytan, A., Wortmann, U.G., 2012. Early Aptian carbon and sulphur isotope signatures at ODP Site 765. Isotopes Environ. Health Stud. 48, 180–194.

Dehairs, F., Goeyens, L., Stroobants, N., Bernard, P., Goyet, C., Poisson, A., Chesselet, R., 1990. On the suspended barite and the oxygen minimum in the Southern Ocean. Global Biogeochem. Cycles 4, 85–102.

Dymond, J., Suess, E., Lyle, M., 1992. Barium in deep-sea sediments: a geochemical proxy for paleoproductivity. Paleoceanography 7, 163–181.

Eckert, T., Brunner, B., Edwards, E.A., Wortmann, U.G., 2011. Microbially mediated re-oxidation of sulfide during dissimilatory sulfate reduction by desulfobacter latus. Geochim. Cosmochim. Acta 75, 3469–3485.

Edmond, J.M., Measures, C., McDuff, R.M., Chan, L.H., Collier, R., Grant, B., Gordon, L.J., Corliss, J.B., 1979. Ridge crest hydrothermal activity and the balances of the major and minor elements in the ocean: the Galapagos data. Earth Planet. Sci. Lett. 46, 1–18.

Farquhar, J., Wing, B., 2003. Multiple sulfur isotopes and the evolution of the Earth's atmosphere. Earth Planet. Sci. Lett. 213, 1–13.

Farquhar, J., Bao, H., Thiemens, M., 2000. Atmospheric influence of Earth's earliest sulfur cycle. Science 289, 756–758.

Fox, J.S., Videtich, P.E., 1997. Revised estimate of $\delta^{34}S$ for marine sulfates from the Upper Ordovician: data from the Williston Basin, North Dakota, USA. Appl. Geochem. 12, 97–103.

Ganeshram, R.S., Francois, R., Commeau, J., Brown-Leger, S.L., 2003. An experimental investigation of Barite formation in seawater. Geochim. Cosmochim. Acta 67, 2599–2605.

Garrels, R.M., Lerman, A., 1984. Couling of the sedimentary sulfur and carbon cycles—an improved model. Am. J. Sci. 284, 989–1007.

Gill, B.C., Lyons, T.W., Saltzman, M.R., 2007. Parallel, high-resolution carbon and sulfur isotope records of the evolving Paleozoic marine sulfur reservoir. Palaeogeogr. Palaeoclimatol. Palaeoecol. 256, 156–173.

Gill, B.C., Lyons, T.W., Jenkyns, H.C., 2011. A global perturbation to the sulfur cycle during the Toarcian Oceanic Anoxic Event. Earth Planet. Sci. Lett. 312, 484–496.

Goldberg, T., Poulton, S.W., Strauss, H., 2005. Sulphur and oxygen isotope signatures of late Neoproterozoic to early Cambrian sulphate, Yangtze Platform, China: diagenetic constraints and seawater evolution. Precambrian Res. 137, 223–241.

Gradstein, F.M., Ogg, J.G., Schmitz, M., Ogg, G., 2012. The Geologic Time Scale 2012. Elsevier B.V, p. 1176.

Habicht, K.S., Gade, M., Thamdrup, B., Berg, P., Canfield, D.E., 2002. Calibration of sulfate levels in the Archean Ocean. Science 298, 2372–2374.

Hansen, K.W., Wallman, K., 2003. Cretaceous and Cenozoic evolution of seawater composition, atmospheric O_2 and CO_2—a model perspective. Am. J. Sci. 303, 94–148.

Holler, T., Wegener, G., Niemann, H., Deusner, C., Ferdelman, T.G., Boetius, A., Brunner, B., Widdel, F., 2011. Carbon and sulfur back flux during anaerobic microbial oxidation of methane and coupled sulfate reduction. Proc. Natl. Acad. Sci. U. S. A. 108, 1484–1490.

Holser, W.T., Kaplan, I.R., 1966. Isotope geochemistry of sedimentary sulfates. Chem. Geol. 1, 93–135.

Horita, J., Zimmermann, H., Holland, H.D., 2002. Chemical evolution of seawater during the Phanerozoic: implications from the record of marine evaporites. Geochim. Cosmochim. Acta 66, 3733–3756.

Hough, M.L., Shields, G.A., Evins, L.Z., Strauss, H., Henderson, R.A., Mackenzie, S., 2006. A major sulphur isotope event at c. 510 Ma: a possible anoxia-extinction-volcanism connection during the Early–Middle Cambrian transition? Terra Nova 18, 257–263.

Hurtgen, M.T., 2012. The marine sulfur cycle, revisited. Science 337, 305–306.

Hurtgen, M.T., Arthur, M.A., Suits, N.S., Kaufman, A.J., 2002. The sulfur isotopic composition of Neoproterozoic seawater sulfate: implications for a snowball Earth? Earth Planet. Sci. Lett. 203, 413–429.

Hurtgen, M.T., Arthur, M.A., Halverson, G.P., 2005. Neoproterozoic sulfur isotopes, the evolution of microbial sulfur species, and the burial efficiency of sulfide as sedimentary pyrite. Geology 33, 41–44.
Hurtgen, M.T., Pruss, S.B., Knoll, A.H., 2009. Evaluating the relationship between the carbon and sulfur cycles in the later Cambrian ocean: an example from the Port au Port Group, western Newfoundland, Canada. Earth Planet. Sci. Lett. 281, 288–297.
Jenkyns, H.C., 1980. Cretaceous anoxic events: from continents to oceans. J. Geol. Soc. London 137, 171–188.
Jenkyns, H.C., 1988. The early Toarcian (Jurassic) anoxic event—stratigraphic, sedimentary, and geochemical evidence. Am. J. Sci. 288, 101–151.
Jenkyns, H.C., 2010. Geochemistry of oceanic anoxic events. Geochem. Geophys. Geosyst. 11, Q03004.
Jenkyns, H.C., 2018. Transient cooling episodes during Cretaceous Oceanic Anoxic Events with special reference to OAE 1a (Early Aptian). Phil. Trans. R. Soc. A 376, 20170073.
Johnston, D.T., 2011. Multiple sulfur isotopes and the evolution of Earth's surface sulfur cycle. Earth Sci. Rev. 106, 161–183.
Jørgensen, B.B., 1982. Mineralization of organic-matter in the sea bed—the role of sulfate reduction. Nature 296, 643–645.
Jørgensen, B.B., 1983. Processes at the sediment-water interface. In: Bolin, B., Cook, R.B. (Eds.), The Major Biogeochemical Cycles and Their Interactions. Scientific Committee on Problems of the Environment, Stockholm, Sweden, pp. 477–515.
Jørgensen, B.B., Kasten, S., 2006. Sulfur cycling and methane oxidation. In: Schulz, H.D., Zabel, M. (Eds.), Marine Geochemistry, second ed. Springer, Berlin, Germany, pp. 271–309.
Kah, L.C., Lyons, T.W., Frank, T.D., 2004. Low marine sulphate and protracted oxygenation of the Proterozoic biospheres. Nature 431, 834–838.
Kah, L.C., Thompson, C.K., Henderson, M.A., Zhan, R., 2016. Behavior of marine sulfur in the Ordovician. Palaeogeogr. Palaeoclimatol. Palaeoecol. 458, 133–153.
Kampschulte, A., Strauss, H., 2004. The sulfur isotopic evolution of Phanerozoic seawater based on the analysis of structurally substituted sulfate in carbonates. Chem. Geol. 204, 255–286.
Kampschulte, A., Buhl, D., Strauss, H., 1998. The sulfur and strontium isotopic compositions of Permian evaporites from the Zechstein basin, northern Germany. Geogr. Rundsch. 87, 192–199.
Kampschulte, A., Bruckschen, P., Strauss, H., 2001. The sulphur isotopic composition of trace sulphates in Carboniferous brachiopods: implications for coeval seawater, correlation with other geochemical cycles and isotope stratigraphy. Chem. Geol. 175, 149–173.
Kaplan, I.R., Emery, K.O., Rittenberg, S.C., 1963. The distribution and isotopic abundance of sulphur in recent marine sediments off southern California. Geochim. Cosmochim. Acta 27, 297–331.
Kump, L.R., Garrels, R.M., 1986. Modeling atmospheric O_2 in the global sedimentary redox cycle. Am. J. Sci. 286, 337–360.
Kurtz, A.C., Kump, L.R., Arthur, M.A., Zachos, J., Paytan, A., 2003. Early Cenozoic decoupling of the global carbon and sulfur cycles. Paleoceanography 18, 1090–1103.
Kusakabe, M., Robinson, B.W., 1977. Oxygen and sulfur isotope equilibria in the $BaSO_4$-HSO_4-H_2O system from 110 to 350°C and applications. Geochim. Cosmochim. Acta 41, 1033–1040.
Lancelot, Y., Seibold, E., Cepek, P., Dean, W.E., Eremeev, V., Gardner, J., et al., 1977. Site 366: Sierra Leone Rise. In: Initial Reports of the Deep Sea Drilling Project. vol. 41. United States Government Printing Office, United States, WA, pp. 121–161.

Langton, S.J., Rabideaux, N.M., Borrelli, C., Katz, M.E., 2016. South-eastern Atlantic deep-water evolution during the late-middle Eocene to earliest Oligocene (Ocean Drilling Program Site 1263 and Deep Sea Drilling Project Site 366). Geosphere 12, 103–1047.
Lazarus, D.B., 1994. Neptune: a marine micropaleontology database. Math. Geol. 26, 817–832.
Longinelli, A., Flora, O., 2007. Isotopic composition of gypsum samples of Permian and Triassic age from the north–eastern Italian Alps: palaeoenvironmental implications. Chem. Geol. 245, 275–284.
Lowenstein, T.K., Hardie, L.A., Timofeeff, M.N., Demicco, R.V., 2003. Secular variation in seawater chemistry and the origin of calcium chloride basinal brines. Geology 31, 857–860.
Luo, G., Kump, L.R., Wang, Y., Tong, J., Arthur, M.A., Yang, H., Huang, J., Yin, H., Xie, S., 2010. Isotopic evidence for an anomalously low oceanic sulfate concentration following end-Permian mass extinction. Earth Planet. Sci. Lett. 300, 101–111.
Marenco, P.J., Corsetti, D.H., Hammond, F.A., Kaufman, A.J., Bottjer, D.J., 2008. Oxidation of pyrite during extraction of carbonate associated sulfate. Chem. Geol. 247, 124–132.
Markovic, S., Paytan, A., Wortmann, U.G., 2015. Pleistocene sediment offloading and the global sulfur cycle. Biogeosciences 12, 3043–3060.
Markovic, S., Paytan, A., Li, H., Wortmann, U.G., 2016. A revised seawater sulfate oxygen isotope record for the last 4 Myr. Geochim. Cosmochim. Acta 175, 239–251.
Masterson, A.L., Wing, B.A., Paytan, A., Farquhar, J., Johnston, D.T., 2016. The minor sulfur isotope composition of Cretaceous and Cenozoic seawater sulfate. Paleoceanography 31, 779–788.
Mazumdar, A., Strauss, H., 2006. Sulfur and strontium isotopic compositions of carbonate and evaporite rocks from the late Neoproterozoic–early Cambrian Bilara Group (Nagaur-Ganganagar Basin, India): constraints on intrabasinal correlation and global sulfur cycle. Precambrian Res. 149, 217–230.
McDonough, W.F., 2000. The composition of the Earth. In: Earthquake Thermodynamics and Phase Transformations in the Earth's Interior. Elsevier, pp. 1–674.
Meija, J., Coplen, T.B., Berglund, M., Brand, W.A., De Bièvre, P., Gröning, M., Holden, N.E., Irrgeher, J., Loss, R.D., Walczyk, T., Prohaska, T., 2016. Atomic weights of the elements 2013 (IUPAC Technical Report). Pure Appl. Chem. 88, 265–291.
Newton, R.J., Pevitt, E.L., Wignall, P.B., Bottrell, S.H., 2004. Large shifts in the isotopic composition of seawater sulphate across the Permo–Triassic boundary in northern Italy. Earth Planet. Sci. Lett. 218, 331–345.
Newton, R.J., Reeves, E.P., Kafousia, N., Wignall, P.B., Bottrell, S.H., Sha, J.-G., 2011. Low marine sulfate concentrations and the isolation of the European epicontinental sea during the Early Jurassic. Geology 39, 7–10.
Nunes, F., Norris, R.D., 2005. Data report: high-resolution stable isotope records across the Paleocene/Eocene boundary, ODP Sites 1220 and 1221. In: Wilson, P.A., Lyle, M., Firth, J.V. (Eds.), Proceedings of the Ocean Drilling Program, Scientific Results. vol. 199, pp. 1–12. Texas A&M University, Ocean Drilling Program, College Station, TX.
Orti, F., Rosell, L., Anadón, P., 2010. Diagenetic gypsum related to sulfur deposits in evaporites (Libros Gypsum, Miocene, NE Spain). Sediment. Geol. 228, 304–318.
Owens, J.D., Gill, B.C., Jenkyns, H.C., Bates, S.M., Severmann, S., Kuypers, M.M.M., Woodfine, R.G., Lyons, T.W., 2013. Sulfur isotopes track the global extent and dynamics of euxinia during Cretaceous Oceanic Anoxic Event 2. Proc. Natl. Acad. Sci. 110, 18407–18412.

Paris, G., Fehrenbacher, J.S., Sessions, A.L., Spero, H.J., Adkins, J.F., 2014. Experimental determination of carbonate-associated sulfate $\delta^{34}S$ in planktonic foraminifera shells. Geochem. Geophys. Geosyst. 15, 1452–1461.

Pavlov, A.A., Kasting, J.F., 2002. Mass-independent fractionation of sulfur isotopes in Archean sediments: strong evidence for an anoxic Archean atmosphere. Astrobiology 2, 27–41.

Paytan, A., Kastner, M., Martin, E.E., Macdougall, J.D., Herbert, T., 1993. Marine barite as a monitor of seawater strontium isotope composition. Nature 366, 445–449.

Paytan, A., Kastner, M., Campbell, D., Thiemens, M.H., 1998. Sulfur isotopic composition of Cenozoic seawater sulfate. Science 282, 1459–1462.

Paytan, A., Mearon, S., Cobb, K., Kastner, M., 2002. Origin of marine barite deposits: Sr and S isotope characterization. Geology 30, 747–750.

Paytan, A., Kastner, M., Campbell, D., Thiemens, M.H., 2004. Seawater sulfur isotope fluctuations in the cretaceous. Science 304, 1663–1665.

Paytan, A., Gray, E.T., Ma, Z., Erhardt, A., Faul, A., 2011. Application of sulphur isotopes for stratigraphic correlation. Isotopes Environ. Health Stud. 48, 195–206.

Peryt, T.M., Halas, S., Kovalevych, V.M., Petrychenko, Y., Dzhinoridze, N.M., 2005. The sulphur and oxygen isotopic composition of Lower Cambrian anhydrites in East Siberia. Geol. Q. 49, 235–242.

Petsch, S.T., Berner, R.A., 1998. Coupling the geochemical cycles of C, P, Fe, and S: the effect on atmospheric O_2 and the isotopic records of carbon and sulfur. Am. J. Sci. 298, 246–262.

Pierre, C., Rouchy, J.M., 1986. Oxygen and sulfur isotopes in anhydrites from Givetian and Visean evaporites of Northern France and Belgium. Chem. Geol. Isot. Geosci. Sect. 58, 245–252.

Planavsky, N.J., Bekker, A., Hofmann, A., Owens, J.D., Lyons, T.W., 2012. Sulfur record of rising and falling marine oxygen and sulfate levels during the Lomagundi event. Proc. Natl. Acad. Sci. 109, 18300–18305.

Present, T.M., Paris, G., Burke, A., Fischer, W.W., Adkins, J.F., 2015. Large carbonate associated sulfate isotopic variability between brachiopods, micrite, and other sedimentary components in Late Ordovician strata. Earth Planet. Sci. Lett. 432, 187–198.

Raab, M., Spiro, B., 1991. Sulfur isotopic variations during seawater evaporation with fractional crystallization. Chem. Geol. 86, 323–333.

Reeburgh, W.S., 1983. Rates of biogeochemical processes in anoxic sediments. Annu. Rev. Earth Planet. Sci. 11, 269–298.

Rees, C.E., 1973. A steady-state model for sulphur isotope fractionation in bacterial reduction processes. Geochim. Cosmochim. Acta 37, 1141–1162.

Rennie, V.C.F., Turchyn, A.V., 2014. The preservation of $\delta^{34}S_{so4}$ and $\delta^{18}O_{so4}$ in carbonate-associated sulfate during marine diagenesis; a 25 Myr test case using marine sediments. Earth Planet. Sci. Lett. 395, 13–23.

Rennie, V.C.F., Paris, G., Sessions, A.L., Abramovich, A., Turchyn, A.V., Adkins, J.F., 2018. Cenozoic record of $\delta^{34}S$ in foraminiferal calcite implies an early Eocene shift to deep-ocean sulfide burial. Nat. Geosci. 11, 761–765.

Rick, B., 1990. Sulphur and oxygen isotopic composition of Swiss Gipskeuper (upper Triassic). Chem. Geol. Isot. Geosci. Sect. 80, 243–250.

Rudnicki, M.D., Elderfield, H., Spiro, B., 2001. Fractionation of sulfur isotopes during bacterial sulfate reduction in deep ocean sediments at elevated temperatures. Geochim. Cosmochim. Acta 65, 777–789.

Sakai, H., 1972. Oxygen isotopic ratios of some evaporites from Precambrian to recent ages. Earth Planet. Sci. Lett. 15, 201–205.

Sakai, H., Casadevall, T.J., Moore, J.G., 1982. Chemistry and isotope ratios of sulfur in basalts and volcanic gases at Kilauea volcano, Hawaii. Geochim. Cosmochim. Acta 46, 729–738.

Schlanger, S.O., Jenkyns, H.C., 1976. Cretaceous oceanic anoxic events: causes and consequences. Geol. Mijnbouw 55, 179–184.
Schobben, M., Stebbins, A., Algeo, T.J., Strauss, H., Leda, L., Haas, J., Struck, U., Korn, D., Korte, C., 2017. Volatile earliest Triassic sulfur cycle: a consequence of persistent low seawater sulfate concentrations and a high sulfur cycle turnover rate? Palaeogeogr. Palaeoclimatol. Palaeoecol. 486, 74–85.
Schröder, S., Bekker, A., Beukes, N.J., Strauss, H., Van Niekerk, H.S., 2008. Rise in seawater sulphate concentration associated with the Paleoproterozoic positive carbon isotope excursion: evidence from sulphate evaporites in the ~2.2–2.1 Gyr shallow-marine Lucknow formation, South Africa. Terra Nova 20, 108–117.
Scott, C., Wing, B.A., Bekker, A., Planavsky, N.J., Medvedev, P., Bates, S.M., Yun, M., Lyons, T.W., 2014. Pyrite multiple-sulfur isotope evidence for rapid expansion and contraction of the early Paleoproterozoic seawater sulfate reservoir. Earth Planet. Sci. Lett. 389, 95–104.
Sim, M.S., Bosak, T., Ono, S., 2011. Large sulfur isotope fractionation does not require disproportionation. Science 333, 74–77.
Sim, M.S., Ono, S., Hurtgen, M.T., 2015. Sulfur isotope evidence for low and fluctuating sulfate levels in the late Devonian ocean and the potential link with the mass extinction event. Earth Planet. Sci. Lett. 419, 52–62.
Smith, L., Oseid, D., Adelman, I., Broderius, S., 1976. Effect of hydrogen sulfide on fish and invertebrates. Part I—acute and chronic toxicity studies. In: Ecological Research Technical Report Series No.3. vol. 8. U.S. Environmental Protection Agency, Duluth, MN, p. Q02009.
Song, H.Y., Tong, J.N., Algeo, T.J., Song, H.J., Qiu, H.O., Zhu, Y.Y., Tian, L., Bates, S., Lyons, T.W., Luo, G.M., Kump, L.R., 2014. Early Triassic seawater sulfate drawdown. Geochim. Cosmochim. Acta 128, 95–113.
Spencer-Cervato, C., 1999. The Cenozoic deep sea microfossil record: explorations of the DSDP/ODP sample set using the Neptune database. Palaeontol. Electron. 2, 1–270.
Strauss, H., 1997. The isotopic composition of sedimentary sulfur through time. Palaeogeogr. Palaeoclimatol. Palaeoecol. 132, 97–118.
Strauss, H., 1999. Geological evolution from isotope proxy signals: sulfur. Chem. Geol. 161, 89–101.
Strauss, H., Banerjee, D.M., Kumar, V., 2001. The sulfur isotopic composition of Neoproterozoic to early Cambrian seawater—evidence from the cyclic Hanseran evaporites, NW India. Chem. Geol. 175, 17–28.
Surakotra, N., Promkotra, S., Charusiri, P., Maruoka, T., Hisada, K.I., 2018. Sulfur, strontium, carbon, and oxygen isotopes of calcium sulfate deposits in late carboniferous rocks of the Loei-Wang Saphung (LWS) Area, Loei Province, Thailand. Geosciences 8, 229–239.
Tedeschi, L.R., Jenkyns, H.C., Robinso, S.A., Sanjinés, A.E.S., Viviers, M.C., Quintaes, C.M.S.P., Vazquez, J.C., 2017. New age constraints on Aptian evaporites and carbonates from the South Atlantic: implications for Oceanic Anoxic event 1a. Geology 45, 543–546.
Thompson, C.K., Kah, L.C., 2012. Sulfur isotope evidence for widespread euxinia and a fluctuating oxycline in early to middle Ordovician greenhouse oceans. Palaeogeogr. Palaeoclimatol. Palaeoecol. 313, 189–214.
Tostevin, R., Turchyn, A.V., Farquhar, J., Johnston, D.T., Eldridge, D.L., Bishop, J.K., McIlvin, M., 2014. Multiple sulfur isotope constraints on the modern sulfur cycle. Earth Planet. Sci. Lett. 396, 14–21.
Turchyn, A.V., Schrag, D.P., Coccioni, R., Montanari, A., 2009. Stable isotope analysis of the Cretaceous sulfur cycle. Earth Planet. Sci. Lett. 285, 115–123.

Ueda, A., Campbell, F.A., Krouse, H.R., Spencer, R.J., 1987. $^{34}S/^{32}S$ variations in trace sulphide and sulphate in carbonate rocks of a Devonian reef, Alberta, Canada, and the Precambrian Siyeh Formation, Montana, USA. Chem. Geol. Isot. Geosci.Sect. 65, 383–390.

Utrilla, R., Pierre, C., Orti, F., Pueyo, J.J., 1992. Oxygen and sulphur isotope compositions as indicators of the origin of Mesozoic and Cenozoic evaporites from Spain. Chem. Geol. 102, 229–244.

Valentine, D.L., 2002. Biogeochemistry and microbial ecology of methane oxidation in anoxic environments: a review. Antonie Van Leeuwenhoek 81, 271–282.

Walker, J.C.G., 1986. Global geochemical cycles of carbon, sulfur, and oxygen. Mar. Geol. 70, 159–174.

Williford, K.H., Foriel, J., Ward, P.D., Steig, E.J., 2009. Major perturbation in sulfur cycling at the Triassic–Jurassic boundary. Geology 37, 835–838.

Worden, R.H., Smalley, P.C., Fallick, A.E., 1997. Sulfur cycle in buried evaporites. Geology 25, 643–646.

Wortmann, U.G., Chernyavsky, B.M., 2007. Effect of evaporite deposition on early Cretaceous carbon and sulphur cycling. Nature 446, 654–656.

Wortmann, U.G., Paytan, A., 2012. Rapid variability of seawater chemistry over the past 130 million years. Science 337, 334–336.

Wortmann, U.G., Bernasconi, S.M., Böttcher, M.E., 2001. Hypersulfidic deep biosphere indicates extreme sulfur isotope fractionation during single-step microbial sulfate reduction. Geology 29, 647–650.

Wortmann, U.G., Chernyavsky, B.M., Bernasconi, S.M., Brunner, B., Böttcher, M.E., Swart, P.K., 2007. Oxygen isotope biogeochemistry of pore water sulfate in the deep biosphere: dominance of isotope exchange reactions with ambient water during microbial sulfate reduction (ODP Site 1130). Geochim. Cosmochim. Acta 71, 4221–4232.

Wotte, T., Strauss, H., Fugmann, A., Garbe-Schönberg, D., 2012. Paired $\delta^{34}S$ data from carbonate-associated sulfate and chromium-reducible sulfur across the traditional lower–middle Cambrian boundary of W-Gondwana. Geochim. Cosmochim. Acta 85, 228–253.

Wu, N., Farquhar, J., Strauss, H., Kim, S.T., Canfield, D.E., 2010. Evaluating the S-isotope fractionation associated with Phanerozoic pyrite burial. Geochim. Cosmochim. Acta 74, 2053–2071.

Wu, N., Farquhar, J., Strauss, H., 2014. $\delta^{34}S$ and $\Delta^{33}S$ records of Paleozoic seawater sulfate based on the analysis of carbonate associated sulfate. Earth Planet. Sci. Lett. 399, 44–51.

Yao, W., Paytan, A., Wortmann, U.G., 2018. Large-scale ocean deoxygenation during the Paleocene–Eocene thermal maximum. Science 361, 804–806.

Yao, W., Paytan, A., Griffith, E.M., Martínez-Ruiz, F., Markovic, S., Wortmann, U.G., 2019. A revised seawater sulfate S-isotope curve for the Eocene. Chem. Geol. under review.

Further reading

Turchyn, A.V., Brüchert, V., Lyons, T.W., Engel, G.S., Balci, N., Schrag, D.P., Brunner, B., 2010. Kinetic oxygen isotope effects during dissimilatory sulfate reduction: a combined theoretical and experimental approach. Geochim. Cosmochim. Acta 74, 2011–2024.

Van Stempvoort, D.R., Krouse, H.R., 1994. Controls of d18O in sulfate—review of experimental data and application to specific environments. In: Alpers, C.N., Blowes, D.W. (Eds.), Environmental Geochemistry of Sulfide Oxidation. vol. 550. American Chemical Society, Washington, D.C., pp. 446–480.

CHAPTER TWO

Sr-isotope chronology of carbonate rocks: Quantifying the uncertainty of inversion

Ruoqi Wan[a,b], Wenjian Liu[c], John M. McArthur[d], Zhengrong Wang[b,*]
[a]State Key Laboratory of Geological Process and Mineral Resources, China University of Geosciences, Beijing, China
[b]Department of Earth and Atmospheric Sciences, CCNY, CUNY, New York, NY, United States
[c]Department of Mathematics and Computer Sciences, Queensborough Community College, CUNY, New York, NY, United States
[d]Department of Earth Sciences, University College London, London, United Kingdom
*Corresponding author: e-mail address: zwang1@ccny.cuny.edu

Contents

Abstract

The ages of carbonate rocks can be determined by comparing their $^{87}Sr/^{86}Sr$ ratios with the well-established seawater $^{87}Sr/^{86}Sr$ curve. Current models and databases provide no explicit solution to the problem of accurately assigning uncertainties to derived ages when the $^{87}Sr/^{86}Sr$ ratios of carbonate rocks contain analytical uncertainties. This is particularly concerning as significant amount of $^{87}Sr/^{86}Sr$ ratios are being measured by multi-collector inductively-coupled-plasma mass-spectrometry (MC-ICP-MS) which typically has a lower precision (1σ is typically several tens of ppm) compared with thermo-ionization mass-spectrometry (TIMs, 1σ can be as low as several ppm) used for calibrating the seawater $^{87}Sr/^{86}Sr$ curve. Deriving ages from carbonate $^{87}Sr/^{86}Sr$ ratios

Stratigraphy & Timescales, Volume 4
ISSN 2468-5178
https://doi.org/10.1016/bs.sats.2019.08.003

with large analytical uncertainties could introduce significant, sometimes asymmetric uncertainties in age estimates following a non-Gaussian distribution. More importantly, uncertainties in age estimates around the local maximum/minimum and saddle points on the Phanerozoic seawater $^{87}Sr/^{86}Sr$ curve are difficult to examine using conventional methods. In this study, we developed a rigorous mathematical framework (including perturbation and probability density methods) to estimate the 95% confidence interval (or any confidence interval of interest) of the derived age with any given analytical uncertainty. We applied this method to estimating uncertainties in derived ages for one version of the Phanerozoic seawater $^{87}Sr/^{86}Sr$ curve, assessing their sources and amplitudes. Our results indicates this method has the potential to be widely applied to estimating the uncertainties in ages derived from inversions using an empirical relationship. A Matlab program is presented that can calculate the age of carbonate rocks (based on their $^{87}Sr/^{86}Sr$ ratios and analytical uncertainties), the duration of intervals between ages, and the confidence interval of choice over the last 509 Ma (Version1 is available from zwang1@ccny.cuny.edu upon request).

1. Introduction

The trace metal concentrations (e.g., Sr, Mg, Ba, Mo, B, U) and stable isotope compositions of carbonates (e.g., $^{18}O/^{16}O$, $^{13}C/^{12}C$, $^{26}Mg/^{24}Mg$, $^{87}Sr/^{86}Sr$, $^{44}Ca/^{40}Ca$) have been widely used as paleoproxies, providing constraints on paleoclimates, chemistry of paleo-oceans, diagenetic processes, and evolution and biomineralization processes of marine organisms (e.g., reviews by Andersson, 2014; Halverson and Théou-Hubert, 2013; Gussone and Dietzel, 2016; Lea, 2014; Lowenstein et al., 2014; McArthur et al., 2012; Ravizza and Zachos, 2014; Saenger and Wang, 2014; Veizer and Prokoph, 2015). Knowing the precise age when carbonate rocks formed is essential to putting these proxies into geological context. However, most conventional radiometric dating methods (including Rb-Sr, Sm-Nd, U-Th-Pb) are not applicable to carbonate rocks, due mostly to the low concentration of these elements in seawater. Ages of carbonate rocks were typically inferred from biostratigraphy, magnetostratigraphy, astronomically tuned age models, and/or deposition-age models anchored by reliable radiometric dates of interbedded volcanic ashes (e.g., reviews by McArthur et al., 2012; Halverson and Théou-Hubert, 2013). These ages could contain large uncertainties due to the relatively lower resolution of biostratigraphy and magnetostratigraphy, and the non-linear deposition histories of carbonate rocks.

Because the $^{87}Sr/^{86}Sr$ ratio of modern seawater is, when measured conventionally without spike, identical to that of marine carbonates (e.g., Asmerom et al., 1991; Burke et al., 1982; Depaolo and Ingram, 1985;

Hess et al., 1986), a significant amount of efforts have been made to reveal the evolution of seawater over geological time according to the $^{87}Sr/^{86}Sr$ ratio of marine carbonates. During the process, the consensus has been reached that the seawater $^{87}Sr/^{86}Sr$ ratio is affected by continental river flux, hydrothermal alteration/volcanic activity, and precipitation or dissolution of carbonate and clay minerals, and the high-precision seawater $^{87}Sr/^{86}Sr$ curve with reasonably fast-paced variation rate could be used as a geologic time scale (e.g., comprehensive reviews by Halverson and Théou-Hubert, 2013; McArthur et al., 2012; Veizer, 1989).

Several versions of seawater $^{87}Sr/^{86}Sr$ curve have been established with the ever expanding database of $^{87}Sr/^{86}Sr$ ratios of marine carbonates (e.g., Burke et al., 1982; McArthur and Howarth, 2004; McArthur et al., 2001, 2012; Shields, 1999; Shields and Veizer, 2002; Veizer et al., 1999). Among them, the version by McArthur et al. (2001) and its updates (McArthur et al., 2012; McArthur and Howarth, 2004) are widely used since the large dataset were fitted based on a rigorous statistic method (i.e., nonparametric LOcally WEighted Scatterplot Smoother or LOWESS statistical regression function). The model also provided the 95% confidence envelope for the seawater $^{87}Sr/^{86}Sr$ curve, and offered a table to look up Phanerozoic ages and their 95% confidence intervals at a given $^{87}Sr/^{86}Sr$ ratio.

These work improved our capacity to date carbonate rocks. However, these models provide no explicit solution to the problem of accurately assigning uncertainties to derived ages when the $^{87}Sr/^{86}Sr$ ratios of carbonate rocks contain significant analytical uncertainties. This is particularly concerning as a significant amount of $^{87}Sr/^{86}Sr$ ratios are being measured by multi-collector inductively-coupled-plasma mass-spectrometry (MC-ICP-MS) which typically has a lower precision (1σ is typically several tens of ppm) compared with thermo-ionization mass-spectrometry (TIMs, 1σ can be as low as several ppm) used for calibrating the seawater $^{87}Sr/^{86}Sr$ curve. Deriving ages from carbonate $^{87}Sr/^{86}Sr$ ratios with large analytical uncertainties could introduce significant, sometimes asymmetric uncertainties following a non-Gaussian distribution. More importantly, as previous studies show, the seawater $^{87}Sr/^{86}Sr$ curve is surjection, i.e., a $^{87}Sr/^{86}Sr$ value could correspond to more than one age. Uncertainties in age estimates around the local maximum/minimum and saddle points are difficult to determine. Because an accurate uncertainty estimate is essential for correlating carbonate stratigraphy occurring at different places, making rational judgement on the duration of geological events, and understanding the meaning of time gaps, we developed a rigorous mathematical framework in this study (including perturbation and probability density method) to estimate the

95% confidence interval of the derived age with any analytical uncertainty. We applied this method to estimating age uncertainties under three conditions (1 sigma standard deviation of measurement $\sigma_m = 0$, 1 ppm and 10 ppm) for one version of the Phanerozoic seawater $^{87}Sr/^{86}Sr$ curve (McArthur and Howarth, 2004), assessing their sources and amplitudes.

2. Method

2.1 Perturbation method

One of the most commonly-used methods for error propagation is the Taylor series expansion (e.g., Anderson, 2011; Heuvelink, 1998; Taylor, 1997), in which, the variation in system input is propagated to the variation in system output through the derivative of the function. Here we refer this method as 'perturbation method.' This method is widely-applied due to its simplicity and low cost of computation.

Assuming y is a function of t, written as $y = f(t)$, then the inverse function, if it exists, can be written as $t = f^{-1}(y)$. When there is an uncertainty (Δy, which can be positive or negative) in y, the uncertainty (Δt) in t can be written as follows:

$$t + \Delta t = f^{-1}(y + \Delta y) \tag{1}$$

The Taylor series of the inverse function can be written as follows:

$$t + \Delta t = f^{-1}(y) + \frac{\Delta y}{f'(t)} + \frac{(\Delta y)^2}{2} \cdot \left[-\frac{f''(t)}{(f'(t))^3} \right] + \frac{(\Delta y)^3}{6} \cdot \left[\frac{3(f''(t))^2 - f'''(t) \cdot f'(t)}{(f'(t))^5} \right] + \ldots \tag{2}$$

where $f'(t)$, $f''(t)$, and $f'''(t)$ represent the first, second and third derivatives of y, respectively, i.e., $\frac{dy}{dt} = f'(t)$, $\frac{d^2y}{dt^2} = f''(t)$ and $\frac{d^3y}{dt^3} = f'''(t)$. The detailed derivation is shown in the Appendix. Thus, the uncertainty can be written, according to the perturbation method, as follows:

$$\Delta t|_t = \frac{\Delta y}{f'(t)} + \frac{(\Delta y)^2}{2} \cdot \left[-\frac{f''(t)}{(f'(t))^3} \right] + \frac{(\Delta y)^3}{6} \cdot \left[\frac{3(f''(t))^2 - f'''(t) \cdot f'(t)}{(f'(t))^5} \right] \ldots \tag{3}$$

In most cases, when the leading term is dominating the uncertainty (i.e., the terms higher than the second order are much smaller than the first term), the uncertainty can be simplified as follows:

$$\Delta t|_t = \frac{\Delta y}{f'(t)}$$
$$\sigma_t|_t = \frac{\sigma_y}{f'(t)} \quad (4)$$

where σ_t and σ_y represent one standard deviations in t and y, respectively. This method can be applied directly to estimating the uncertainty associated with dating carbonate rocks using the seawater $^{87}Sr/^{86}Sr$ curve by letting $y = {}^{87}Sr/^{86}Sr$ and t=geologic time in mega-annum (Ma). As depicted in Fig. 1, for any $^{87}Sr/^{86}Sr$ varying with time t, any variation (i.e., σ_y) in $^{87}Sr/^{86}Sr$ can be linked with variation in time t (i.e., σ_t) through the local tangent of the function (or the first derivative of the curve) at time t.

The uncertainty in age estimate when dating carbonate rocks using the seawater $^{87}Sr/^{86}Sr$ curve is typically derived from two sources: (1) the uncertainties (σ_f) arising from fitting the LOWESS regression to experimental data, which have been provided by McArthur et al. (2001); McArthur et al. (2012); and McArthur and Howarth (2004) based on their statistical regression of experimental data using Locally Weighted Scatterplot Smoother (LOWESS) (e.g., Cleveland, 1979; Cleveland et al., 1992; Howarth and McArthur, 1997). This uncertainty also includes the analytical uncertainty, the density of experimental data at a specific time, and the age model, used for calibrating the seawater $^{87}Sr/^{86}Sr$ curve; and (2) the analytical uncertainty (σ_m) due to the method and instruments used, and the number of measurements made (e.g., McArthur et al., 2001, 2012; McArthur and Howarth, 2004 and references therein). For example, the thermo-ionization mass-spectrometer (TIMS) typically has a precision for $^{87}Sr/^{86}Sr$ as low as a few ppm (i.e., $1\sigma_m$ is on the order of 10^{-6}) whereas multi-collector inductively coupled plasma mass-spectrometer (MC-ICP-MS) has around a few tens of ppm (i.e., $1\sigma_m$ is on the order of 10^{-5}) (e.g., Balcaen et al., 2005; Liu et al., 2013; McArthur et al., 2001, 2012; Veizer, 1989). When more measurements were made, the uncertainty, expressed as standard error, would be reduced in proportion to the square root of the number of analysis.

If these two uncertainties are independent, their sum should also be a normal distribution according to the statistic principle, and the total σ_y can be

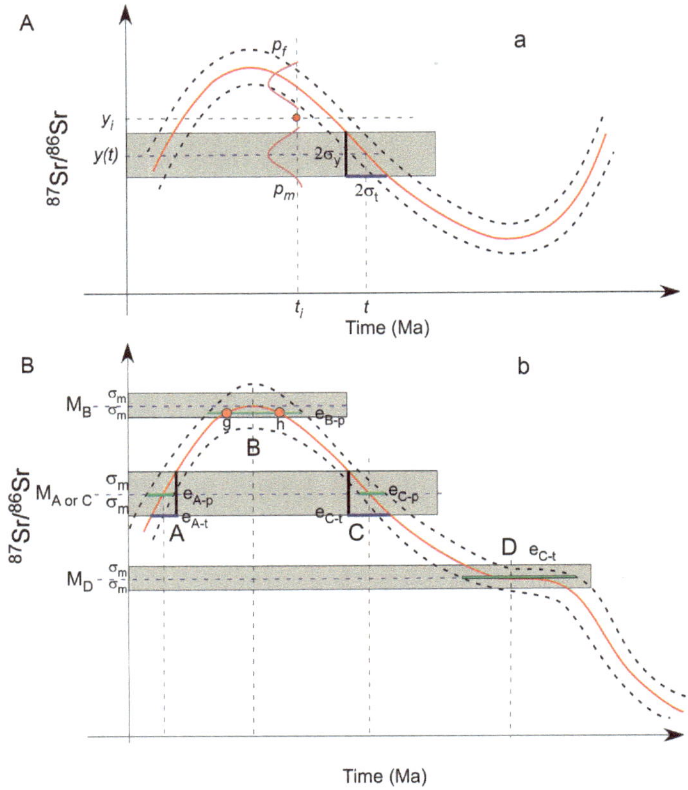

Fig. 1 The sketches of assumed $^{87}Sr/^{86}Sr$ ($y(t)$) variations with time (t in Ma). (A): a segment of the seawater $^{87}Sr/^{86}Sr$ curve containing two critical points (a maximum and a minimum). When the 95% confidence envelope is parallel to the seawater $^{87}Sr/^{86}Sr$ curve, a variation in y ($2\sigma_y$) can be projected to a variation in time ($2\sigma_t$) through the slope of the curve at the time of interest, as described by the perturbation method in the text; at a given position (y_i,t_i), the probability density subject to the seawater $^{87}Sr/^{86}Sr$ curve is p_f, subject to the measurement that obeys Gaussian distribution is p_m; (B): a segment of the seawater $^{87}Sr/^{86}Sr$ curve containing two special points (a local maximum in region B and a saddle point in region D). When the 95% confidence envelope is not parallel to the seawater $^{87}Sr/^{86}Sr$ curve (in regions A and C), the upper and lower bounds of the 95% confidence interval (e_{A-p} and e_{C-p}) may be asymmetric around the age, which is also different from the 95% confidence interval calculated using our perturbation method (e_{A-t} and e_{C-t}); at the local maximum (region B), two ages (g and h) can be found on both sides of the critical point, while at the saddle point (region D), one age can be found; in both regions, the sketch show uncertainties are bounded, which cannot be described by the perturbation method due to the almost-zero first derivative of the seawater curve.

written as follows (e.g., Anderson, 2011; Blachman and Machol, 1987; Heuvelink, 1998; Taylor, 1997, see also Section 3.1 for a numerical test): Substituting the σ_y into Eq. (3) or (4) can yield the uncertainty estimate for time (σ_t).

$$\sigma_y = \sqrt{\sigma_m^2 + \sigma_f^2} \tag{5}$$

This method can be readily implemented since the first derivative of the seawater $^{87}Sr/^{86}Sr$ can be numerically calculated with economic calculation cost. The only attention that needs to be paid is to conduct a bit data-smoothing before the derivative is taken to avoid calculation artefact.

However, as shown in Fig. 1B, this method has several shortcomings which are not related to data-smoothing. First, this method requires the existence of an inverse function for $y=f(t)$, i.e., $y=f(t)$ is a bijection or the $^{87}Sr/^{86}Sr$ and time are one-to-one correspondence. The Figs. 7 through 13 clearly show that the seawater $^{87}Sr/^{86}Sr$ curve during the Phanerozoic era is a surjection, i.e., a $^{87}Sr/^{86}Sr$ value could correspond to more than one age (e.g., McArthur et al., 2001, 2006, 2012; Veizer, 1989). Although this issue can be resolved by defining $y=f(t)$ as a piece-wise function, within each piece the $y=f(t)$ is a bijection, when the data is closer to the boundary of the piece-wise function, Eq. 3 is unlikely to converge and yet Fig. 1B in area B show the uncertainty close to the boundary (the critical point) is finite. Second, the perturbation method assumes the slope of the function at time t equals the slope of the 95% confidence envelop. Fig. 1B in regions A and C exhibits, as an example, that upper and lower bounds of the 95% confidence envelope are not parallel to the function y, resulting an asymmetric uncertainty estimate (i.e., $\sigma_t^+ \neq \sigma_t^-$). This phenomenon can be observed frequently in the seawater $^{87}Sr/^{86}Sr$ curve (Figs. 6 through 13). Under those conditions, age uncertainty cannot be correctly estimated using Eq. (4) which would yield a symmetric one. Although Eq. (3) could produce asymmetric estimates of uncertainty after considering high order terms, it takes no account of the first derivative of the 95% confidence envelope. Perhaps the biggest issue using perturbation method is around the critical point when the first derivative of the seawater $^{87}Sr/^{86}Sr$ ratio approaches zero (i.e., $f'(t) \to 0$). Using Eqs. (3) or (4) will yield extremely high estimates of uncertainty whereas they should still be bounded according to the graphical representation (e.g., areas B and D in Fig. 1B). In the next sections, a numerical method is developed to calculate the probability density, the confidence interval of ages and age differences.

2.2 Probability density method

Because the seawater $^{87}Sr/^{86}Sr$ curve and its 95% confidence envelope lack explicit function forms, (i.e., they are essentially empirical functions which are statistically best fits to the experimental data), numerical methods have their advantages to make reasonable estimates of uncertainty under complex conditions. In this study, a numerical method is developed to estimate the confidence interval by directly calculating the probability density.

As shown in Fig. 1A, since both the analytical uncertainty of $^{87}Sr/^{86}Sr$ and that of model-fit obey Gaussian distribution (or normal distribution), the probability density at a point (t_i, y_i) subject to those two Gaussian distributions can be calculated as p_f for the model fit and p_m for the analytical uncertainty as follows:

$$\begin{aligned} p_f &= f\left(y_i \mid y(t_i), \sigma_{t_i}^2\right) = \frac{1}{\sqrt{2\pi\sigma_{t_i}^2}} e^{-\frac{[y_i - y(t_i)]^2}{2\sigma_{t_i}^2}} \\ p_m &= f\left(y_i \mid y(t), \sigma_m^2\right) = \frac{1}{\sqrt{2\pi\sigma_m^2}} e^{-\frac{[y_i - y(t)]^2}{2\sigma_m^2}} \end{aligned} \tag{6}$$

where σ_{t_i} is one standard deviation of seawater $^{87}Sr/^{86}Sr$ curve at time t_i; σ_m is one standard deviation of $^{87}Sr/^{86}Sr$ ratio measurement; function $f(y_i \mid y(t_i), \sigma_{t_i}^2)$ represents the probability density at (t_i, y_i) which follows a Gaussian distribution with an average of $y(t_i)$ and 1 standard deviation of σ_{t_i}; similarly function $f(y_i \mid y(t), \sigma_m^2)$ represents the probability density at (t_i, y_i) which follows a Gaussian distribution with an average of $y(t)$ and 1 standard deviation of σ_m.

If these two uncertainties are independent (in most cases this is true), the total probability can be written as the product of the probability density:

$$p_t\big|_{t_i, y_i} = p_f \cdot p_m \tag{7}$$

Integrating the probability density over y axis from $-\infty$ to ∞ and over age axis (t) from $t - \Delta^-$ to $t + \Delta^+$ shall yield a given confidence interval (here a 95% level as default) at time t as follows:

$$P(t - \Delta^- < t < t + \Delta^+) = \int_{t-\Delta^-}^{t+\Delta^+} dt_i \int_{-\infty}^{\infty} \left\{ \frac{1}{\sqrt{2\pi\sigma_{t_i}^2}} e^{-\frac{[y_i - y(t_i)]^2}{2\sigma_{t_i}^2}} \cdot \frac{1}{\sqrt{2\pi\sigma_m^2}} e^{-\frac{[y_i - y(t)]^2}{2\sigma_m^2}} \right\} dy_i = 0.95 \tag{8}$$

From this equation, the 95% confidence interval (Δ^-, Δ^+) representing the cumulative probability between 2.5% and 97.5% can be obtained, and the age can be represented as $t_{\Delta^-}^{\Delta^+}$. Due to the lack of explicit expression of the seawater $^{87}Sr/^{86}Sr$ curve (y) and its 95% confidence envelope, Eq. (8) needs to be calculated numerically.

Compared with the perturbation method, this numerical method is more versatile and can obtain confidence interval under a variety of conditions, overcoming all of those shortcomings for the perturbation method mentioned in Section 2.1. However, because each calculation involves making a dense grid around the target age (t), evaluating their probability density, and integrating over a large number of points, calculation using this method typically comes at a higher computational cost.

2.3 Model implementation

The two aforementioned methods are sufficient to constrain 95% confidence intervals of output of any empirical function with known 95% confidence envelope of the function and input. In this study, we apply these methods to the Phanerozoic seawater $^{87}Sr/^{86}Sr$ curve (version V4B_08_04) from McArthur and Howarth (2004), an update version of McArthur et al. (2001).

First, the Phanerozoic seawater $^{87}Sr/^{86}Sr$ curve is parameterized via a cubic smoothing spline (e.g., csaps in Matlab, Copyright 1987–2010, The Mathworks. Inc.), from which 26 critical points, where the first derivative is zero (i.e., $f'(t)=0$) but the second and third derivatives are finite, are solved and listed in Table 1. These critical points reflect a shift in the prevailing mechanism that controls the $^{87}Sr/^{86}Sr$ ratio of seawater. Along with the starting and ending time of 0 and 509 Ma, respectively, the Phanerozoic seawater $^{87}Sr/^{86}Sr$ curve is divided into 27 segments, within which the seawater $^{87}Sr/^{86}Sr$ curve is a bijection, and can be calculated using both perturbation and probability density methods, while at around critical points (i.e., $f'(t) \rightarrow 0$) only probability density method yields meaningful results. For the perturbation method, the second and third derivatives are also calculated numerically to evaluate if Eq. (3) is converging and if the upper and lower bounds of the 95% confidence envelope are symmetric. For the probability density method, the 95% confidence envelopes of the Phanerozoic seawater $^{87}Sr/^{86}Sr$ curve are also parameterized with a cubic spline. A ± 5 Ma year around the target time (t) and $\pm 6\ \sigma_m$ window around the measured $^{87}Sr/^{86}Sr$ ratio were typically chosen which were equally

Table 1 A list of 26 critical points in the seawater $^{87}Sr/^{86}Sr$ curve.

ID	Ages[a] (Ma)	ID	Ages (Ma)	ID	Ages (Ma)	ID	Ages (Ma)	ID	Ages (Ma)	ID	Ages (Ma)	ID	Ages (Ma)
Start	0.0000												
1	39.0888	5	90.1065	9	159.4705	15	226.9583	18	287.0951	22	365.6542	25	451.4087
2	44.3710	6	102.1483	10	172.3445	16	245.0218	19	297.0983	23	388.5858	26	496.3604
3	50.6265	7	114.1751	11	183.3602	17	261.3440	20	309.8635	24	416.0911	end	509.8000
4	65.7334	8	127.8850	12	197.9107			21	334.9635				
				13	200.3089								
				14	205.9762								

[a]The critical points were calculated based on cubic smoothing spline of interpolation of seawater $^{87}Sr/^{86}Sr$ curve (version V4B-08-04, McArthur et al., 2001; McArthur and Howarth, 2004).

divided into 1000×300 grids for calculating probability densities using Eq. (7). Integrating these probability density over $^{87}Sr/^{86}Sr$ axis yields a probability density distribution over time. One can fit this distribution using Gaussian distribution, or solve Eq. (8) for the cumulative probability density between 2.5% and 97.5% which is very effective if the uncertainty estimate follows a non-Gaussian distribution. Whenever possible, calculation using the probability density method is preferred. But around the starting (0 Ma) and ending (509 Ma) time when there is no enough grid to make accurate estimates of the confidence interval, the perturbation method is used.

In the following section, the two models are first compared between each other and with previous results. Detailed discussions are made on uncertainty estimations around critical points, and of age differences. The 95% confidence interval is presented for all Phanerozoic ages at an interval of 0.01 Ma given three different analytical precisions: $1\sigma_m = 0$, $1\sigma_m = 1$ ppm, and $1\sigma_m = 10$ ppm.

3. Result and discussion

3.1 Comparison between the two methods

In most cases where $^{87}Sr/^{86}Sr$ ratios are far away from critical points, these two methods yield exactly the same result. Fig. 2 shows an example of uncertainty estimation at $t = 29.26$ Ma using both methods. Fig. 2A through C show results calculated assuming $1\sigma_m = 0$. The perturbation method estimates the 95% confidence interval at this time is ± 0.21 Ma, using Eq. (3) or (4). The probability density is calculated based on the numerical integration over all grids for $^{87}Sr/^{86}Sr$ at a given time, and shown in Fig. 2B. An almost perfect fit for the Gaussian distribution yields a mean age of 29.25 Ma with a 95% confidence interval of ± 0.20 Ma. The numeric solution of Eq. (8) found the 95% confidence interval is ± 0.21 Ma. Similarly, when $1\sigma_m = 10$ ppm, the perturbation method estimates the 95% confidence interval is between -0.53 and 0.56 Ma, in which case the second term in Eq. (3) contribute slightly (Fig. 2D and E). The Gaussian fit of the probability density (Fig. 2E) shows a little deviation, with a mean age of 29.25 Ma and 95% confidence interval of ± 0.54 Ma, and the integration of probability density finds the 95% confidence interval is between -0.53 and 0.57 Ma (Fig. 2F), almost exactly the same as those calculated by the perturbation method (Fig. 2D). This exercise demonstrates that under regular conditions when

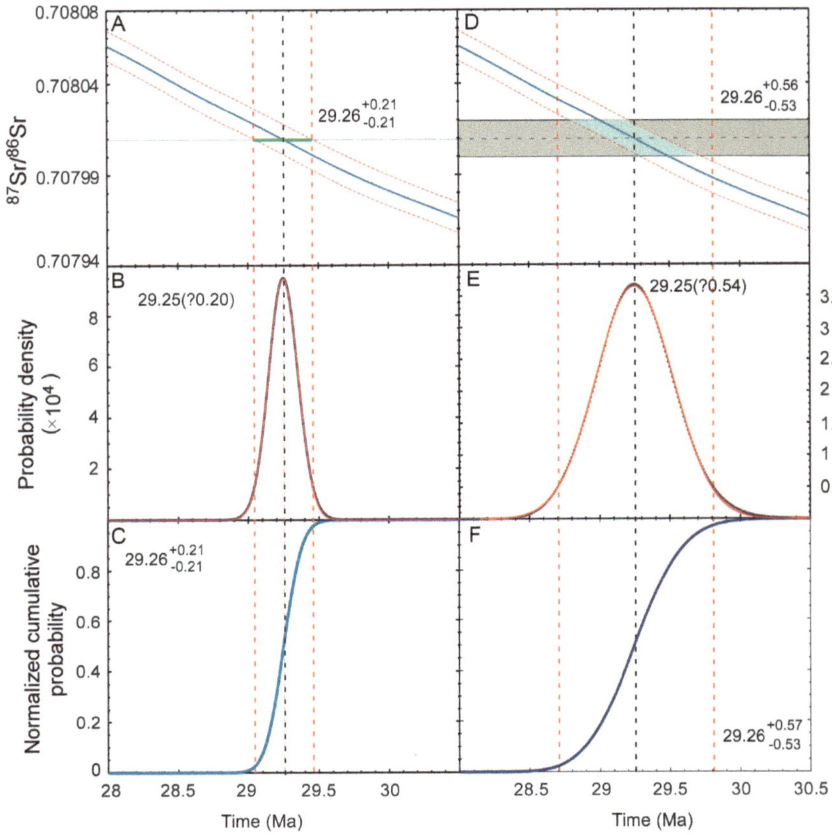

Fig. 2 Comparison between the perturbation method and probability density method. The calculation is made at 29.26 Ma for σ_m (1 standard deviation of the measurement) = 0 in subplots (A–C), and σ_m = 10 ppm in subplots (D–F). Subplots (A and D) show the seawater $^{87}Sr/^{86}Sr$ curve between 28 and 30.5 Ma based on the V4B-08-04 database (McArthur et al., 2001, 2006, 2012), along with the confidence intervals calculated by our perturbation method. Subplots (B and E) show the calculated probability densities and the results of fitting them to a Gaussian distribution. Subplots (C and F) show normalized cumulative probability integrated over the range and the calculated 95% confidence interval (between 2.5% and 97.5%). In this case, all three methods agree well with each other, while when σ_m is higher, the upper and lower bounds of the 95% confidence interval tend to be asymmetric.

the age estimate is away from critical point and the 95% confidence envelope is almost parallel to the seawater $^{87}Sr/^{86}Sr$ curve, both methods yield identical uncertainty estimates. Moreover, when $1\sigma_m = 1$ ppm, the 95% confidence interval calculated using both methods are very similar to the case when $1\sigma_m = 0$.

3.2 Estimating uncertainty near a critical point

When the first derivative of the seawater $^{87}Sr/^{86}Sr$ curve is approaching zero (i.e., $f'(t) \to 0$), the perturbation method is unable to yield a bounded confidence interval, and therefore, the probability density method is a better choice to estimate the confidence interval despite its higher computation cost. When $f'(t)=0$, two cases need to be discussed separately: (1) critical points when $f'(t)=0$ but $f''(t) \neq 0$; and (2) stationary point of inflection (or saddle-point) when $f'(t)=0$ and $f''(t)=0$. At the saddle-point (e.g., region D in Fig. 1B; 218.12 or 276.82 Ma of the Phanerozoic seawater $^{87}Sr/^{86}Sr$ curve in Fig. 10), the seawater $^{87}Sr/^{86}Sr$ curve is still close to a bijection, and therefore, it can be treated using the probability density method directly. However, at the critical point, the seawater $^{87}Sr/^{86}Sr$ curve reaches a local maximum or minimum. Thus on both sides of the critical point, the curve becomes a surjection. Because at a given $^{87}Sr/^{86}Sr$ value one age may be obtained on one side of a critical point, each of the two solutions will have its own confidence interval. Thus, the 26 critical points are also used as boundaries for 95% confidence intervals.

Fig. 3 depicts an example of calculating the 95% confidence interval around a critical point (102.148 Ma). Around this time, a carbonate sample with a $^{87}Sr/^{86}Sr$ ratio of 0.707423 corresponds to two possible ages (or 'twin ages,' 98.0 Ma and 104.45 Ma). Assuming analytical uncertainty is zero, ($1\sigma_m=0$, Fig. 3A through C), using the perturbation method will yield 98.0 ± 2.9 Ma and 104.6 ± 2.6 Ma if only considering the leading term (Eq. 4), but will yield $98.0^{7.2}_{-2.8}$ and $104.5^{1.6}_{-0.7}$ if considering the first three terms (Eq. 3). Since the second and third terms are as large as the first term, the Taylor expansion in Eq. (2) is technically non-converging. Therefore, the perturbation method is unreliable here. A poor fit of the probability density to the Gaussian distribution is noticeable in Fig. 3B, and it yields solutions of 98.8 ± 3.3 Ma and 104.5 ± 2.6 Ma, while the cumulative probability between 2.5% and 97.5% yields solutions of $98.0^{3.9}_{-1.5}$Ma and $104.5^{2.6}_{-2.0}$Ma which are the best estimates of uncertainty for these two ages since their 95% confidence intervals are asymmetric.

Similar exercises can be conducted for $1\sigma_m=10$ ppm. The results are shown in Fig. 3D through F. Solutions of 98^{27}_{-11}Ma and 104^{10}_{-6}Ma can be obtained using the perturbation method considering the first three terms. Solutions of 99.3 ± 5.0 Ma and 104.4 ± 4.5 Ma are yielded for fitting the probability density to a Gaussian distribution (Fig. 3E). The confidence intervals with the cumulative probability density between 2.5% and 97.5% are $98.0^{4.0}_{-2.3}$ Ma and $104.5^{3.5}_{-2.2}$ Ma. This exercise shows that using

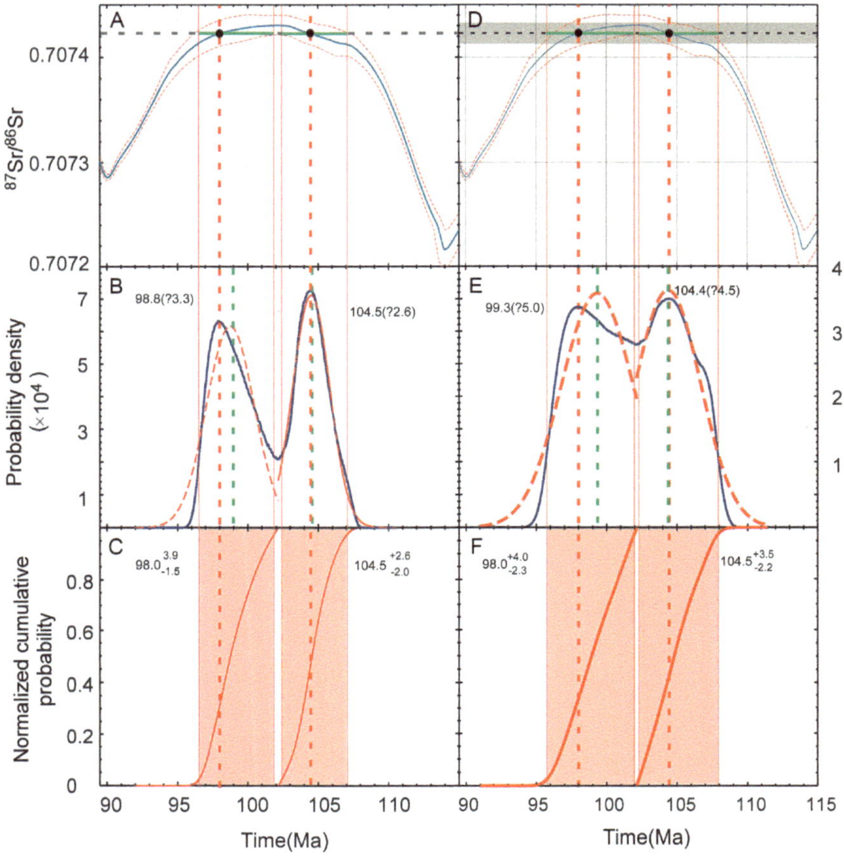

Fig. 3 An example of uncertainty estimation around a critical point (102.148 Ma). Near 100 Ma, a carbonate sample with a $^{87}Sr/^{86}Sr$ ratio of 0.707423 corresponds to two possible ages (98.0 and 104.45 Ma), each of which has its own 95% confidence interval. Subplots (A–C) show the calculation results for $1\sigma_m = 0$, and subplots (D–F) show the calculation results for $1\sigma_m = 10$ ppm. Subplots A and D show the seawater Ma based on the V4B-08-04 database from McArthur and Howarth (2004). In this case, the calculation by our perturbation method yields extremely wide confidence intervals. Subplots (B and E) show the calculated probability densities and the results of fitting them to a Gaussian distribution. Subplots (C and F) show the normalized cumulative probability integrated over the range and the calculated 95% confidence interval (between 2.5% and 97.5%). In this case, the upper and lower bounds of the confidence intervals are asymmetric.

the cumulative probability density between 2.5% and 97.5% will typically yield a narrower confidence interval compared with those obtained using perturbation method due to $f'(t) \rightarrow 0$, or the fitting of probability density to a Gaussian distribution because the lower or the upper bound of the confidence interval is limited by the critical point.

3.3 Comparison with previous methods

McArthur and Howarth (2004) provided tables to look up Phanerozoic ages and their associated 95% confidence interval for a given $^{87}Sr/^{86}Sr$ ratio. As illustrated in Fig. 4, their results (black lines based on the look-up table version V4B_08_04) are overlapping with this study when $\sigma_m = 0$ (the red line), indicating our probability density method is consistent with those used in McArthur et al. (2001) and McArthur and Howarth (2004). Moreover, Fig. 4 shows that even when the seawater $^{87}Sr/^{86}Sr$ curve is rather smooth and increases quasi-linearly over the last 30 Ma, the upper and lower bounds of the 95% confidence intervals can be asymmetric, particularly when uncertainty is larger than ~0.3 Ma or smaller than ~ −0.3 Ma (i.e., comparing the black line and the dotted black line, which is the mirror image of lower bounds of the 95% confidence interval from McArthur and Howarth, 2004). The larger uncertainties with significant asymmetry occur where the first derivative ($f'(t)$) is closest to zero (i.e., within the first 15 Ma and around

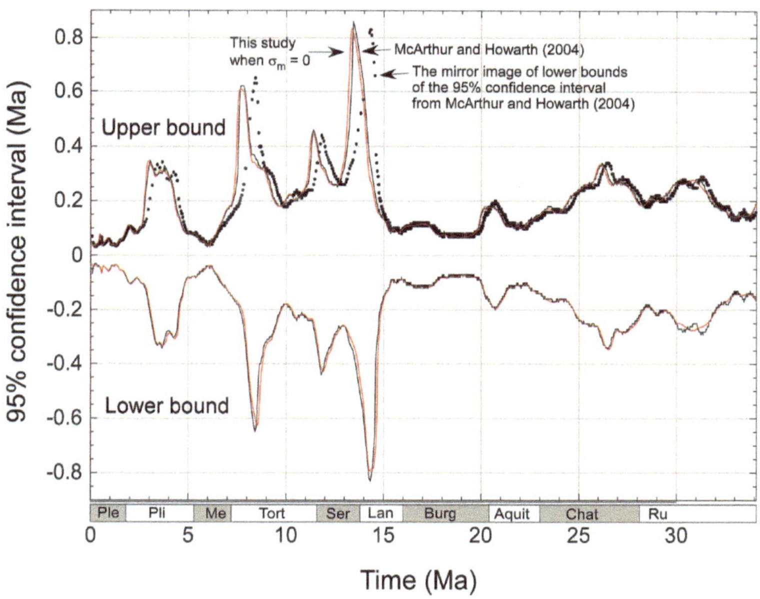

Fig. 4 Comparison between our results with the look-up table (version V4B_08_04) by McArthur and Howarth (2004) within 0–35 Ma. Red lines are from this study; black lines are from the Look-up table; dotted lines is the mirror image of lower bounds of the 95% confidence interval from the lookup table. As shown here, the 95% confidence interval provided by the lookup table is overlapping with this study when $\sigma_m = 0$. It also shows that the upper and lower bounds of the confidence interval can be asymmetric when uncertainties are high. The geologic time scale is based on Walker et al. (2013), and the abbreviated symbols are listed in Table 3.

Oligocene Chattian, Figs. 4 and 6). Under those conditions, our probability density method works better. The results of 95% confidence interval at $\sigma_m = 1$ ppm and $\sigma_m = 10$ ppm are also shown in Fig. 6, which indicates this asymmetry can extend in both amplitude and range with increasing σ_m values.

3.4 Estimating uncertainty for time-difference

One of the most vital applications of Sr isotope stratigraphy is to assess the duration between two numerical ages derived from LOWESS. Assuming the ages of two carbonate rocks (t_1 and t_2) were determined using their $^{87}Sr/^{86}Sr$ ratios, if their uncertainties obey Gaussian distribution with one standard deviation of σ_{t1} and σ_{t2}, respectively, the one standard deviation of their difference can be calculate readily as follows (e.g., Anderson, 2011; McArthur et al., 2001, 2012; McArthur and Howarth, 2004; Oberkampf et al., 2002; Taylor, 1997):

$$\sigma_{t_2-t_1} = \sqrt{\sigma_{t1}^2 + \sigma_{t2}^2} \tag{9}$$

However, as mentioned in previous sections, in many cases, the lower and upper bounds of the 95% confidence intervals are asymmetric, particularly when σ_m is large. Under those conditions, Eq. (9) is not applicable, and our probability method can provide a better estimate.

Combining Eqs. (6) and (7), one can calculate the probability densities of t_i subject to the distribution of t_1 and t_2 as follows:

$$\begin{aligned} p_{t_1}\big|_{t_i} &= \int_{-\infty}^{\infty} \left\{ \frac{1}{\sqrt{2\pi\sigma_{t_i}^2}} e^{-\frac{[y_i - y(t_i)]^2}{2\sigma_{t_i}^2}} \cdot \frac{1}{\sqrt{2\pi\sigma_m^2}} e^{-\frac{[y_i - y(t_1)]^2}{2\sigma_m^2}} \right\} dy_i \\ p_{t_2}\big|_{t_i} &= \int_{-\infty}^{\infty} \left\{ \frac{1}{\sqrt{2\pi\sigma_{t_i}^2}} e^{-\frac{[y_i - y(t_i)]^2}{2\sigma_{t_i}^2}} \cdot \frac{1}{\sqrt{2\pi\sigma_m^2}} e^{-\frac{[y_i - y(t_2)]^2}{2\sigma_m^2}} \right\} dy_i \end{aligned} \tag{10}$$

Thus, assuming t_1 and t_2 are independent, the probability density of $t_2 - t_1$ can be written as follows:

$$p_{t_2-t_1}\big|_{t_i} = (p_{t_2} \cdot p_{t_1})\big|_{t_i} \tag{11}$$

Similar to Eq. (8), the 95% confidence interval (Δ^-, Δ^+) between 2.5% and 97.5% for t_2 - t_1 can be written as follows:

$$P(t_2 - t_1 - \Delta^- < t_2 - t_1 < t_2 - t_1 + \Delta^+) = \int_{t_2 - t_1 - \Delta^-}^{t_2 - t_1 + \Delta^+} (p_{t_2} \cdot p_{t_1}) \Big|_{t_i} dt_i = 0.95 \quad (12)$$

Six cases were calculated using this method assuming $\sigma_m = 5$ ppm, as represented in Table 2, two of which are depicted in Fig. 5. For example, in case 2, the uncertainties for t_1 are symmetric and obey the Gaussian distribution (Fig. 5A, $t_1 = 17.98 \pm 0.15$ Ma, 95% confidence interval or 1.96 times of one standard deviation for Gaussian distribution), whereas the uncertainties for t_2 slightly deviate from a Gaussian distribution (Fig. 5A, $t_2 = 18.68^{+0.14}_{-0.15}$ Ma, 95% confidence interval). The difference between them (i.e., $t_2 - t_1$) is $0.70^{+0.20}_{-0.21}$, which is almost the same as the solution by fitting the probability density to a Gaussian distribution (Fig. 5B, $t_2 - t_1 = 0.70 \pm 0.21$ Ma, 95% confidence interval), or as that calculated using Eq. (9) (Table 2, method-I). This exercise demonstrates that our probability density method is capable of predicting confidence interval precisely. Similar conclusions can be drawn from cases 3 and 4 in Table 2.

When the lower and upper bounds of 95% confidence intervals are asymmetric, our probability density method has more advantages over others (e.g., Eq. 9 or perturbation method). For example, in case 6 (again assuming $\sigma_m = 5$ ppm), $t_1 = 7.50^{+0.85}_{-0.38}$ and $t_2 = 13.00^{+1.1}_{-0.49}$, both of which have highly asymmetric uncertainties. Our probability density method yields an age difference of $t_2 - t_1 = 5.50^{+1.1}_{-0.97}$ (95% confidence interval). In comparison, the Gaussian fit of the probability density yields $t_2 - t_1 = 5.48 \pm 0.92$ Ma which is close to, but not a precise description of the 95% confidence interval of the age difference. Such deviation will be even more severe when the analytical uncertainty (σ_m) or the age difference is large (e.g., case 5).

Table 2 also lists two different treatments on propagating asymmetric uncertainties. Assuming the 95% confidence intervals for t_1 and t_2 are (Δ^-_{t1}, Δ^+_{t1}) and (Δ^-_{t2}, Δ^+_{t2}), respectively. In method-I, the lower and higher bounds of 95% confidence interval is estimated, in analogy to the Gaussian distribution, as follows:

$$\begin{aligned} \Delta^-_{t_2 - t_1} &\approx -\sqrt{\left(\Delta^-_{t_1}\right)^2 + \left(\Delta^-_{t_2}\right)^2} \\ \Delta^+_{t_2 - t_1} &\approx \sqrt{\left(\Delta^+_{t_1}\right)^2 + \left(\Delta^+_{t_2}\right)^2} \end{aligned} \quad (13)$$

Table 2 Case studies of age differences with 95% confidence intervals between age pairs during the last 100 Ma assuming analytical uncertainty $\sigma_m = 5$ ppm.

Case		Probability density method[a]			Gaussian fit[b]		Probability density method[c]			Method-I[d]		Model-II[e]	
		Age			95% confidence interval		95% confidence interval						
		(Ma)	lower bound	upper bound	Difference t_2-t_1(Ma)	$\pm 1.96\sigma$[f]	Difference (Ma)	lower bound	upper bound	lower bound	upper bound	lower bound	upper bound
1	t_1	13.59	−0.61	0.98									
	t_2	16.44	−0.20	0.21	2.68	0.51	2.85	−1.01	0.63	−0.64	1.00	−1.00	0.64
2	t_1	17.98	−0.15	0.15									
	t_2	18.68	−0.15	0.14	0.70	0.21	0.70	−0.21	0.20	−0.21	0.20	−0.21	0.21
3	t_1	16.79	−0.21	0.21									
	t_2	18.04	−0.15	0.15	1.25	0.26	1.25	−0.26	0.25	−0.26	0.25	−0.25	0.26
4	t_1	15.64	−0.18	0.17									
	t_2	17.66	−0.17	0.14	2.02	0.24	2.02	−0.24	0.23	−0.25	0.22	−0.24	0.23
5	t_1	78.00	−0.53	0.69									
	t_2	98.00	−1.73	3.95	20.91	3.83	20.00	−1.87	3.99	−1.81	4.01	−1.86	3.99
6	t_1	7.50	−0.38	0.85									
	t_2	13.00	−0.49	1.14	5.48	0.92	5.50	−0.97	1.11	−0.62	1.43	−0.98	1.20

The cases dicussed in the text were highlighted with color.
[a] Calculated using our probability density method in Eq. (8).
[b] Fitting the probability density of $t_2 - t_1$ to a Gaussian distribution, calculated using Eq. (11).
[c] Calculated using the probability density method in Eq. (12).
[d] Calculated using empirical equation for error propagation for Gaussian distribution in Eq. (13).
[e] Calculated using empirical equation for error propagation for Gaussian distribution in Eq. (14), which has a better match with the estimates using probability density method.
[f] For Gaussian distribution, $\pm 1.96\sigma$ represents 95% confidence interval.

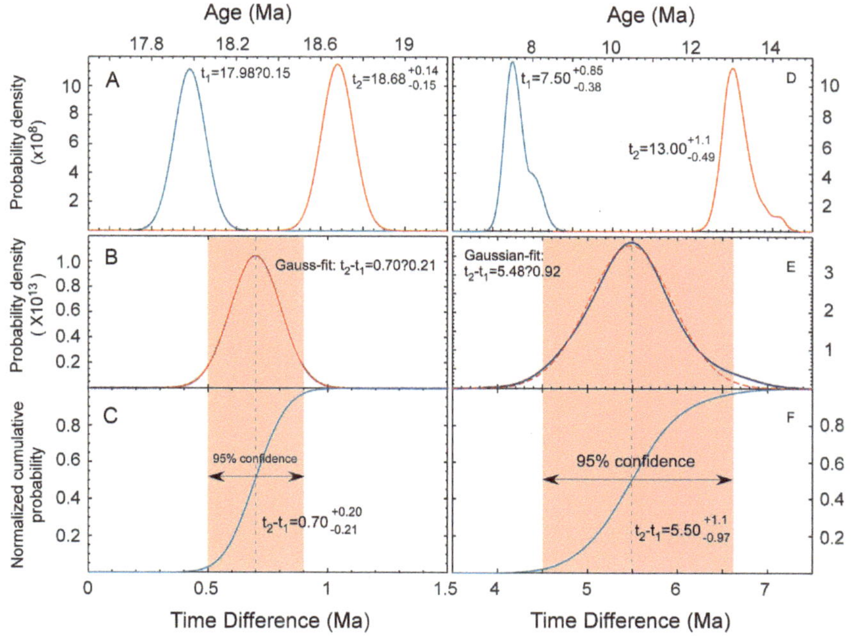

Fig. 5 Two examples of assessing the uncertainties of age differences. Subplots (A–C) depict case 2 in Table 2. In subplot 5A, two age estimates are assumed ($\sigma_m = 5$ ppm): $t_1 = 17.98 \pm 0.15$ Ma with symmetric uncertainties and obeying the Gaussian distribution, and $t_2 = 18.68^{+0.14}_{-0.15}$ Ma deviating slightly from the Gaussian distribution. The difference between them (i.e., $t_2 - t_1$) is $0.70^{+0.20}_{-0.21}$ Ma, which is almost the same as the solution by fitting the probability density to a Gaussian distribution (Fig. 5B, $t_2 - t_1 = 0.70 \pm 0.21$ Ma, 95% confidence interval), or that calculated using Eq. (9) (Table 2, method-I). Subplots (D–F) depict case 6 in Table 2. In subplot 5D, two age estimates are assumed ($\sigma_m = 5$ ppm): $t_1 = 7.50^{+0.85}_{-0.38}$ Ma and $t_2 = 13.00^{+1.1}_{-0.49}$ Ma, both of which have highly asymmetric uncertainties. Our probability density method yields an age difference of $t_2 - t_1 = 5.50^{+1.1}_{-0.97}$ (95% confidence interval, subplot 5F). In comparison, the Gaussian fit of the probability density yields $t_2 - t_1 = 5.48 \pm 0.92$ Ma (subplot 5E).

Calculation results in Table 2 show that this method produced larger errors compared with using method-II when the high and low bounds of the confidence interval are asymmetric. Method-II is calculated as follows:

$$\Delta^-_{t_2-t_1} \approx -\sqrt{\left(\Delta^-_{t_2}\right)^2 + \left(\Delta^+_{t_1}\right)^2} \qquad (14)$$
$$\Delta^+_{t_2-t_1} \approx \sqrt{\left(\Delta^+_{t_2}\right)^2 + \left(\Delta^-_{t_1}\right)^2}$$

This method empirically has a better agreement with the results from our probability density method, and therefore, it represents a better description of confidence interval, as particularly shown in general cases (e.g., cases 1, 5 and 6). However, case 6 suggests that sometimes even Method-II is not that accurate. In any event, our probability density method is always a better choice despite a slightly higher computation cost.

3.5 Uncertainty in age analyses using the seawater $^{87}Sr/^{86}Sr$ curve during the Phanerozoic era

As depicted in Figs. 6 through 13, the 95% confidence interval is presented for all Phanerozoic ages at an interval of 0.01 Ma given three different analytical precisions: $1\sigma_m = 0$, $1\sigma_m = 1$ ppm, and $1\sigma_m = 10$ ppm. The maximum and minimum of the upper and lower bounds of these uncertainties for

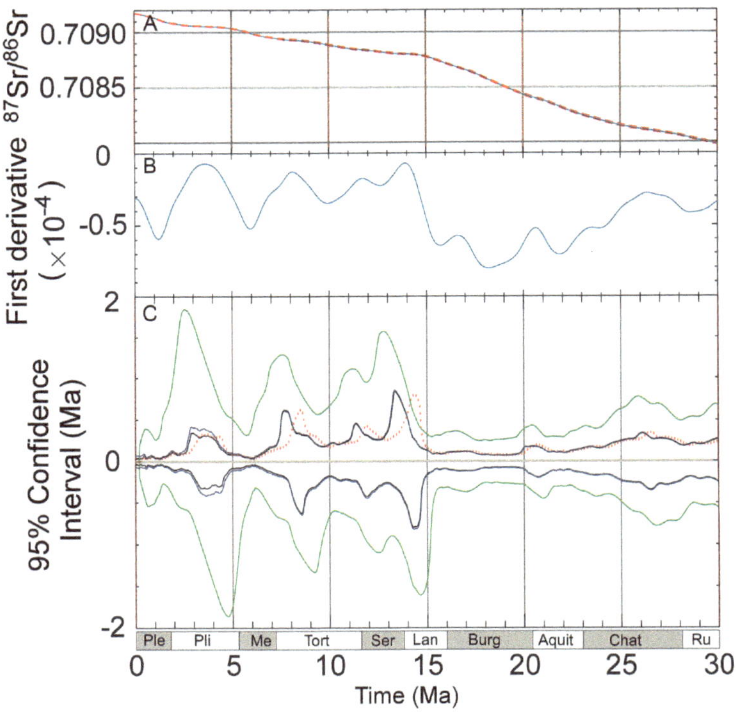

Fig. 6 The 95% confidence interval of any given age estimate between 0 and 30 Ma with known one standard deviation of $^{87}Sr/^{86}Sr$ determination ($1\sigma_m$). Subplot (A) shows the seawater $^{87}Sr/^{86}Sr$ curve and its 95% confidence envelope based on the V4B-08-04 database by McArthur and Howarth (2004); subplot (B) shows the first derivative of the seawater $^{87}Sr/^{86}Sr$ curve; subplot (C) shows the 95% confidence intervals calculated using our probability density method: black lines are for $\sigma_m = 0$, blue lines are for $\sigma_m = 1$ ppm, and green lines are for $\sigma_m = 10$ ppm, and the red dotted line is the mirror image of the lower bound for $\sigma_m = 1$ ppm (i.e., the negative blue line). The geologic time scale is based on Walker et al. (2013), and the abbreviated symbols are listed in Table 3.

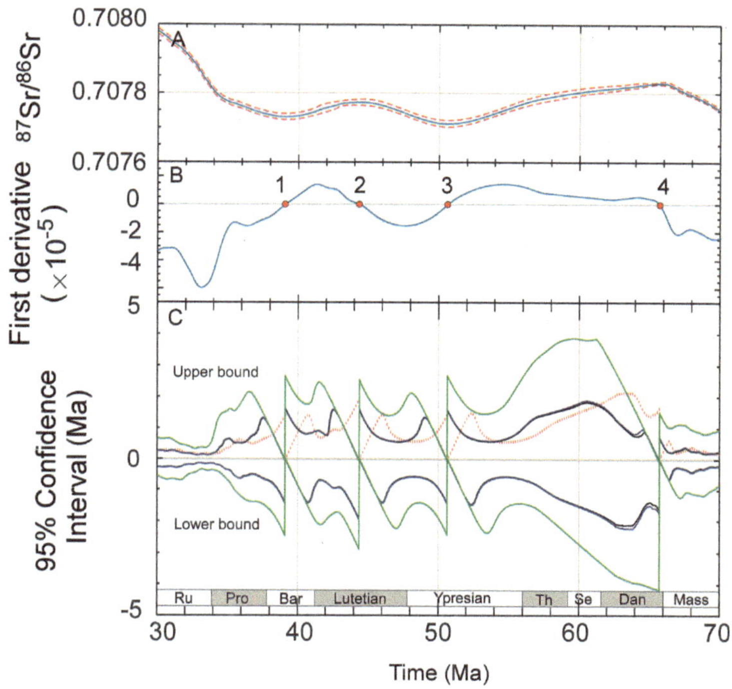

Fig. 7 The 95% confidence interval of any given age estimate between 30 and 70 Ma with known one standard deviation of $^{87}Sr/^{86}Sr$ determination ($1\sigma_m$). Subplot (A) show the seawater $^{87}Sr/^{86}Sr$ curve and its 95% confidence envelope based on the V4B-08-04 database by McArthur and Howarth (2004); subplot (B) shows the first derivative of the seawater $^{87}Sr/^{86}Sr$ curve, number indices #1 through #4 are 4 critical points listed in the first columns of Table 1; subplot (C) shows the 95% confidence intervals calculated using our probability density method: black lines are for $\sigma_m = 0$, blue lines are for $\sigma_m = 1$ ppm, green lines are for $\sigma_m = 10$ ppm, and the red dotted line is the mirror image of the lower bound for $\sigma_m = 1$ ppm (i.e., the negative blue line). The geologic time scale is based on Walker et al. (2013), and the abbreviated symbols are listed in Table 3.

each stage/epoch of geologic time scale based on Walker et al. (2013) is summarized in Table 3. Following McArthur et al. (2001), uncertainties for the Phanerozoic time in the last 509 Ma are divided into six time periods: 0–30 Ma (in Fig. 6), 30–70 Ma (in Fig. 7), 70–130 Ma (in Fig. 8), 130–210 Ma (in Fig. 9), 210–280 Ma (in Fig. 10), 280–350 Ma (in Fig. 11), 350–430 Ma (in Fig. 12), and 430–509 Ma (in Fig. 13). They are discussed separately as follows.

Between 30 and 0 Ma (Fig. 6, Table 3), the seawater $^{87}Sr/^{86}Sr$ ratio increases (Fig. 6A). There is no critical point or saddle point within this time range (Fig. 6B). When $1\sigma_m = 0$, the baseline uncertainty is $\Delta^+_{max} = \sim 0.07–0.18$ Ma for the upper bound. Bigger and asymmetric uncertainties

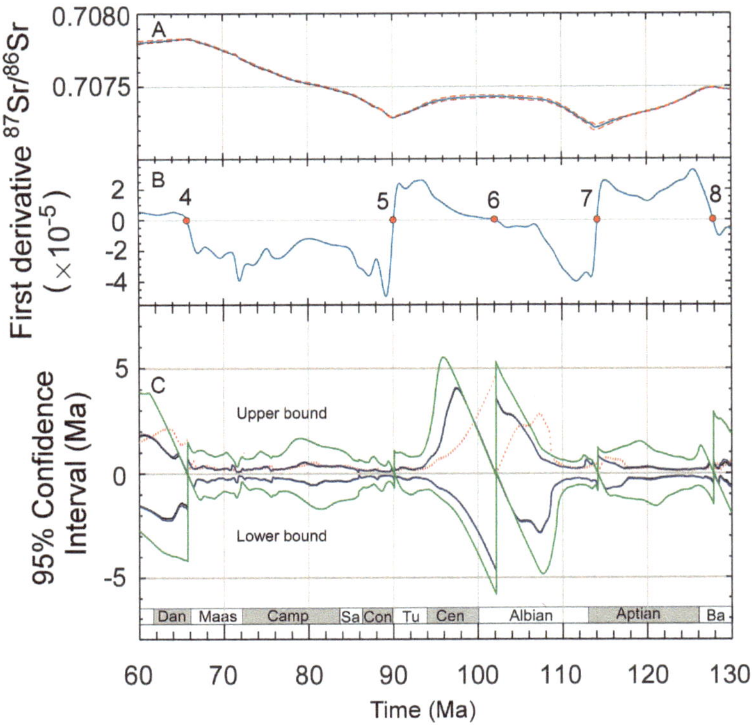

Fig. 8 The 95% confidence interval of any given age estimate between 60 and 130 Ma with known one standard deviation of $^{87}Sr/^{86}Sr$ determination ($1\sigma_m$). Subplot (A) shows the seawater $^{87}Sr/^{86}Sr$ curve and its 95% confidence envelope based on the V4B-08-04 database by McArthur and Howarth (2004); subplot (B) shows the first derivative of the seawater $^{87}Sr/^{86}Sr$ curve, number indices #5 through #8 are 4 critical points listed in the second columns of Table 1; subplot (C) shows the 95% confidence interval calculated using our probability density method: black lines are for $\sigma_m = 0$, blue lines are for $\sigma_m = 1$ ppm, green lines are for $\sigma_m = 10$ ppm, and the red dotted line is the mirror image of the lower bound for $\sigma_m = 1$ ppm (i.e., the negative blue line). The geologic time scale is based on Walker et al. (2013), and the abbreviated symbols are listed in Table 3.

occur at Pliocene, Tortonoian, Serravallian, Langhian, and Chattian (Fig. 6C) as the first derivative of the seawater curve is closer to zero (Fig. 6B). The upper and lower bounds for $1\sigma_m = 0$ is almost exactly overlapping with those for $1\sigma_m = 1$ ppm, except where the uncertainties are bigger than ~0.3 or smaller than ~−0.3 (Table 3). When $1\sigma_m = 10$ ppm, the maximum of upper bound (Δ^{+}_{max}) can be as large as ~2.0 Ma at Gelasian (Table 3). Also note, as $1\sigma_m$ increases from 0 to 10 ppm, both the range and the magnitude of the upper and lower bounds are increasing, and the asymmetry of uncertainties intensifies.

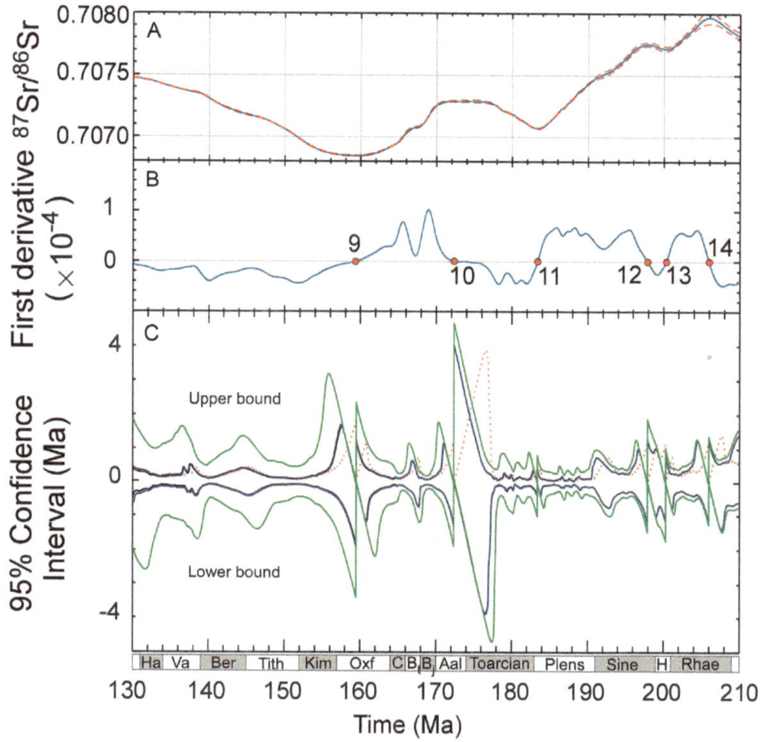

Fig. 9 The 95% confidence interval of any given age estimate between 130 and 210 Ma with known one standard deviation of $^{87}Sr/^{86}Sr$ determination ($1\sigma_m$). Subplot (A) shows the seawater $^{87}Sr/^{86}Sr$ curve and its 95% confidence envelope based on the V4B-08-04 database by McArthur and Howarth (2004); subplot (B) shows the first derivative of the seawater $^{87}Sr/^{86}Sr$ curve, number indices #9 through #14 are 6 critical points listed in the third columns of Table 1; subplot (C) shows the 95% confidence interval calculated using our probability density method: black lines are for $\sigma_m=0$, blue lines are for $\sigma_m=1$ ppm, green lines are for $\sigma_m=10$ ppm, and the red dotted line is the mirror image of the lower bound for $\sigma_m=1$ ppm (i.e., the negative blue line). The geologic time scale is based on Walker et al. (2013), and the abbreviated symbols are listed in Table 3.

Between 30 and 70 Ma (Fig. 7, Table 3), the seawater $^{87}Sr/^{86}Sr$ ratio shows two local maximum and two local minimum (Fig. 7A), having four critical points (from #1 to #4) at Bartonian, Lutetian, Ypresian, and Danian (Fig. 7B), respectively, where the upper and lower bounds of the 95% confidence intervals show a 'butterfly' pattern. This occurs because the critical point is defined as the boundary for uncertainties. For example, as time gets closer to the critical point #1 (around 39.089 Ma), the upper bound (Δ^+) of the 95% confidence interval for one of the 'twin ages' on the left diminishes to zero and its lower bound (Δ^-) reaches the minimum, whereas the lower

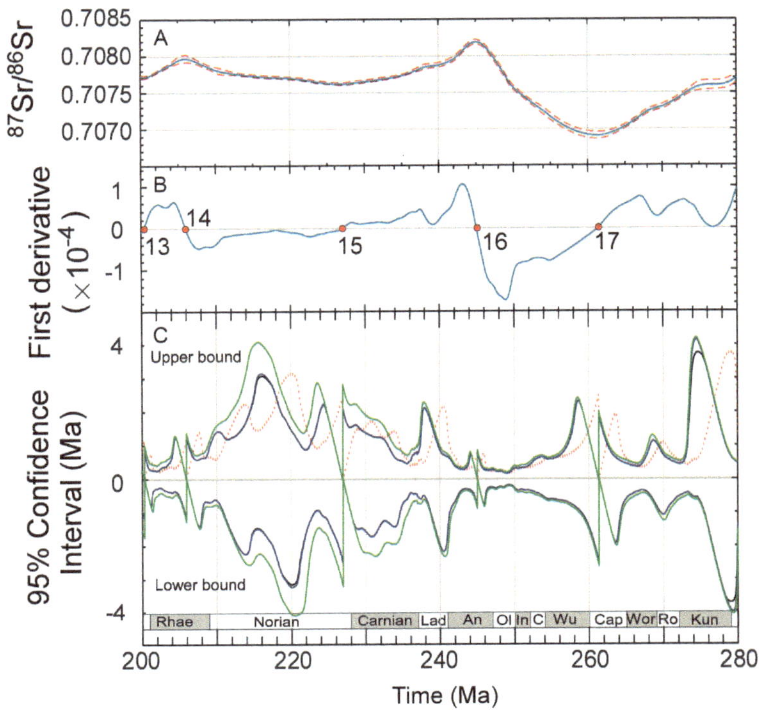

Fig. 10 The 95% confidence interval of any given age estimate between 200 and 280 Ma with known one standard deviation of $^{87}Sr/^{86}Sr$ determination ($1\sigma_m$). Subplot (A) shows the seawater $^{87}Sr/^{86}Sr$ curve and its 95% confidence envelope based on the V4B-08-04 database by McArthur and Howarth (2004); subplot (B) shows the first derivative of the seawater $^{87}Sr/^{86}Sr$ curve, number indices #13 through #17 are 5 critical points listed in the fourth columns of Table 1; subplot (C) shows the 95% confidence interval calculated using our probability density method: black lines are for $\sigma_m = 0$, blue lines are for $\sigma_m = 1$ ppm, green lines are for $\sigma_m = 10$ ppm, and the red dotted line is the mirror image of the lower bound for $\sigma_m = 1$ ppm (i.e., the negative blue line). The geologic time scale is based on Walker et al. (2013), and the abbreviated symbols are listed in Table 3.

bound (Δ^-) of the 95% confidence interval for the other one of the 'twin ages' on the right diminishes to zero and the upper bound reaches the maximum. Right at the critical point where the twin ages are identical, Δ^- reaches the minimum and Δ^+ reaches the maximum. Again, the upper and lower bounds for $1\sigma_m = 0$ is almost exactly overlapping with those for $1\sigma_m = 1$ ppm, except where the uncertainties are high. When $1\sigma_m = 10$ ppm, the maximum of upper bounds can be as big as 2.7 Ma at critical point #3 (Ypresian). But the biggest uncertainty in this range occurs at Thanetian and Selandian where the seawater $^{87}Sr/^{86}Sr$ curve is

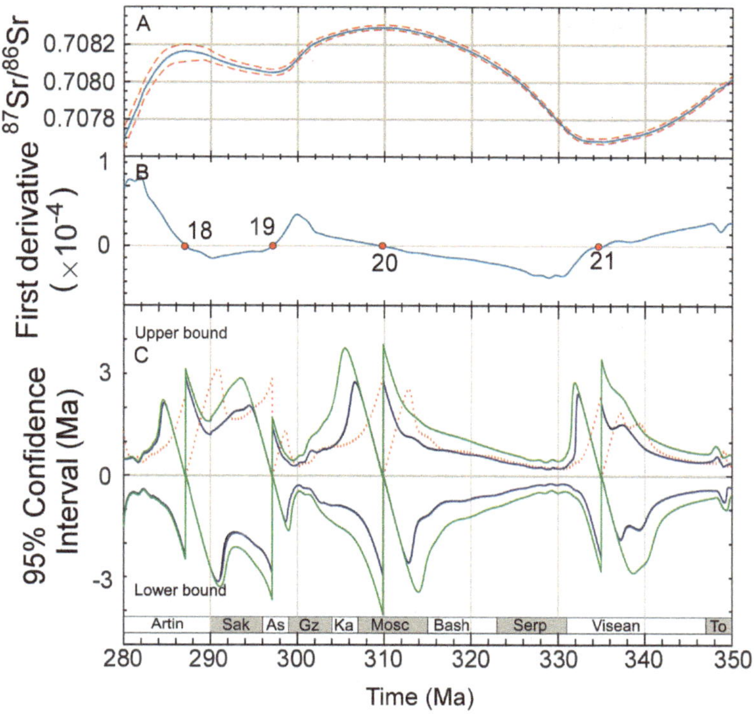

Fig. 11 The 95% confidence interval of any given age estimate between 280 and 350 Ma with known one standard deviation of $^{87}Sr/^{86}Sr$ determination ($1\sigma_m$). Subplot (A) shows the seawater $^{87}Sr/^{86}Sr$ curve and its 95% confidence envelope based on the V4B-08-04 database by McArthur and Howarth (2004); subplot (B) shows the first derivative of the seawater $^{87}Sr/^{86}Sr$ curve, number indices #18 through #21 are 4 critical points listed in the fifth columns of Table 1; subplot (C) shows the 95% confidence interval calculated using our probability density method: black lines are for $\sigma_m = 0$, blue lines are for $\sigma_m = 1$ ppm, green lines are for $\sigma_m = 10$ ppm, and red dotted line is the mirror image of the lower bound for $\sigma_m = 1$ ppm (i.e., the negative blue line). The geologic time scale is based on Walker et al. (2013), and the abbreviated symbols are listed in Table 3.

close to be flat (Fig. 7C). Between 34 and 66 Ma, the upper and lower bounds of the confidence interval are strongly asymmetric.

Between 70 and 130 Ma (Fig. 8, Table 3), the seawater $^{87}Sr/^{86}Sr$ ratio shows two local maximum and two local minimum (Fig. 8A), having four critical points (from #5 through #8) at Turonian, Albian, Aptianand, and Barremian (Fig. 8B), respectively, where the upper and lower bounds of the 95% confidence intervals show a 'butterfly' pattern. The smallest 95% confidence interval occurs between 66 and 90 Ma with $\Delta^- = -1.7 \sim -0.24$ Ma and $\Delta^+ = 0.29$–1.7 Ma (Table 3), showing a roughly symmetric upper and lower bounds.

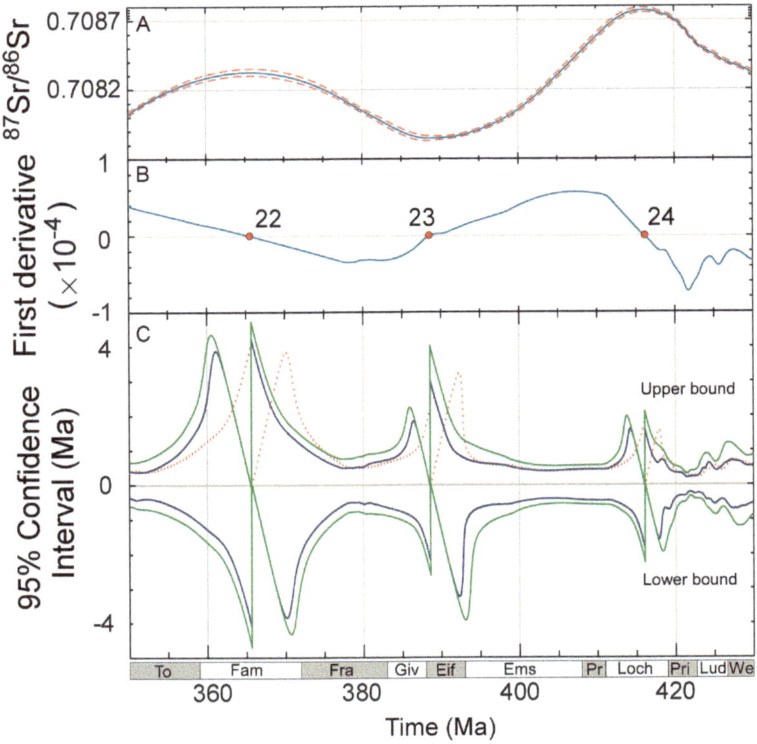

Fig. 12 The 95% confidence interval of any given age estimate between 350 and 430 Ma with known one standard deviation of $^{87}Sr/^{86}Sr$ determination ($1\sigma_m$). Subplot (A) shows the seawater $^{87}Sr/^{86}Sr$ curve and its 95% confidence envelope based on the V4B-08-04 database by McArthur and Howarth (2004); subplot (B) shows the first derivative of the seawater $^{87}Sr/^{86}Sr$ curve, number indices #22 through #24 are 3 critical points listed in the sixth columns of Table 1; subplot (C) shows the 95% confidence interval calculated using our probability density method: black lines are for $\sigma_m = 0$, blue lines are for $\sigma_m = 1$ ppm, green lines are for $\sigma_m = 10$ ppm, and the red dotted line is the mirror image of the lower bound for $\sigma_m = 1$ ppm (i.e., the negative blue line). The geologic time scale is based on Walker et al. (2013), and the abbreviated symbols are listed in Table 3.

Between 130 and 210 Ma (Fig. 9, Table 3), the seawater $^{87}Sr/^{86}Sr$ ratio shows three local maximum and three local minimum (Fig. 8A), having six critical points (from #9 to #14) at Oxfordian, Aalenian, Plensbachian, Sinemurian, Hettangian and Rhaetian (Fig. 9B), respectively, where the upper and lower bounds of the 95% confidence intervals show a 'butterfly' pattern. The largest 95% confidence interval occurs around critical point #10, where $\Delta^+_{max} = 4.7$ and $\Delta^-_{min} = -4.7$ when $1\sigma_m = 10$ ppm (Table 3). This happens because the second order derivative is also closer to zero around the critical point. Between Kimmeridgian and Toarcian, uncertainties are in general larger.

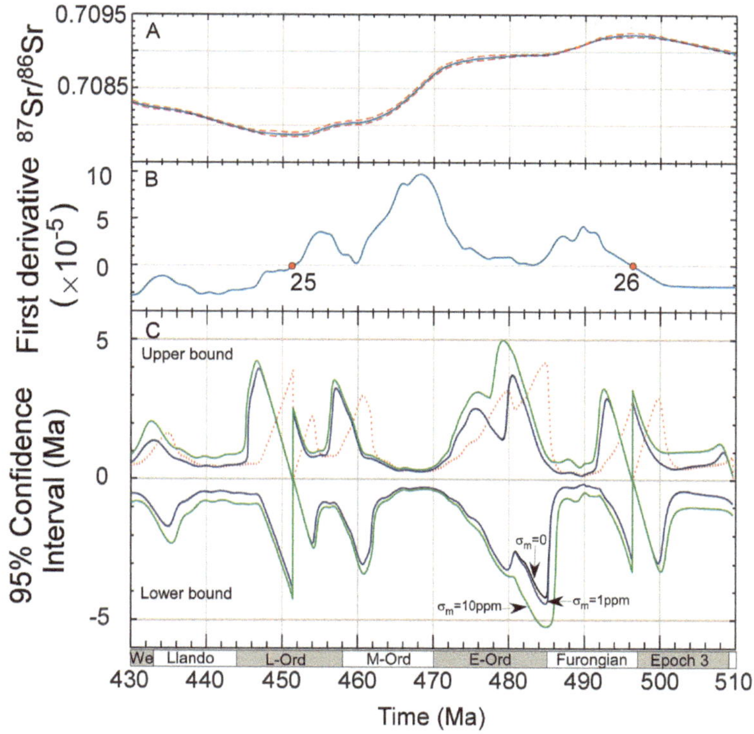

Fig. 13 The 95% confidence interval of any given age estimate between 430 and 509 Ma with known one standard deviation of $^{87}Sr/^{86}Sr$ determination ($1\sigma_m$). Subplot (A) shows the seawater $^{87}Sr/^{86}Sr$ curve and its 95% confidence envelope based on the V4B-08-04 database by McArthur and Howarth (2004); subplot (B) shows the first derivative of the seawater $^{87}Sr/^{86}Sr$ curve, number indices #25 through #26 are 2 critical points listed in the seventh columns of Table 1; subplot (C) shows the 95% confidence interval calculated using our probability density method: black lines are for $\sigma_m = 0$, blue lines are for $\sigma_m = 1$ ppm, green lines are for $\sigma_m = 10$ ppm, and the red dotted line is the mirror image of the lower bound for $\sigma_m = 1$ ppm (i.e., the negative blue line). The geologic time scale is based on Walker et al. (2013), and the abbreviated symbols are listed in Table 3.

Between 210 and 280 Ma (Fig. 10, Table 3), the seawater $^{87}Sr/^{86}Sr$ ratio shows one local maximum and two local minimum (Fig. 10A), having three critical points (from #15 through #17) at Norian, Anisian, and Capitanian (Fig. 10B), respectively, where the upper and lower bounds of the 95% confidence intervals show a 'butterfly' pattern. There are also two saddle points: one at Norian and the other at Kunguri (Fig. 10B), around which the 95% confidence interval are the widest ($\Delta^-_{min} = -3.8$ Ma and $\Delta^+_{max} = 4.2$ Ma when $1\sigma_m = 10$ ppm, Table 3). It can also be noticed that when the first derivative is high (e.g., higher than 5×10^{-5} at Olenekian), the 95%

Table 3 Compilation of range of low and high bounds of 95% confidence intervals over the last 509 Ma (Walker et al., 2013).

Era	Period_1	Period_2	Epoch	Age_1	Symbol[a]	Starting time (Ma)[b]	$\sigma_m = 0$ LB[c] Min[e]	$\sigma_m = 0$ LB[c] Max[f]	$\sigma_m = 0$ UB[d] Min	$\sigma_m = 0$ UB[d] Max	$\sigma_m = 1$ ppm LB Min	$\sigma_m = 1$ ppm LB Max	$\sigma_m = 1$ ppm UB Min	$\sigma_m = 1$ ppm UB Max	$\sigma_m = 10$ ppm LB Min	$\sigma_m = 10$ ppm LB Max	$\sigma_m = 10$ ppm UB Min	$\sigma_m = 10$ ppm UB Max
Cenozoic	Quaternary		Holocene			0.00												
			Pleistocene	Calabrian	Ple	0.01	−0.07	−0.03	0.03	0.07	−0.09	−0.04	0.04	0.12	−0.54	−0.19	0.19	0.78
				Gelasian	Ple	1.8	−0.11	−0.07	0.07	0.12	−0.13	−0.09	0.10	0.16	−0.58	−0.36	0.78	2.0
	Tertiary	Neogene	Pliocene	Piacenzian	Pli	2.6	−0.32	−0.11	0.12	0.39	−0.41	−0.13	0.16	0.42	−1.2	−0.58	1.2	1.8
				Zanclean	Pli	3.6	−0.36	−0.07	0.07	0.32	−0.42	−0.10	0.06	0.10	−1.9	−1.1	0.43	1.2
			Miocene	Messinian	Me	5.3	−0.17	−0.04	0.04	0.18	−0.17	−0.05	0.04	0.20	−1.1	−0.32	0.32	1.2
				Tortonian	Tort	7.2	−0.58	−0.17	0.18	0.60	−0.64	−0.17	0.19	0.62	−1.3	−0.59	0.57	1.3
				Serravallian	Ser	11.6	−0.52	−0.06	0.27	1.2	−0.48	−0.27	0.27	0.85	−1.1	−0.82	0.95	1.6
				Langhian	Lan	13.8	−1.4	0.00	0.00	0.76	−0.82	−0.09	0.09	0.66	−1.6	−0.32	0.30	1.0
				Burbigalian	Burg	16.0	−0.17	−0.07	0.07	0.18	−0.17	−0.08	0.07	0.18	−0.37	−0.26	0.25	0.4
				Aquitanian	Aquit	20.4	−0.18	−0.11	0.11	0.18	−0.20	−0.11	0.10	0.18	−0.45	−0.29	0.29	0.4
		Paleogene	Oligocene	Chattian	Chat	23.0	−0.31	−0.15	0.15	0.31	−0.35	−0.14	0.15	0.34	−0.78	−0.36	0.40	0.78
				Rupelian	Ru	28.1	−0.29	−0.13	0.13	0.28	−0.30	−0.14	0.13	0.29	−0.70	−0.34	0.34	0.69
			Eocene	Probiabonian	Pro	33.9	−0.69	−0.14	0.16	1.3	−0.70	−0.15	0.17	1.3	−1.6	−0.36	0.58	2.2
				Bartonian	Bar	37.8	−1.4	0.00	0.00	1.6	−1.5	0.00	0.00	1.7	−2.5	0.00	0.00	2.7
				Lutetian	Lutetian	41.2	−1.9	0.00	0.00	1.6	−2.0	0.00	0.00	1.6	−2.9	0.00	0.00	2.6
				Ypresian	Ypresian	47.8	−1.5	0.00	0.00	1.6	−1.5	0.00	0.00	1.6	−2.4	0.00	0.00	2.7
			Paleocene	Thanetian	Th	56.0	−1.4	−0.64	1.0	1.6	−1.4	−0.7	1.0	1.7	−2.4	−1.5	2.7	3.9
				Selandian	Se	59.2	−1.8	−1.4	1.6	1.9	−1.8	−1.4	1.7	1.9	−3.3	−2.4	3.6	3.9
				Danian	Dan	61.6	−2.1	0.00	0.00	1.7	−2.2	0.00	0.00	1.7	−4.2	0.00	0.00	3.6

Mesozoic	Cretaceous	Late	Maastrichtian	Maas	66.0	−0.61	−0.09	0.10	0.48	−0.64	−0.10	0.10	0.50	−1.2	−0.24	0.35	1.3
			Campanian	Camp	72.1	−0.53	−0.12	0.13	0.53	−0.56	−0.13	0.14	0.56	−1.7	−0.38	0.70	1.7
			Santonian	Sa	83.6	−0.37	−0.08	0.08	0.37	−0.39	−0.09	0.09	0.39	−1.3	−0.70	0.59	1.2
			Coniacian	Con	86.3	−0.26	−0.10	0.09	0.24	−0.28	−0.10	0.10	0.27	−0.83	−0.37	0.29	0.81
			Turonian	Tu	89.8	−0.30	0.00	0.00	0.48	−0.32	0.00	0.00	0.49	−1.0	0.00	0.00	1.4
			Cenomanian	Cen	93.9	−2.8	−0.26	0.48	4.0	−2.0	−0.27	0.49	4.1	−3.8	−0.76	1.4	5.5
		Early	Albian	Albian	100	−4.7	0.00	0.00	3.6	−4.7	0.00	−0.08	3.6	−5.8	0.00	0.00	5.3
			Aptian	Aptian	113	−0.76	0.00	0.00	0.85	−0.77	0.00	0.00	0.86	−1.6	0.00	0.00	1.5
			Barremian	Ba	126	−0.55	0.00	0.00	0.51	−0.65	0.00	0.00	0.62	−2.4	0.00	0.00	2.9
			Hauterivian	Ha	131	−0.32	−0.12	0.11	0.27	−0.38	−0.15	0.15	0.33	−2.6	−1.0	0.99	1.5
			Valanginian	Va	134	−0.41	−0.12	0.11	0.48	−0.45	−0.15	0.16	0.51	−1.7	−1.0	0.56	1.7
			Berriasian	Ber	139	−0.41	−0.09	0.08	0.39	−0.43	−0.10	0.09	0.41	−1.5	−0.49	0.48	1.4
	Jurassic	Late	Tithonian	Tith	145	−0.42	−0.08	0.08	0.37	−0.44	−0.09	0.09	0.39	−1.4	−0.46	0.45	1.3
			Kimmeridgian	Kim	152	−0.48	−0.08	0.08	1.3	−0.51	−0.10	0.09	1.4	−1.5	−0.46	0.47	3.2
			Oxfordian	Oxf	15	−1.9	0.00	0.00	1.7	−1.9	0.00	0.00	1.7	−3.4	0.00	0.00	2.4
		Middle	Callovian	C	164	−0.16	−0.08	0.08	0.20	−0.16	−0.09	0.10	0.21	−0.67	−0.25	0.26	0.58
			Bathonian	B_t	166	−0.74	−0.14	0.12	0.64	−0.76	−0.15	0.12	0.66	−1.2	−0.28	0.31	1.2
			Bajocian	B_j	168	−0.19	−0.08	0.08	0.21	−0.19	−0.08	0.08	0.21	−1.2	−0.20	0.20	0.96
			Aalenian	Aal	170	−1.6	0.00	0.00	4.0	−1.6	0.00	0.00	4.1	−1.9	0.00	0.00	4.7
		Early	Toarcian	Toarcian	174	−3.9	−0.05	0.05	2.4	−3.9	−0.06	0.06	2.4	−4.7	−0.38	0.34	3.0
			Plensbachian	Plens	183	−0.41	0.00	0.00	0.70	−0.42	0.00	0.00	0.70	−1.0	0.00	0.00	0.99
			Sinemurian	Sine	191	−1.2	0.00	0.00	1.3	−1.2	0.00	0.00	1.3	−1.5	0.00	0.00	1.9
			Hettangian	H	199	−1.2	0.00	0.00	1.0	−1.2	0.00	0.00	1.0	−1.7	0.00	0.00	1.2

Continued

Table 3 Compilation of range of low and high bounds of 95% confidence intervals over the last 509 Ma (Walker et al., 2013).—Cont'd

Era	Period_1	Period_2				Starting time	$\sigma_m = 0$				$\sigma_m = 1$ ppm				$\sigma_m = 10$ ppm			
							LB[c]		UB[d]		LB		UB		LB		UB	
			Epoch	Age_1	Symbol[a]	(Ma)[b]	Min[e]	Max[f]	Min	Max	Min	Max	Min	Max	Min	Max	Min	Max
Mesozoic			Late	Rhaetian	Rhae	201	−1.0	0.00	0.00	1.3	−1.4	0.00	0.00	1.3	−1.5	0.00	0.00	1.4
				Norian	Norian	209	−3.4	0.00	0.00	3.1	−3.2	0.00	0.00	3.2	−4.1	0.00	0.00	4.1
				Carnian	Carnian	228	−1.1	−0.40	0.46	1.6	−1.7	−0.41	0.46	1.6	−2.3	−0.64	0.70	2.3
			Middle	Ladinian	Lad	237	−2.2	−0.49	0.67	2.1	−2.2	−0.49	0.67	2.1	−2.3	−0.61	0.75	2.3
				Anisian	An	241	−1.7	0.00	0.00	0.79	−1.7	0.00	0.00	0.80	−2.2	0.00	0.00	0.86
			Early	Olenekian	Olen	247	−0.28	−0.18	0.17	0.32	−0.28	−0.18	0.17	0.32	−0.31	−0.21	0.20	0.39
				Induan	In	250	−0.34	−0.26	0.30	0.40	−0.34	−0.26	0.30	0.40	−0.42	−0.30	0.37	0.49
Paleozoic		Permian	Lopingian	Changhsingian	C	252	−0.55	−0.34	0.40	0.65	−0.55	−0.34	0.40	0.65	−0.60	−0.41	0.49	0.70
				Wuchiapingian	Wu	254	−1.4	−0.55	0.58	2.3	−1.4	−0.55	0.58	2.3	−1.5	−0.60	0.63	2.4
			Guadalupian	Capitanian	Cap	260	−2.5	0.00	0.00	1.9	−2.5	0.00	0.00	1.9	−2.6	0.00	0.00	2.0
				Wordian	Wor	265	−0.80	−0.34	0.35	1.1	−0.80	−0.35	0.35	1.1	−0.88	−0.43	0.44	1.3
				Roadian	Ro	269	−1.1	−0.53	0.44	1.1	−1.1	−0.54	0.44	1.1	−1.3	−0.64	0.54	1.2
			Cisuralian	Kungurian	Kun	272	−3.8	−0.41	0.44	3.8	−4.0	−0.41	0.44	4.2	−4.0	−0.49	0.53	4.2
				Artinskian	Artin	27	−3.8	0.00	0.00	2.9	−4.0	0.00	0.00	2.9	−4.1	0.00	0.00	3.1
				Sakmarian	Sak	290	−3.1	−1.7	1.0	2.1	−3.1	−1.7	1.0	2.1	−3.3	−2.1	1.0	2.9
				Asselian	As	296	−2.9	0.00	0.00	1.5	−2.9	0.00	0.00	1.5	−3.7	0.00	0.00	1.7
	Carboniferous	Pennsylvanian	Late	Gzhelian	Gz	299	−0.88	−0.28	0.29	0.86	−0.89	−0.28	0.29	0.87	−1.6	−0.45	0.45	2.1
				Kasimovian	Ka	304	−1.2	−0.87	0.86	2.8	−1.2	−0.88	0.87	2.8	−2.0	−1.4	2.1	3.8
			Middle	Moscovian	Mosc	307	−2.9	0.00	0.00	2.8	−3.0	0.00	0.00	2.8	−4.1	0.00	0.00	3.9
			Early	Bashkirian	Bash	315	−1.2	−0.55	0.47	0.84	−1.2	−0.56	0.48	0.85	−2.4	−0.90	0.78	1.5
		Mississippian	Late	Serpukhovian	Serp	323	−0.55	−0.22	0.21	0.47	−0.56	−0.22	0.22	0.48	−0.90	−0.42	0.42	0.78
			Middle	Visean	Visean	331	−2.4	0.00	0.00	2.4	−2.4	0.00	0.00	2.4	−2.9	0.00	0.00	3.4
			Early	Tournasian	To	347	−1.1	−0.32	0.31	1.73	−1.1	−0.32	0.31	1.7	−1.4	−0.59	0.57	2.6

Paleozoic	Devonian	Late	Famennian	Fam	359	−4.1	0.00	0.00	4.3	−4.1	0.00	0.00	4.3	−4.7	0.00	0.00	4.8
			Frasnian	Fra	372	−2.0	−0.50	0.49	1.2	−2.0	−0.50	0.49	1.2	−2.8	−0.76	0.75	1.5
		Middle	Givetian	Giv	383	−1.7	−0.65	0.54	1.9	−1.7	−0.65	0.54	1.9	−2.1	−0.88	0.53	2.3
			Eifelian	Eif	388	−3.2	0.00	0.00	3.0	−3.3	0.00	0.00	3.1	−3.9	0.00	0.00	4.0
		Early	Emsian	Ems	393	−1.3	−0.40	0.40	0.76	−1.3	−0.40	0.40	0.77	−3.9	−0.54	0.54	1.3
			Pragian	Pr	408	−0.45	−0.43	0.44	0.4	−0.45	−0.44	0.44	0.45	−0.58	−0.55	0.56	0.59
			Lochkovian	Loch	411	−1.9	0.00	0.00	1.6	−1.9	0.00	0.00	1.6	−2.2	0.00	0.00	2.1
	Silurian	Pridoli		Pri	419	−0.72	−0.21	0.20	0.48	−0.72	−0.21	0.20	0.49	−1.5	−0.33	0.32	0.72
		Ludlow	Ludfordian	Lud	423	−0.56	−0.24	0.28	0.62	−0.56	−0.24	0.28	0.63	−0.87	−0.41	0.57	0.97
			Gorstian	Lud	426	−0.65	−0.42	0.52	0.68	−0.65	−0.42	0.53	0.69	−0.89	−0.65	0.97	1.2
		Wenlock	Homerian	We	427	−0.70	−0.52	0.53	0.70	−0.71	−0.53	0.53	0.71	−1.2	−0.85	0.81	1.2
			Sheinwoodian	We	430	−1.1	−0.51	0.58	1.4	−1.1	−0.52	0.58	1.4	−1.3	−0.79	0.88	2.1
		Llando-very	Telychian	Llando	433	−1.7	−0.48	0.44	1.4	−1.7	−0.49	0.44	1.4	−2.3	−0.87	0.76	2.1
			Aeronian	Llando	439	−0.48	−0.45	0.43	0.48	−0.49	−0.45	0.44	0.48	−0.87	−0.78	0.76	0.81
			Rhuddanian	Llando	441	−0.51	−0.43	0.43	0.51	−0.51	−0.44	0.43	0.52	−0.89	−0.76	0.76	0.92
	Ordovician	Late	Hirnantian	L-Ord	444	−0.52	−0.50	0.51	0.60	−0.52	−0.51	0.52	0.61	−0.89	−0.88	0.92	1.2
			Katian	L-Ord	445	−2.3	−0.49	0.54	1.0	−0.71	−0.50	0.55	1.0	−1.1	0.00	0.00	4.3
			Sandvian	L-Ord	453	−2.3	−0.82	0.80	3.3	−2.3	−0.82	0.81	3.3	−2.4	−0.96	0.35	3.6
		Middle	Darriwilian	M-Ord	458	−3.0	−0.31	0.31	2.8	−3.0	−0.31	0.31	2.8	−3.4	−0.38	0.38	3.0
			Dapingian	M-Ord	467	−0.34	−0.29	0.28	0.38	−0.34	−0.29	0.28	0.38	−0.41	−0.35	0.35	0.47
		Early	Floian	E-Ord	470	−0.33	−0.33	0.38	2.6	−2.6	−0.33	0.38	2.6	−2.8	−0.39	0.47	3.7
			Tremadocian	E-Ord	478	−4.2	−2.5	0.61	3.7	−4.4	−2.6	0.61	3.8	−5.2	−2.8	0.89	5.0

Continued

Table 3 Compilation of range of low and high bounds of 95% confidence intervals over the last 509 Ma (Walker et al., 2013).—Cont'd

Era	Period_1	Period_2					$\sigma_m = 0$				$\sigma_m = 1$ ppm				$\sigma_m = 10$ ppm			
						Starting time	LB[c]		UB[d]		LB		UB		LB		UB	
			Epoch	Age_1	Symbol[a]	(Ma)[b]	Min[e]	Max[f]	Min	Max	Min	Max	Min	Max	Min	Max	Min	Max
Paleozoic		Cambrian	Furongian	Age 10	Furongian	485	−4.1	−0.15	0.16	0.61	−4.4	−0.15	0.16	0.61	−5.2	−0.46	0.46	0.89
				Jiangshanian	Furongian	490	−1.1	−0.16	0.21	2.9	−1.1	−0.16	0.22	2.9	−1.5	−0.46	0.61	3.3
				Paibian	Furongian	494	−2.8	0.00	0.00	2.8	−2.8	0.00	0.00	2.8	−3.3	0.00	0.00	3.2
			Epoch 3	Guzhangian	Epoch 3	497	−2.9	−0.58	0.47	2.2	−2.9	−0.58	0.48	2.2	−3.2	−0.57	1.0	2.6
				Drumian	Epoch 3	501	−1.1	−0.41	0.41	0.54	−1.1	−0.42	0.42	0.55	−2.1	−0.99	1.0	1.0
				Age 5	Epoch 3	505	−0.71	−0.49	0.54	1.0	−0.71	−0.50	0.55	1.0	−1.1	−1.0	0.73	1.3
			Epoch 2	Age 4		509												

[a]All symbols listed here are based on America geologic time scale (Walker et al., 2013).
[b]Numerical age of the starting time estimate in of each mega-annum (Ma) based on Walker et al. (2013).
[c]LB: lower bound.
[d]UB: upper bound.
[e]Minimum values with two significant numbers.
[f]Maximum values with two significant numbers.

confidence interval for $1\sigma_m = 1$ ppm differ slightly from that for $1\sigma_m = 10$ ppm, suggesting analytical uncertainty is less important under those conditions.

Between 280 and 350 Ma (Fig. 11, Table 3), the seawater $^{87}Sr/^{86}Sr$ ratio shows two local maximum and two local minimum (Fig. 11A), having four critical points (from #18 to #21) at Artinskian, Asselian, Moscovian, and Visean (Fig. 11B), respectively, where the upper and lower bounds of the 95% confidence intervals show a 'butterfly' pattern. The highest age uncertainty occurs at around critical point #20 (e.g., $\Delta^-_{min} = -3.8$ Ma and $\Delta^+_{max} = 4.2$ Ma when $1\sigma_m = 10$ ppm, Table 3), where the second derivative is closer to zero. Moreover, the range of the 95% interval around critical point #18 (e.g., $\Delta^-_{min} = -2.4$ Ma and $\Delta^+_{max} = 3.1$ Ma when $1\sigma_m = 10$ ppm, Table 3) is similar to that around #19 (e.g., $\Delta^-_{min} = -3.7$ Ma and $\Delta^+_{max} = 1.7$ Ma when $1\sigma_m = 10$ ppm, Table 3), despite a noticeable wider 95% confidence envelope around critical point #18 than around critical point #19 (Fig. 11A).

Between 350 and 430 Ma (Fig. 12, Table 3), the seawater $^{87}Sr/^{86}Sr$ ratio shows two local maximum and one local minimum (Fig. 12A), having three critical points (from #22 to #24) at Famennian, Eifelian, and Lochkovian (Fig. 12B), respectively, where the upper and lower bounds of the 95% confidence intervals show a 'butterfly' pattern. The highest age uncertainty occurs at around critical point #22 (e.g., $\Delta^-_{min} = -4.7$ Ma and $\Delta^+_{max} = 4.8$ Ma when $1\sigma_m = 10$ ppm, Table 3), where the second derivative is closer to zero. Between 395 and 410 Ma, the uncertainties are smallest (e.g., $\Delta^-_{min} = -0.58$ Ma and $\Delta^+_{max} = 0.59$ Ma when $1\sigma_m = 10$ ppm, Table 3).

Between 430 and 509 Ma (Fig. 13, Table 3), the seawater $^{87}Sr/^{86}Sr$ ratio shows one local maximum and one local minimum (Fig. 13A), having two critical points (from #25 to #26) at Late-Ordovician and Furongian (Fig. 13B), respectively, where the upper and lower bounds of the 95% confidence intervals show a 'butterfly' pattern. There are also two saddle points: one at Middle-Ordovician and the other at Early-Ordovician (Fig. 13B), around which the 95% confidence intervals are the widest ($\Delta^-_{min} = -5.2$ Ma and $\Delta^+_{max} = 5.0$ Ma when $1\sigma_m = 10$ ppm, Table 3). During Dapingian, the uncertainties are smallest ($\Delta^-_{min} = -0.41$ Ma and $\Delta^+_{max} = 0.47$ Ma when $1\sigma_m = 10$ ppm, Table 3).

These exercises demonstrate that the slope of the seawater $^{87}Sr/^{86}Sr$ curve is the most important factor to control the magnitude of uncertainty at a given inverted age. When the slope is zero, the age uncertainties around the saddle points are in general larger than around critical points, both of which are much bigger than when the slope is non-zero. When the slope is close to zero, the uncertainties are also affected by analytical uncertainty and the 95% confidence interval of the seawater $^{87}Sr/^{86}Sr$ curve. Under these conditions, the 95% confidence interval of inverted age is likely to be asymmetric, and will be best treated by our probability density method. When the slope of the seawater $^{87}Sr/^{86}Sr$ curve is high, the uncertainties tend to be symmetric following Gaussian distribution and can also be treated by the perturbation method.

These results could be very useful for making a decision regarding whether a specific geological problem can be solved by dating carbonate rocks using seawater $^{87}Sr/^{86}Sr$ curve. For example, around critical points or saddle points when uncertainies are high, increasing analytical precision is not able to reduce the uncertainty by much. However, even for old carbonate rocks (e.g., in Pragian around 400 Ma), high precision date with small uncertainties in age can be obtained when $^{87}Sr/^{86}Sr$ of carbonates are measured with reasonably small analytical uncertainty. These results can also be used as a guideline to determine the minimum analytical uncertainty required for solving a given problem using Sr-isotope stratigraphy.

To facilitate dating by comparing $^{87}Sr/^{86}Sr$ ratio of carbonate sample with that of Phanerozoic seawater, a Matlab program is developed using the V4B-08-04 database by McArthur and Howarth (2004). For any given $^{87}Sr/^{86}Sr$ ratio and its analytical uncertainty, the program can find all age solutions between 0 and 509 Ma and their confidence intervals of choice (the default is 95%). When given two ages and analytical uncertainties, it can also calculate the confidence interval of the age difference. The version 1.0 of this program is available from j.mcarthur@ucl.ac.uk or zwang1@ccny.cuny.edu upon request.

4. Conclusion

In this study, we developed two approaches (perturbation and probability density methods) to calculate 95% confidence intervals of ages inverted from the well-established seawater $^{87}Sr/^{86}Sr$ curve. While the perturbation method provides fast solution with low computation cost, we demonstrate the probability density method is a more versatile tool to estimate confidence intervals under a variety of conditions, including constraining uncertainy in age around critical points and saddle points, assessing the asymmetric uncertainties where the slope of seawater $^{87}Sr/^{86}Sr$ curve is close to zero and the analytical uncertainty is high, and calculating the confidence interval of age difference when one or both ages contain asymmetric uncertainties that follow a non-Gaussian distribution. Applying these methods to calculate the 95% confidence intervals of ages inverted from one version of the seawater $^{87}Sr/^{86}Sr$ curve show that the magnitude uncertainties in age are mostly controlled by the slope of the seawater $^{87}Sr/^{86}Sr$ curve. These results indicates our method may have the potential to be widely applied to estimates of uncertainty involving inversions using an empirical relationship.

Acknowledgments

This research is partly supported by Open Research Project (GPMR201814) from the State Key Laboratory of Geological Processes and Mineral Resources, China University of Geosciences, and start-up funds to Z.W. from CCNY. Z.W. would like to thank Prof. Mark Pagani for his friendship and continuous encouragement for finishing this work.

Appendix.

Assuming a function $y=f(t)$, whose inverse function also exists (i.e., $y=f(t)$ is a bijection) and can be written as $t=f^{-1}(y)$. As shown in Eq. (1), when there is an uncertainty (Δy) in y, the uncertainty (Δt) in t can be written as follows:

$$t+\Delta t=f^{-1}(y+\Delta y)$$

Expanding this function around y as a Taylor series will yield the following equation:

$$f^{-1}(y+\Delta y)=f^{-1}(y)+\frac{[f^{-1}(y)]'}{1!}\cdot\Delta y+\frac{[f^{-1}(y)]''}{2!}(\Delta y)^2 +\frac{[f^{-1}(y)]'''}{3!}(\Delta y)^3\ldots \quad (15)$$

Thus, the Taylor expansion requires the calculation of the first, second and third derivatives of the inverse function.

From the definition, the first, second and third derivative of the original function $y=f(t)$ can be written as:

$$\begin{aligned}\frac{dy}{dt}&=f'(t)\\ \frac{d^2y}{dt^2}&=f''(t)\\ \frac{d^3y}{dt^3}&=f'''(t)\end{aligned} \tag{16}$$

Thus the first derivative of the inverse function $t=f^{-1}(y)$ can be written as follows:

$$\frac{dt}{dy}=\left(f^{-1}(y)\right)'=\frac{1}{f'(t)} \tag{17}$$

Similarly, the second derivative of the inversion function can be written as follows:

$$\begin{aligned}\frac{d^2t}{dy^2}&=\left(f^{-1}(y)\right)''=\frac{d\left(f^{-1}(y)\right)'}{dy}=\frac{d\left(\frac{1}{f'(t)}\right)}{dy}=-\frac{f''(t)}{\left(f'(t)\right)^2}\cdot\frac{dt}{dy}\\ &=-\frac{f''(t)}{\left(f'(t)\right)^3}\end{aligned} \tag{18}$$

and the third derivative of the inversion function can be written as follows:

$$\begin{aligned}\frac{d^3t}{dy^3}&=\left(f^{-1}(y)\right)'''=\frac{d\left(f^{-1}(t)\right)''}{dy}=\frac{d\left(-\frac{f''(t)}{\left(f'(t)\right)^3}\right)}{dy}\\ &=\left(\frac{3\left(f''(t)\right)^2}{\left(f'(t)\right)^4}-\frac{f'''(t)}{\left(f'(t)\right)^3}\right)\cdot\frac{dt}{dy}=\frac{3\left(f''(t)\right)^2-f'''(t)\cdot f'(t)}{\left(f'(t)\right)^5}\end{aligned} \tag{19}$$

Substituting Eqs. (17) through (19) into Eq. (15) will yield Eq. (2).

References

Anderson, T.V., 2011. Efficient, Accurate, and Non-Gaussian Error Propagation Through Nonlinear, Closed-Form, Analytical System Models. Department of Mechanical Engineering. Brigham Young University.

Andersson, A.J., 2014. 19—the oceanic CaCO3 cycle. In: Holland, H.D., Turekian, K.K. (Eds.), Treatise on Geochemistry, Second ed. vol. 8. Elsevier, Oxford, pp. 519–542.

Asmerom, Y., Jacobsen, S.B., Knoll, A.H., Butterfield, N.J., Swett, K., 1991. Strontium isotopic variations of Neoproterozoic seawater: implication for crustal evolution. Geochim. Cosmochim. Acta 55, 2883–2894.

Balcaen, L., Schrijver, I.D., Moens, L., Vanhaecke, F., 2005. Determination of the 87Sr/86Sr isotope ratio in USGS silicate reference materials by multi-collector ICP–mass spectrometry. Int. J. Mass Spectrom. 242, 251–255.

Blachman, N.M., Machol, R.E., 1987. Confidence intervals based on one or more observations. IEEE Trans. Inf. Theory 33, 373–382.

Burke, W.H., Denison, R.E., Hetherington, E.A., Koepnick, R.B., Nelson, H.F., Otto, J.B., 1982. Variation of seawater 87Sr/86Sr throughout Phanerozoic time. Geology 10, 516–519.

Cleveland, W.S., 1979. Robust locally weighted regression and smoothing scatterplots. J. Am. Stat. Assoc. 74, 829–836.

Cleveland, W.S., Grosse, E., Shyu, W.M., 1992. Local regression models. In: Chambers, J.M., Hastie, T. (Eds.), Statistical Models. S. Wadsworth and Brooks/Cole, Pacific Grove (CA), pp. 309–376.

Depaolo, D.J., Ingram, B.L., 1985. High-resolution stratigraphy with strontium isotopes. Science 227, 938.

Gussone, N., Dietzel, M., 2016. Calcium isotope fractionation during mineral precipitation from aqueous solution. In: Gussone, N., Schmitt, A.-D., Heuser, A., Wombacher, F., Dietzel, M., Tipper, E., Schiller, M. (Eds.), Calcium Stable Isotope Geochemistry. Springer Berlin Heidelberg, Berlin, Heidelberg, pp. 75–110.

Halverson, G.P., Théou-Hubert, L., 2013. Seawater Sr curve. In: Rink, W.J., Thompson, J. (Eds.), Encyclopedia of Scientific Dating Methods. Springer, Netherlands, Dordrecht, pp. 1–10.

Hess, J., Bender, M.L., Schilling, J.-G., 1986. Evolution of the ratio of strontium-87 to strontium-86 in seawater from Cretaceous to present. Science 231, 979–984.

Heuvelink, G.B.M., 1998. Error Propagation in Enviromental Modelling with GIS. Taylor & Francis.

Howarth, R.J., McArthur, J.M., 1997. Statistics for strontium isotope stratigraphy: a Robust LOWESS Fit to the Marine Sr-Isotope Curve for 0 to 206 Ma, with look-up table for derivation of numeric age. J. Geol. 105, 441–456.

Lea, D.W., 2014. 14—elemental and isotopic proxies of past ocean temperatures. In: Holland, H.D., Turekian, K.K. (Eds.), Treatise on Geochemistry, Second ed. vol. 8. Elsevier, Oxford, pp. 373–397.

Liu, C., Wang, Z., Raub, T.D., 2013. Geochemical constraints on the origin of Marinoan cap dolostones from Nuccaleena Formation, South Australia. Chem. Geol. 351, 95–104.

Lowenstein, T.K., Kendall, B., Anbar, A.D., 2014. 21—the geologic history of seawater. In: Holland, H.D., Turekian, K.K. (Eds.), Treatise on Geochemistry, second ed. vol. 8. Elsevier, Oxford, pp. 569–622.

McArthur, J.M., Howarth, R.J., 2004. Sr-isotope stratigraphy. In: Gradstein, F.M., Ogg, J.G., ans Smith, A.G. (Eds.), A Geological Timescale 2004. Cambridge University Press, Cambridge, p. 589.

McArthur, J.M., Howarth, R.J., Bailey, T.R., 2001. Strontium isotope stratigraphy: LOWESS version 3: best fit to the Marine Sr-Isotope curve for 0-509 Ma and accompanying look-up table for deriving numerical age. J. Geol. 109, 155–170.

McArthur, J.M., Rio, D., Massari, F., Castradori, D., Bailey, T.R., Thirlwall, M., Houghton, S., 2006. A revised Pliocene record for marine-87Sr/86Sr used to date an interglacial event recorded in the Cockburn Island Formation, Antarctic Peninsula. Palaeogeogr. Palaeoclimatol. Palaeoecol. 242, 126–136.

McArthur, J., Howarth, R.J., Shields, G.A., 2012. Strontium isotope stratigraphy. In: The Geological Time Scale. Elsevier, pp. 127–144.

Oberkampf, W.L., DeLand, S.M., Rutherford, B.M., Diegert, K.V., Alvin, K.F., 2002. Error and uncertainty in modeling and simulation. Reliab. Eng. Syst. Saf. 75, 333–357.
Ravizza, G.E., Zachos, J.C., 2014. 20—records of cenozoic ocean chemistry. In: Holland, H.D., Turekian, K.K. (Eds.), Treatise on Geochemistry, Second ed. vol. 8. Elsevier, Oxford, pp. 543–568.
Saenger, C., Wang, Z., 2014. Magnesium isotope fractionation in biogenic and abiogenic carbonates: implications for paleoenvironmental proxies. Quat. Sci. Rev. 90, 1–21.
Shields, G.A., 1999. Working towards a new stratigraphic calibration scheme for the Neoproterozoic-Cambrian. Eclogae Geol. Helv. 92, 221–233.
Shields, G., Veizer, J., 2002. Precambrian marine carbonate isotope database: version 1.1. Geochem. Geophys. Geosyst. 3, 1–12.
Taylor, J.R., 1997. An Introduction to Error Analysis: The Study of Uncertainties in Physical Measurements. University Science Books, Sausalito, CA.
Veizer, J., 1989. Strontium isotopes in seawater through time. Annu. Rev. Earth Planet. Sci. 17, 141–167.
Veizer, J., Prokoph, A., 2015. Temperatures and oxygen isotopic composition of Phanerozoic oceans. Earth Sci. Rev. 146, 92–104.
Veizer, J., Ala, D., Azmy, K., Bruckschen, P., Buhl, D., Bruhn, F., Carden, G.A.F., Diener, A., Ebneth, S., Godderis, Y., Jasper, T., Korte, C., Pawellek, F., Podlaha, O.G., Strauss, H., 1999. 87Sr/86Sr, δ13C and δ18O evolution of Phanerozoic seawater. Chem. Geol. 161, 59–88.
Walker, J.D., Geissman, J.W., Bowring, S.A., Babcock, L.E., 2013. The geological society of America geologic time scale. GSA Bull. 125, 259–272.

CHAPTER THREE

Isotope stratigraphy of Precambrian sedimentary rocks from Brazil: Keys to unlock Earth's hydrosphere, biosphere, tectonic, and climate evolution

Fabrício de A. Caxito[a,*], Gabriel J. Uhlein[a], Alexandre Uhlein[a], Antônio Carlos Pedrosa-Soares[a], Matheus Kuchenbecker[a,b], Humberto Reis[c], Alcides N. Sial[d], Valderez P. Ferreira[d], Carlos José Souza Alvarenga[e], Roberto Ventura Santos[e], Lucieth Cruz Vieira[e], Elton Dantas[e], Marly Babinski[f], Ricardo Trindade[g], Paulo César Boggiani[f], Lucas Warren[h], João Pedro Hippertt[a,c], Marcus Paulo Sotero[a,i], Janaína Rodrigues de Paula[a,j]

[a]Centro de Pesquisas Manoel Teixeira da Costa, Instituto de Geociências, Universidade Federal de Minas Gerais (CPMTC-IGC-UFMG), Campus Pampulha, Belo Horizonte, Brazil
[b]Laboratório de Estudos Tectônicos/CeGeo/ICT/UFVJM, Diamantina, Brazil
[c]Laboratório de Modelagem Tectônica (LabMod), Departamento de Geologia, Universidade Federal de Ouro Preto (UFOP), Campus Morro do Cruzeiro, Ouro Preto, Brazil
[d]NEG-LABISE, Departamento de Geologia, Universidade Federal de Pernambuco, Recife, Brazil
[e]Instituto de Geociências, Universidade de Brasília, Campus Darcy Ribeiro, Brasília, Brazil
[f]Instituto de Geociências, Universidade de São Paulo, São Paulo, Brazil
[g]Instituto de Astronomia, Geofísica e Ciências Atmosféricas, Universidade de São Paulo, São Paulo, Brazil
[h]Departamento de Geologia Aplicada, Instituto de Geociências e Ciências Exatas, Universidade Estadual Paulista, Rio Claro, Brazil
[i]CPRM—Serviço Geológico do Brasil (Brazilian Geological Survey), SUREG/BH, Belo Horizonte, Brazil
[j]VALE S/A, Rio de Janeiro, Brazil
*Corresponding author: e-mail address: caxito@ufmg.br

Contents

Stratigraphy & Timescales, Volume 4
ISSN 2468-5178
https://doi.org/10.1016/bs.sats.2019.08.002

Abstract

Brazil is a natural laboratory for the application of isotope stratigraphy tools on the study of Precambrian sedimentary successions, with excellent registers of Archean, Paleo, Meso, and Neoproterozoic age. Those successions witnessed the most extreme and unidirectional events in Earth's history, such as the Paleoproterozoic and Neoproterozoic oxygenation events (GOE and NOE), the main stages of iron formation deposition, the global Cryogenian glaciations and the explosive diversification of metazoans near the Precambrian/Cambrian border. Chemical sedimentary rocks such as carbonates, iron formations and evaporites that can act as natural recorders of seawater composition are present in all of those successions and furnish important pieces of evidence in order to reconstruct the chemostratigraphic record. The field of isotope stratigraphy keeps expanding in Brazil and novel information from developing isotopic systems will surely add to the growing database and provide important information on Earth's ancient biogeochemical cycles in the years to come.

1. Introduction

Stable and radiogenic isotopes are used as natural tracers of origin and processes in the most varied fields, such as ecology, forensic sciences, agronomy, archeology, and food science. In the Earth Sciences, isotopes have found immense niches in both geochronology and as proxies for the sources and ancient chemical conditions of the environment where rocks were formed. The analysis of punctual samples in comprehensive stratigraphic packages permits the possibility of assessing vertical variations of the isotopic proxies, which can reflect ancient Earth's atmosphere, hydrosphere and biosphere geochemical conditions. Those geochemical conditions are in turn complexly interlinked to tectonic, climate and biological changes that occurred during the evolution of the planet. Chemical sedimentary rocks (carbonates, Banded Iron Formations—BIFs, evaporites, phosphorites, cherts) are especially useful in this respect, as (when free of detrital particles and post-depositional fluid-rock interaction) they are thought to broadly represent the chemical composition of seawater during sedimentary deposition.

Isotope stratigraphy of ancient sedimentary rocks is thus used to unravel: (1) seawater chemistry during deposition and its historical evolutionary patterns; (2) sedimentary provenance patterns; (3) establishing of local, regional and global stratigraphic correlations based on similar isotope values or breaks in the record; (4) causal or inter-relational links with biological, tectonic and climate conditions. In Precambrian sedimentary rocks, isotope stratigraphy is a crucial tool because the lack of abundant fossiliferous strata hampers the use of biostratigraphic tools (e.g., Kah et al., 2001; Knoll and Walter, 1992). Due to the current state of understanding of fractionation processes, natural cycles, and analytical development, some isotope systems have been more widely used than others in chemostratigraphic studies of Precambrian successions (e.g., C, O, Sr, S, etc.), but the field keeps growing and other, non-traditional systems have been put to test, some of them yielding promising results (e.g., Li, Mg, Ca, Ba, Cr, Mo, Fe, Cu, Zn, Cd, Hg, B, Se, N, triple O isotopes, etc.).

Brazil is a privileged place for the study of Precambrian sedimentary successions and a natural laboratory for the application of isotope stratigraphy tools, with important outcropping rock units of Archean, Paleo, Meso, and Neoproterozoic age (Fig. 1). Those successions span the most extreme and unidirectional events in Earth's history, such as the first Global Oxygenation Event (GOE) and the main stage of BIF deposition, the widespread Cryogenian glaciations and the Neoproterozoic Oxygenation Event (NOE) followed by the explosive diversification of metazoans. Carbonates, BIFs and evaporites are present in Brazilian sedimentary successions throughout geological time, furnishing important pieces of evidence in order to reconstruct the chemostratigraphic record, ancient geochemical and biochemical conditions.

In this chapter, we will focus on practical descriptions of the main isotope systems used in the last years as proxies to unravel the isotope stratigraphy of Brazilian Precambrian sedimentary sequences, followed by a description of the main results and interpretations put forward for each sequence through the gathered dataset. Detailed descriptions of the fundamental concepts of

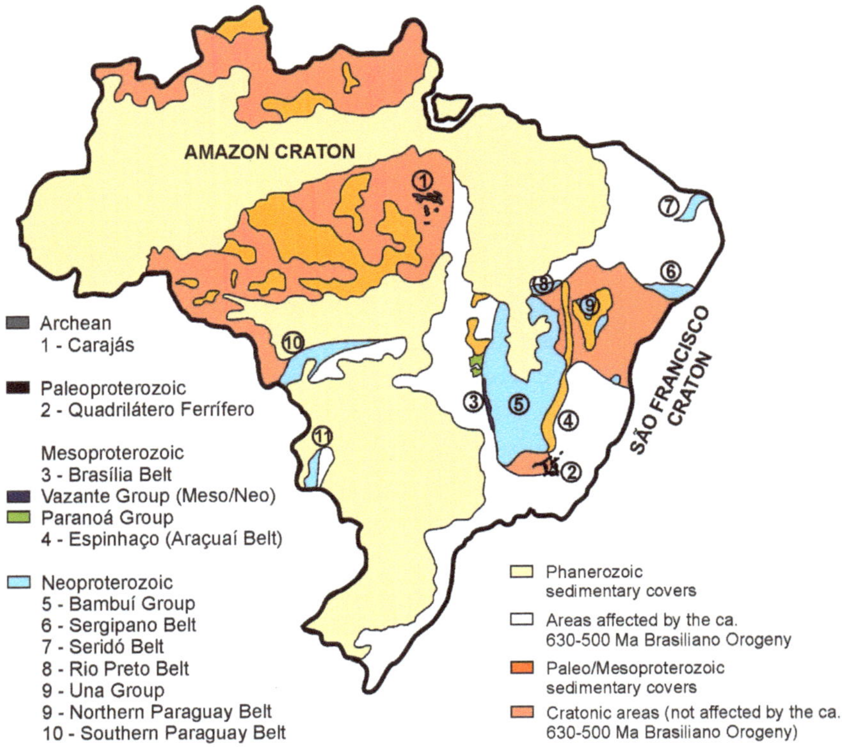

Fig. 1 Location of the discussed Precambrian sedimentary successions in the Brazilian territory.

fractionation, mass spectrometry and representation of isotope ratios of each system can be found elsewhere (e.g., Faure and Mensing, 2005; Hoefs, 2009) and will not be discussed here.

2. Isotope systems as proxies for Earth's hydrosphere, biosphere, tectonic and climate evolution

Stable isotopes of C (both in carbonates and organic matter), O, S (both in pyrite and carbonate-associated sulfate—CAS) and N are the most widely applied systems in isotope stratigraphy, although the latter has not yet been applied to Brazilian Precambrian sedimentary successions and interpretation of the $\delta^{15}N$ signals as tracers of ancient seawater composition is still controversial (Halverson et al., 2010; Sial et al., 2015a, 2018). Radiogenic isotopes of Sr and Nd follow closely. The non-traditional isotope systems of Li, Mg, Ca, Ba, Cr, Mo, Fe, Cu, Zn, Cd, Hg, B, Se, and triple O isotopes have all been proposed as possible tracers of ancient biogeochemical conditions, with some of them already put to test and proven useful in Precambrian sedimentary successions (e.g., Fe, Cr, Zn, Cd, Ba, Hg, B and triple O). We will only discuss below the systems that have been applied to Brazilian sequences. Information about the other systems can be found elsewhere (e.g. Faure and Mensing, 2005; Halverson et al., 2010; Sial et al., 2015a, 2018; Teng et al., 2017).

2.1 Stable isotopes

The natural distribution of stable isotopes of light elements is the most widely used isotope stratigraphy tool. C, O, and S isotopes are the champions in this respect, although studies using the first two surpass by far studies using S isotopes, due to its more time-consuming analytical settings. Other, non-traditional stable isotope systems are being constantly applied, albeit some are yet in the early stages of analytical development and/or their natural cycles and fractionation processes are not yet fully understood.

2.1.1 Carbon isotopes

By far, stable carbon isotopes are the most widely used tools in the chemostratigraphy of Brazilian Precambrian sedimentary rocks (e.g., reviews in Misi et al., 2007; Sial et al., 2016). This is mostly due to (1) the fast acquisition of high volumes of data in gas-source Isotope Ratio Mass Spectrometry (IRMS) machines, which are readily available in various Brazilian isotope geochemistry labs; (2) the relatively easiness of collection and preparation of samples, allowing to obtain large datasets with the number of analyzed samples depending mostly on the sampling strategies, required stratigraphic resolution and costs involved; (3) the relatively low cost of individual analysis or analysis of sample batches; (4) the level of development and understanding of the carbon cycle and isotope fingerprinting of distinct processes and reservoirs, which, albeit far from completely understood, are definitely on a more advanced level of comprehension than other isotope systems used in isotope stratigraphy.

The rationale behind using stable carbon isotopes to unravel ancient biogeochemical conditions is that unaltered carbon-rich sedimentary rocks (such as carbonates and carbonate-facies BIF) precipitated in equilibrium with seawater should reflect the composition of the Dissolved Inorganic Carbon (DIC) pool, which in turn depends on variations on the global carbon sources and sinks (Halverson et al., 2010 and references therein). There are, however, intense debates as whether carbon isotope ratios in

Precambrian carbonates represent faithful records of original seawater composition, especially for those that have suffered post-depositional alteration during diagenesis, metamorphism, surficial weathering and/or hydrothermalism (Jacobsen and Kaufman, 1999). It should also be remarked that clastic, reworked carbonate rocks (i.e., grainstone, calcarenite), despite sometimes being analyzed as chemical sedimentary rocks, could represent a mixed isotope signal from the source carbonate rocks; and that only marine carbonates deposited in open-sea settings should be regarded as possible recorders of global marine signals. Nevertheless, the extreme fluctuations recorded in the $\delta^{13}C$ record, especially for the Neoproterozoic and during the Paleoproterozoic Lomagundi excursion, are widely regarded as produced by real seawater variations that testify extreme fluctuations in the Precambrian carbon cycle. A number of petrographic and elemental/isotopic (e.g., Mn/Sr, Mg/Ca versus $\delta^{13}C$, $\delta^{13}C$ versus $\delta^{18}O$) screening proxies can be used to assess the extent of secondary alteration and thus how dependable the obtained $\delta^{13}C$ values are for the reconstruction of ancient seawater signals.

Carbon isotope studies have been conducted in order to address a number of issues (Halverson et al., 2010) including: (i) understanding and modeling the global carbon cycle and oxygen generation (Derry et al., 1992); (ii) constraining the mechanisms and causes of Neoproterozoic glaciations and associated cap carbonates (Halverson et al., 2002; Higgins and Schrag, 2003; Hoffman et al., 1998; Kaufman et al., 1997; Kennedy et al., 2001; Sansjofre et al., 2011; Schrag et al., 2002); (iii) understanding basin architecture and sequence stratigraphy (e.g., Caetano-Filho et al., 2019; Cozzi et al., 2004; Martins and Lemos, 2007; Perrella Júnior et al., 2017; Schrag et al., 2002); (iv) tracking changes in bioproductivity and sedimentary provenance both locally and basinwide (e.g., Hayes et al., 1999; Hippertt et al., 2019; Uhlein et al., 2019); (v) local and global correlation of carbonate-rich sections through stratigraphic comparison of $\delta^{13}C$ anomalies (e.g., Caxito et al., 2012; Halverson et al., 2005).

The isotope composition of carbon in sedimentary carbonates remained fairly constant and close to the VPDB (Vienna Pee Dee Belemnite) standard during most of the Precambrian, which indicates a steady-state of Earth's carbon cycle during this time interval. There are, however, exceptional regional and global anomalies, especially during the Paleoproterozoic (the positive Lomagundi excursion) and during the Neoproterozoic (characterized by large, both positive and negative swings of the $\delta^{13}C$ values). Those variations are most probably related to changes in the atmosphere and biosphere caused by enhanced organic carbon burial and global glaciations, respectively (Bekker et al., 2006; Hoffman et al., 1998).

The Lomagundi excursion was defined in Paleoproterozoic dolomites of the Lomagundi Group of northwestern Zimbabwe, which are characterized by positively fractionated $\delta^{13}C$ values of +2.7‰ to +13.4‰ (Schidlowski et al., 1976). The authors interpreted such excursion as the result of deposition in evaporite basins under ^{13}C-enriched seawater, with the corresponding complementary ^{12}C deposited separately in kerogen-rich shales in other parts of the basin. ^{13}C-enriched carbonates deposited in the same time interval (2.3–2.2 Ga) were further recognized in over 15 formations worldwide, allowing to classify the Lomagundi excursion as the fingerprint of a global-scale perturbation in the carbon cycle (Bekker et al., 2003, 2006; Maheshwari et al., 2010; Melezhik et al., 1999, 2007).

The hypothetical variations of Earth's carbon cycle in response to global glaciation are explained through the biological carbon pump model (e.g., Kump, 1991). According to this model, ^{12}C is preferentially incorporated over ^{13}C by pelagic organisms, which upon death sink to the bottom of the ocean, where most of the organic matter is oxidized to CO_2 (while some of it is buried). This causes partitioning of the marine carbon sink into two distinct reservoirs. The DIC pool in oceanic bottom waters is boosted in ^{12}C due to organic matter settling, while a correspondent enrichment in ^{13}C of DIC occurs in oceanic surface waters. The difference in $\delta^{13}C$ values ($\Delta^{13}C$) of the two reservoirs is proportional to bioproductivity in surface waters, which in turn, is proportional to nutrient availability. Under normal conditions, nutrients are delivered to surface waters through continental weathering of continental crust. If the continents and oceans are covered by perennial ice sheets, then chemical weathering and the flowing of rivers would temporarily be interrupted (or significantly diminished), which along with the fact that the oceans would be isolated from the atmosphere creating widespread anoxia and locking circulation, would cause a bioproductivity crisis. Mantle-derived carbon would then dominate ocean waters, driving down $\delta^{13}C$ values closer to the canonical mantle value of −6‰.

Hoffman et al. (1998) put forward the "Snowball Earth" hypothesis to explain the variations in $\delta^{13}C$ recorded in cap carbonates related to Neoproterozoic glaciations worldwide. According to this model, a fully (or nearly fully) ice-covered Planet Earth (the icehouse condition) would cause an increase of the partial pressure of volcanic-derived CO_2 in the atmosphere due to physical separation from the oceans by a nearly continuous ice sheet and interruption of chemical weathering in the continents (i.e., the major CO_2 sinks would have temporarily been put out of operation). In turn, the elevated ρCO_2 in the atmosphere would eventually cause a greenhouse condition and deglaciation. Post-glacial upwelling of ^{13}C-depleted bottom waters then flooded the continental shelves causing deposition of shallow-water cap carbonates with negatively fractionated $\delta^{13}C$ sitting directly in sharp contact atop glacial diamictites and related rocks, generated the well-known climatic dichotomy of Cryogenian sequences (cold-climate rocks such as glacial diamictites in sharp contact with warm-climate rocks such as carbonates). The periods of greenhouse caused acceleration of continental weathering leading to enhanced delivery of nutrients to the oceans, which in turn boosted productivity and caused deposition of carbonates with anomalously high $\delta^{13}C$ following the ^{13}C-depleted cap carbonates.

The swinging back and forth of $\delta^{13}C$ values during the Neoproterozoic (with major glaciations being both preceded and succeeded by major negative shifts) suggests at least two major periods of global glaciation, the mid-Cryogenian Sturtian (717–663 Ma; Cox et al., 2018; Macdonald et al., 2010; MacLennan et al., 2018) and the end-Cryogenian Marinoan (ca. 650–635 Ma; Condon et al., 2005; Hoffmann et al., 2004, 2017 and references therein), preceded by the Islay and Trezona negative anomalies, respectively. The cap carbonate of the Marinoan glaciation (dated through zircon U-Pb ID-TIMS analysis from volcanic tuff intercalations at 635.2 ± 0.5 Ma in China, 636.4 ± 0.5 Ma in South Australia, 635.5 ± 0.2 Ma in Namibia and at 632.3 ± 5.9 Ma through Re-Os isochronic dating of black shales in Canada; Calver et al., 2013; Condon et al., 2005; Hoffmann et al., 2004; Prave et al., 2016; Rooney et al., 2015) represents a synchronous worldwide depositional event and marks the base of the formally recognized Ediacaran Period (Knoll et al., 2004). Recent developments in the geochronology of Cryogenian sequences around the world suggest that the Sturtian glaciation likewise probably started and ended synchronously everywhere at ca. 717 Ma (MacLennan et al., 2018) and 668 Ma (Cox et al., 2018), respectively. The continuous refining of geochronological constraints on the Cryogenian successions lends strong support to the global nature of isotope signals and regional correlations based on isotope stratigraphy, as most of those sequences present similar $\delta^{13}C$ patterns (e.g., as compiled by Halverson et al., 2005) as well as other distinctive isotope fingerprints (e.g., Crockford et al., 2018). Care must always be taken, however, to evaluate local factors that might mask the global signals (e.g., Paula-Santos et al., 2017; Uhlein et al., 2019) as well as preservation of the original marine signals versus diagenetic, metamorphic, weathering and/or hydrothermal overprints (e.g., Caxito et al., 2018).

Other explanations have been put forward to explain the extreme $\delta^{13}C$ fluctuations in Neoproterozoic carbonates. As the values of $\delta^{13}C$ in biogenic methane are extremely fractionated, reaching values with ranges between −60‰ and −95‰, Kennedy et al. (2001) presented a hypothesis for deposition of shallow-water cap carbonates involving the destabilization of large volumes of solid methane hydrate at the bottom of the oceans. The high volume of gas hydrate necessary to create the signals down to ca. −6‰ in cap carbonate sequences around the world was, however, questioned by Hoffman and Schrag (2002).

2.1.2 Oxygen isotopes: Both double and triple

As the reaction of carbonates with phosphoric acid generates CO_2 gas in the source which is then analyzed in IRMS machines, the $^{18}O/^{16}O$ ratio (expressed as $\delta^{18}O$) is often presented along with $\delta^{13}C$ data. However, although this proxy is widely used in chemostratigraphic studies of Phanerozoic successions, the interpretation of oxygen isotope data in Precambrian chemical sedimentary rocks is a highly controversial topic. Precambrian limestones and calcareous organisms are commonly depleted in ^{18}O relative to Phanerozoic carbonates, and thus interpretation of the corresponding negatively fractionated $\delta^{18}O$ is open to debate—in effect, most workers interpret it as reflecting diagenetic overprints (Halverson et al., 2010). Occasional interpretations regarding depositional environments and evolving climate conditions were put forward for *syn*-glacial and cap carbonate sequences bearing unusually high $\delta^{18}O$ (Fairchild et al., 1989; Halverson et al., 2004; Hoffman and Schrag, 2002; Hoffman et al., 2007; Shields, 2005).

An *avant-garde* application of oxygen isotopes in the chemostratigraphy of ancient sedimentary rocks was presented by Bao et al. (2008, 2009), which measured the mass-independent oxygen isotope fractionation using all three stable oxygen isotopes (^{16}O, ^{17}O and ^{18}O), commonly denoted by the capital delta notation ($\Delta^{17}O = \ln(\delta^{17}O + 1) - 0.5305 \times \ln(\delta^{18}O + 1)$). The authors studied sulfates, including barite layers, deposited during and shortly after the end-Cryogenian (Marinoan) glaciation. The very low $\Delta^{17}O$ values, unmatched in the whole stratigraphic record, are interpreted as the result of extremely high atmospheric ρCO_2 and sluggish photosynthetic O_2 production, as predicted in the Snowball Earth models. Those conditions damaged the ozone layer, creating a mass-independent enrichment of ^{17}O into O_3 and CO_2 molecules during dissociation and reformation of ozone in the stratosphere (Wen and Thiemens, 1993). The resulting negative $\Delta^{17}O$ anomaly is preserved in SO_4 deposited along with cap carbonates. Since anomalous $\Delta^{17}O$ values below −0.4‰ are only known for Marinoan-aged glacial deposits and associated cap carbonate intervals and absent from Sturtian or post-Sturtian sulfates (Bao et al., 2008; Crockford et al., 2018) argue that those anomalous values are unique to the aftermath of the Marinoan glaciation, implicating an extreme global event in Earth's history and representing a key isotope stratigraphic marker akin to the iridium anomaly in the Cretaceous-Paleogene boundary.

2.1.3 Sulfur isotopes

Sulfur isotopes are widely applied tools in resolving the geochemical evolution of Precambrian Earth systems, with increases in the range of $\delta^{34}S$ values as measured in pyrites from shales broadly coinciding with increases in the levels of free O_2 in the atmosphere, as suggested by geological and geochemical data (Canfield and Teske, 1996; Shields-Zhou and Och, 2011). Sulfate-bearing minerals (such as evaporites, barite, phosphorite) and Carbonate-Associated Sulfate (CAS) are interpreted as recorders of seawater sulfur isotope composition, while the isotopic signatures of sulfides (e.g., pyrite) are interpreted as reflecting complex local fractionation effects such as those arising from Bacterial Sulfate Reduction (BRS) and oxidative recycling of sulfide (Canfield and Teske, 1996; Detmers et al., 2001; Hurtgen et al., 2005).

In the Paleoproterozoic, a sharp increase in $\Delta^{34}S$ values (the difference between $\delta^{34}S$ values measured in sulfate and pyrite from the same rocks) is regarded as evidence for a switch from anoxic to euxinic (anoxic with dissolved free H_2S) conditions in the final stages of the main depositional event of Superior-type BIFs (Canfield, 1998; Poulton and Canfield, 2005; Poulton et al., 2004). Although $\Delta^{34}S$ varied significantly through the Neoproterozoic, an overall increase seems to represent a second oxygenation pulse (NOE), possibly leading (or related to) earliest complex animal evolution (Canfield and Teske, 1996; Hurtgen et al., 2005).

Multiple sulfur isotopes yielded some of the cornerstone arguments for the models of evolution of Earth's atmospheric oxygenation. Both positively and negatively fractionated $\Delta^{33}S$ and $\Delta^{36}S$ are widely preserved in Archean metasedimentary rocks (Stage I of Farquhar and Wing, 2003). This suggests Mass-Independent Fractionation (MIF) of multiple sulfur isotopes in the Archean, probably caused by photochemical reactions in the atmosphere, especially SO_2-SO photolysis by UV radiation. This suggests that the

Archean atmosphere was largely transparent to UV radiation, and thus, that the ozone UV shield did not exist before ca. 2.45 Ga. From 2.45 to ca. 2.0 Ga (Stage II of Farquhar and Wing, 2003), $\Delta^{33}S$ largely stabilize around 0, but with some clear slightly positive fractionation (below 0.5‰). This is interpreted as evidence for the first GOE and a consequent development of a weak ozone layer in the atmosphere, with the slightly anomalous values interpreted as caused by either (i) sulfide weathering due to the rise of oxygen levels, generating sulfate which would retain the MIF signature of Archean sulfides; or (ii) atmospheric oxygen levels that were intermediate between the very low Stage I levels and the close-to-modern Stage III (Farquhar and Wing, 2003) levels, allowing photolysis of SO_2 and SO by passing UV radiation but not efficient transfer of anomalous $\Delta^{33}S$ to the sedimentary record. Stage III, from ca. 2.0 Ga to the present, is characterized by near-zero $\Delta^{33}S$.

2.2 Non-traditional stable isotope systems

2.2.1 Calcium and boron

Although the processes affecting calcium isotope fractionation during carbonate precipitation and diagenesis effects are not yet well understood, some authors suggested that carbonates could act as archives for the calcium isotope composition of seawater (Higgins and Schrag, 2003; Kasemann et al., 2005; Silva-Tamayo et al., 2010a,b). For example, variations in $\delta^{44}Ca$ patterns in cap carbonates deposited after the extensive Cryogenian glaciations suggest rapid glacier melting and increase in the calcium input to the ocean due to deglaciation. There are also suggestions that post-Sturtian monotonic $\delta^{44}Ca$ signals might differ from the post-Marinoan oscillating signals, thus implying differences in calcium mass balance due to the two distinct deglaciation events and that calcium isotopes might be useful in the future as an additional proxy for chemostratigraphic correlations (Silva-Tamayo et al., 2010a,b). A comprehensive set of Ca isotope analysis in cap carbonates around the world to support a correlation scheme is, however, not yet available.

Calcium and boron isotopes were used by Kasemann et al. (2005) to investigate the ratio of Ca fluxes and palaeo-seawater pH, respectively, and to estimate atmospheric CO_2 levels in the aftermath of the Marinoan (Ghaub) glaciation in Namibia. Variations in $\delta^{44}Ca$ values (between +0.35‰ and +1.14‰) correlate with carbon isotope ratios implying enhanced postglacial weathering rates, and a negative (2.7‰ to −6.2‰) $\delta^{11}B$ excursion is thought to record a decrease in marine pH during post-glacial cap carbonate deposition. Both effects are assigned to high ρCO_2 levels during cap carbonate deposition, as predicted by the Snowball Earth model; however, boron isotopes do not correlate with carbon isotopes, and thus Kasemann et al. (2005) suggest caution in using carbon isotopes alone for assessing ancient ρCO_2 conditions.

2.2.2 Iron

Iron delivered to the modern oceans may have different sources and predominantly depends on geographic location and climate. Iron-rich dust, dissolved Fe(II) from continental margins and hydrothermal iron are the main sources and account for almost the entire low concentrations of iron in the modern oceans (<1 nmol/L; e.g., Conway and John, 2014). Iron has four naturally occurring stable isotopes: ^{54}Fe (5.84%), ^{56}Fe (91.76%), ^{57}Fe (2.12%), and ^{58}Fe (0.28%), and workers have reported data mainly in terms of ^{57}Fe/^{54}Fe (δ^{57}Fe) or ^{56}Fe/^{54}Fe (δ^{56}Fe) ratios. Variations in the isotope ratios are almost exclusively driven by redox changes between Fe(II) and Fe(III). Inorganic and organic mechanisms can oxidize dissolved ferrous iron in seawater during BIF genesis, including direct oxidation of Fe^{2+} by O_2 across a redoxcline and microbial iron oxidation by chemolithotrophic or photosynthetic microorganisms. However, experiments have shown that Fe isotope ratios cannot be used to discriminate these oxidative pathways due to the similar magnitude and direction of their respective fractionation factors (Johnson and Beard, 2006). Therefore, while not reliable tracers for the oxidative pathway or for the source of iron, δ^{56}Fe and δ^{57}Fe values can be used to track the redox history of oceans and sediments.

Fe(II) oxidation involves hydrolysis and electron transfer steps that convert Fe(II)(aq) species to intermediate Fe(III)(aq) prior to precipitation of ferric oxyhydroxide minerals. The experimentally determined equilibrium isotopic fractionation upon oxidation of Fe(II)(aq) to Fe(III)(aq) (ΔFe(III)(aq)$-$Fe(II)(aq)) is ~+3.0‰ at 22 °C, and the fractionation observed from Fe(III)(aq) to ferric oxyhydroxide (ΔFe(III)(aq)$-Fe(OH)_3$) is ~+2.0‰. Thus, the δ^{56}Fe of ferric minerals is consistently heavier (~1.0‰) than that of the coexisting aqueous Fe (Beard and Johnson, 2004; Bullen et al., 2001).

Late Archean to Early Proterozoic BIFs define a significant range of Fe isotope compositions from −2.5‰ to +1.1‰. The large Fe isotope variations recorded in BIFs are generally correlated with the mineralogy of individual BIF layers (Johnson et al., 2003), where δ^{56}Fe values increase following the order pyrite/Fe-carbonate/oxides. Those isotopic differences may be related to equilibrium iron isotope fractionations that are intrinsic to specific minerals, reflecting precipitation or diagenetic reactions that occurred in relation to dissimilatory iron reduction (DIR) of Fe(III) to Fe(II) (e.g., Konhauser et al., 2005; Teixeira et al., 2017). Iron isotope data from the Neoproterozoic Rapitan iron formations of the northern Canadian Cordillera (Halverson et al., 2011) show a trend of increasing δ^{57}Fe up-section from ~-0.7‰ to 1.2‰, corresponding to a shift from a muddy IF facies to a dominantly jaspilitic IF facies during a sea level rise. This distinct isotopic pattern likely records a steep isotopic gradient across the iron chemocline in Rapitan seawater.

2.2.3 Chromium

Chromium isotopes have been proposed as useful proxies for redox conditions of seawater and, by extension, of the atmosphere during deposition of chemical sedimentary rocks (e.g., Frei et al., 2009, 2011). The oxidized Cr(VI) anion is enriched in ^{53}Cr in respect to the reduced and immobile Cr(III) form, and thus oxidation of chromium on Earth's surface is marked by positive isotope fractionation. Hexavalent chromium is mobile in subaqueous conditions and might be carried into the oceans as a solute. Thus, the δ^{53}Cr signatures of BIF and carbonate are thought to reflect the oxygenation state of the hydrosphere-atmosphere system. Under anoxic conditions, the δ^{53}Cr signature of seawater would reflect the lithogenic sources and plot within the range of igneous silicate reservoirs (-0.12 ± 0.10‰ 2σ; Schoenberg et al., 2008). Positively fractionated values, with a small spike during the first GOE and up to +4.9‰ in the Neoproterozoic, are thought to represent pulses of atmosphere oxygenation (Frei et al., 2009).

2.2.4 Cadmium

Dissolved Cd is thought to mimic the concentration patterns of bionutrients such as phosphorous and zinc in seawater, consumed during primary production and released during biomass degradation (Boyle et al., 1976; Middag et al., 2018). Cadmium isotopes fractionate upon uptake into biomass (Abouchami et al., 2014) and represent a promising proxy for revealing patterns of biogeochemical cycling as recorded in chemical sedimentary rocks, such as changes in marine productivity in surface waters (e.g., Georgiev et al., 2015; Hohl et al., 2017; John et al., 2017), monitored by $\varepsilon^{112/110}$Cd values. In an recent publication, cadmium isotopes were applied to study metal cycling in microbial mats of the Paranoá Mesoproterozoic carbonates of central Brazil (Viehmann et al., 2019).

2.2.5 *Barium*

Barium isotopes have recently been applied on the study of ancient sedimentary rocks to investigate the origins of barite layers interleaved within cap carbonate sequences around the world (Crockford et al., 2019). At this point of development, the technique consists in comparing the $^{138}Ba/^{134}Ba$ ratios ($\delta^{138/134}Ba_{NIST}$) of barite samples from ancient sequences with those of modern environments. Marine (pelagic) Ba represents a will-mixed reservoir with homogeneous $\delta^{138}Ba$ ratios of $+0.04 \pm 0.06‰$. This strongly contrasts with terrestrial ($+0.12 \pm 0.31‰$) and cold seep ($0.05 \pm 0.28‰$) barites which show a much wider variation of isotope ratios, and also with hydrothermal barite which shows negatively fractionated values ($-0.07 \pm 0.02‰$). Post-Marinoan barites show $\delta^{138}Ba$ ratios of $+0.08 \pm 0.06‰$, identical within error to modern barite samples, and thus (Crockford et al., 2019) put forward a model for deposition of post-glacial barite involving a well-mixed Ba-rich marine reservoir accumulated during the glacial interval, which upon deglaciation is transported upward and encounters continental weathering-derived SO_4 causing globally synchronous deposition of barite layers in the cap carbonate successions.

2.3 Radiogenic isotopes

Radiogenic isotope systems are widely applied in geochronology, but are also powerful tools in the investigation of sedimentary provenance, tracking of hydrothermal fluids, variations of continental erosion and silicate weathering. These processes are, in turn, linked to tectonic and climatic factors. Different radiogenic elements have variable residence times, which allow the use of separate isotope systems to investigate sediment supply and patterns of continental weathering in multiple scales of time and space (e.g., Halverson et al., 2010). Strontium and neodymium are the main isotope systems used in this regard, and to some extent they mirror and complement each other. Rubidium is more incompatible than strontium, thus causing enrichment in ^{87}Sr (due to the radioactive decay of the parent isotope ^{87}Rb) and higher $^{87}Sr/^{86}Sr$ ratios in the continental crust relative to the mantle. On the other hand, Nd isotopes show the inverse relation, with enrichment of the parent isotope ^{147}Sm causing higher $^{143}Nd/^{144}Nd$ ratios in mantle-derived rocks and hydrothermal fluxes. Strontium, however, has been much more widely used in isotope stratigraphy, especially using Sr-rich carbonates due to its much greater residence time. The application of Nd isotopes in suitable sedimentary rocks (e.g., BIF-bearing intervals;

Cox et al., 2016a,b) is steadily growing, while its use in sedimentary provenance studies, specially when coupled with detrital zircon U-Pb and Lu-Hf isotope analysis, is one of the most widely tools for unraveling sedimentary provenance and tectonic evolution of orogenic and cratonic areas (e.g., Caxito et al., 2014).

2.3.1 Strontium isotopes

Next to C and O isotopes, Sr isotopes were also widely applied in the study of Brazilian Precambrian sedimentary rocks. The $^{87}Sr/^{86}Sr$ ratio is a useful tracer of tectonic and climate changes because it is thought to represent changes in the input of ^{87}Sr-rich continental riverine fluxes versus ^{87}Sr-poor hydrothermal input (or erosion of juvenile sources such as island arcs). Highly detailed Sr-evolution curves are available for the Cenozoic and Mesozoic oceans (e.g., McArthur et al., 2001; Paytan et al., 1993), but the record becomes scattered and less well-resolved in the Precambrian, especially for rocks older than Neoproterozoic (Shields and Veizer, 2002; Veizer et al., 1999).

The general trend of the Neoproterozoic marine Sr record as detected by various authors based on the most reliable datasets (e.g., Hoffman et al., 2007; Shields, 2007; Thomas et al., 2004) is one of quasi-continuous rising $^{87}Sr/^{86}Sr$ from <0.7055 in the beginning of the Era to >0.7080 toward the end. It appears that $^{87}Sr/^{86}Sr$ rose almost monotonically (i.e., without major swings or fluctuations) during the Neoproterozoic (Halverson et al., 2010) which led some authors to propose that $^{87}Sr/^{86}Sr$ ratios could be useful in establishing relative timing of deposition and for global stratigraphic correlations (Halverson et al., 2007; Melezhik et al., 2001). Local variations might, however, complicate the use of $^{87}Sr/^{86}Sr$ as a relative dating tool (Paula-Santos et al., 2017). Despite the apparent monotonically rise, there are, however, some interesting "peaks" that appear to be global signals, for example, the sharp rise in $^{87}Sr/^{86}Sr$ values observed in early Ediacaran cap carbonates worldwide, from 0.7072 to >0.7078, then falling back to ca. 0.7072–0.7074 (Caxito et al., 2012, 2018; Halverson et al., 2007, 2010). This sharp peak could be related to unusually high ρCO_2 in the aftermath of the end-Cryogenian glaciation (Higgins and Schrag, 2003) which would cause a temporary significant increase in continental weathering and thus, a major input of ^{87}Sr through riverine fluxes.

There are a number of screening proxies for selecting the $^{87}Sr/^{86}Sr$ ratios most likely to represent original seawater values. The most used are Rb/Sr, Mn/Sr, Fe/Sr, Ca/Sr and strontium concentrations (Halverson et al., 2007).

The strontium concentration cut-off value is somewhat arbitrarily defined (e.g., Halverson et al., 2007; Paula-Santos et al., 2017), but as, in general, secondary alteration tends to drive $^{87}Sr/^{86}Sr$ ratios up, the least radiogenic values of a given dataset (provided stratigraphic resolution allows) are commonly regarded as closer to the original seawater values, especially when correlated to the higher strontium concentrations and lower Rb/Sr and Mn/Sr ratios in the dataset.

2.3.2 Neodymium isotopes

Neodymium isotopes can provide useful chemostratigraphic information both on provenance and tectonic evolution of sedimentary basins and source areas (e.g., Barovich and Hand, 2008; McLennan et al., 1993; Wade et al., 2005). The most useful data is commonly obtained from fine-grained sediments (which are though to correspond to a homogenized mixture of source area compositions, while coarse-grained sediments can be biased toward specific source areas, besides commonly containing less Rare Earth Elements than their pelitic counterparts) and chemical precipitates, in which the initial $^{143}Nd/^{144}Nd$ is thought to record local seawater composition due to its lower residence time than strontium. Neodymium isotope ratios are often used in conjunction with detrital zircon U-Pb and Lu-Hf geochronology and isotope systematics, in order to track orogenic events and source areas to sedimentary basins.

Neodymium isotope data are very useful in tracking the extent of contribution or detrital input from the erosion of large igneous provinces such as continental flood basalts and their influence in seawater composition and geochemical evolution (Barovich and Foden, 2000; Cox et al., 2016a). As basalt weathers 5–10 times faster than granite (Dessert et al., 2003), erosion of widespread continental floods might have played a major role on climate modulation and influenced the onset of major Cryogenian glaciations (Cox et al., 2016a; Goddéris et al., 2003).

Global curves for the evolution of seawater Nd ratios, although limited by the lower residence time compared to the ocean mixing time, have been proposed in the same way as the carbon and strontium curves (e.g., Halverson et al., 2010; Keto and Jacobsen, 1988). Those curves, yet representing rough estimates, suggest a pattern of progressively lower εNd(t) values from 800 to 550 Ma, perhaps mirroring the breakup of Rodinia and opening of new passive margins, exposing a larger surface of old continental crust for erosion, an interpretation similar to that of (Halverson et al., 2007) for the mirroring Sr curve.

3. Applications of isotope stratigraphy in Brazilian Precambrian sedimentary successions

3.1 Archean: Isotope stratigraphy of the Carajás formation

Sedimentary carbonates are relatively rare in Archean greenstone belts, with most occurrences interpreted as hydrothermally-influenced levels. When present, however, those rocks are important because they could act as recorders of the chemical conditions of Archean seawater during the time of deposition. The Carajás domain of the southeastern Amazon Craton in northern Brazil is one of the most important metallogenetic provinces of the South American platform (Fig. 2). The BIF-bearing Carajás Formation is part of the ca. 2.76 Ga Grão Pará Group (Machado et al., 1991; Trendall et al., 1998), a meta-volcanosedimentary unit containing major iron and manganese deposits. In the N4E iron mine, the ca. 220 m thick Carajás Formation bears carbonate-rich jaspilites that are thought to represent carbonate-facies BIF. Sial et al. (2000) presented C and O isotope data on these jaspilites, with homogeneous $\delta^{13}C$ predominantly around −5‰ and highly variable $\delta^{18}O$ predominantly around −15‰ to −10‰. Sial et al. (2000) proposed that the Carajás carbonate-facies BIF equilibrated with deep basinal waters influenced by hydrothermal fluid input. In this scenario, the Archean mantle was probably the main source for CO_2 as suggested by the retaining of mantle-like $\delta^{13}C$ values (Fig. 2).

Based on S isotopes from BIF and black shales from the Archean Grão Pará Group and from the Paleoproterozoic (ca. 2.1 Ga) Águas Claras Formation, Fabre et al. (2011) and Cabral et al. (2013) suggest that dissolved sulfate was the limiting factor controlling bacterial sulfate reduction and biogenic pyrite formation during the Neoarchean and Paleoproterozoic in the Carajás domain. Also, the quasi-absence of MIF in S isotope distribution in the ca. 2.1 Ga Águas Claras Formation suggests deposition at a time when the oxygen partial pressure in the atmosphere was already high enough to hinder the photochemical reactions that drive this process (Fabre et al., 2011).

The Fe isotope composition of the Carajás BIFs show positively fractionated $\delta^{57}Fe$ (+2.06 ± 0.34‰) suggesting partial oxidation of FeII(aq) in the surface ocean following upwelling of anoxic deep seawater (Fabre et al., 2011). Cabral et al. (2013) also presented molybdenum isotope ratios for black shales of the Grão Pará Group, suggesting the mixing between a clastic end member with about $\delta^{98/95}Mo = 0.2‰$ (continental input) and

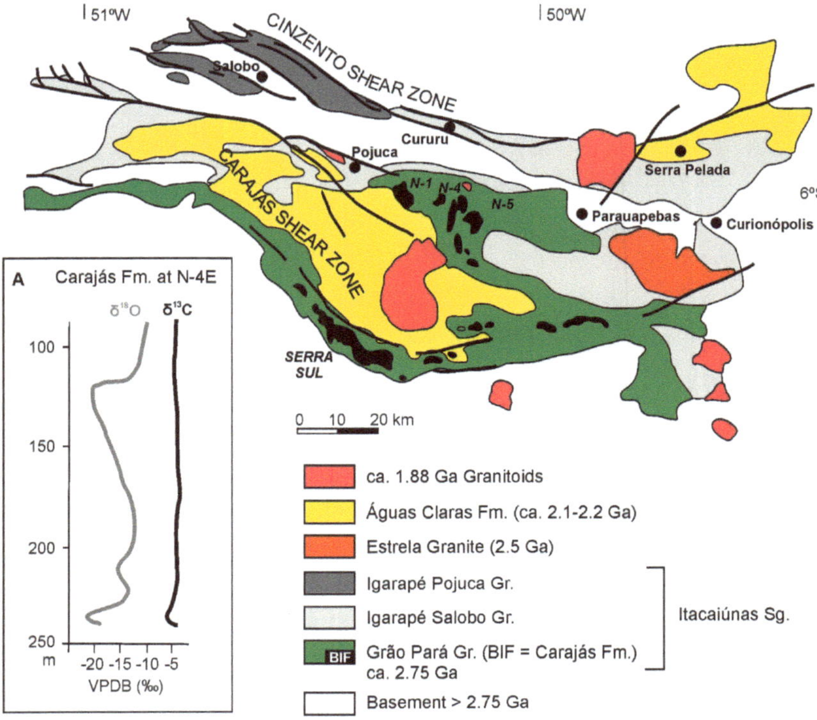

Fig. 2 Simplified geological features of the Carajás area. Inset (A) shows the C and O isotope stratigraphy of BIFs from the Carajás Fm. at N-4E mine. *Compiled and modified from Dardenne, M.A., Schobbenhaus, C., 2001. Metalogênese do Brasil. Brasília: UNB; CPRM, 392 pp. ISBN 8523006478; Fabre, S. et al., 2011. Iron and sulphur isotopes from the Carajás mining province (Pará, Brazil): implications for the oxidation of the ocean and the atmosphere across the Archaean–Proterozoic transition. Chem. Geol. Elsevier, 289 (1–2), 124–139. doi: 10.1016/J.CHEMGEO.2011.07.019. Inset (A): Data from Sial, A.N., et al., 2000. Carbon isotope fluctuations in Precambrian carbonate sequences of several localities in Brazil. An. Acad. Bras. Cienc. Academia Brasileira de Ciências 72 (4), 539–558. doi: 10.1590/S0001-37652000000400006.*

a hydrogenous component characterized by $\delta^{98/95}$Mo = 0.9‰. Black shale samples localized at the bottom of the BIF sequence yielded heavy molybdenum isotopic compositions up to $\delta^{98/95}$Mo = 1.8‰, suggesting either an early and transient "whiff of oxygen" or hydrothermal fluid imprint.

3.2 Paleoproterozoic: Isotope stratigraphy of carbonates and BIFs from the Quadrilátero Ferrífero: The first GOE, microbial influence on BIF deposition and the Lomagundi event

The Quadrilátero Ferrífero (Iron Quadrangle) corresponds to an area of circa 15.000 km^2 to the south of Belo Horizonte, Brazil, in the southern

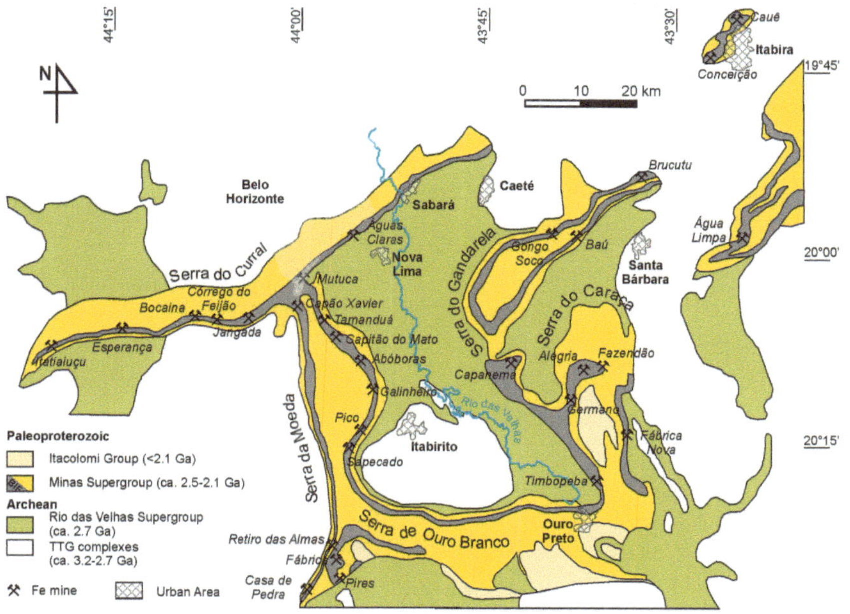

Fig. 3 Simplified geological features of the Quadrilátero Ferrífero area. *Modified from Caxito, F.A., Gonçalves-Dias, T. 2018. Ferro. In: Pedrosa-Soares, A.C., Voll, E., Cunha, E.C. (Eds.) Recursos Minerais de Minas Gerais On Line. Companhia de Desenvolvimento de Minas Gerais, Belo Horizonte.*

São Francisco Craton. This region is characterized by ranges sustained mainly by Paleoproterozoic metasedimentary rocks (Minas Supergroup) and lowlands composed of Archean rocks (crystalline basement and the Rio das Velhas greenstone belt) (Fig. 3).

The Minas Supergroup is a circa 8 km thick metasedimentary sequence metamorphosed dominantly under greenschist facies conditions (Alkmim and Marshak, 1998; Chemale et al., 1994; Dorr, 1969; Renger et al., 1994). It is composed, from bottom to top, of the Caraça Group, representing the transition of a deltaic to a marine system, overlain by the Itabira Group, subdivided in the Cauê (BIF and dolomitic BIF) and Gandarela (dolomite, metapelite, BIF intercalations and breccias) formations. The upper Caraça Group and the Itabira Group record a regional marine transgression (Dorr, 1969) and the development of an extensive continental margin.

The Cauê Formation comprises nearly 200–400 m thick of Lake Superior-type metamorphosed BIFs, with local development of high-grade iron ore bodies. Metadolomites of the overlying Gandarela Formation were dated through the Pb-Pb whole-rock isochron method at 2420 ± 19 Ma (Babinski et al., 1995), coherent with U-Pb detrital zircon data obtained from subjacent units (maximum depositional age of circa 2.58 Ga for the Moeda Formation; Hartmann et al., 2006; Koglin et al., 2014; Machado et al., 1996; Martínez Dopico et al., 2017) (Fig. 4). Stromatolitic dolostones of the Fecho do Funil Formation in the overlying Piracicaba Group yielded a Pb-Pb whole-rock isochron age of 2.11 ± 0.11 Ga (Babinski et al., 1995) interpreted as reflecting fluid mobilization during an important 2.1–2.0 Ga tectono-metamorphic event. This event caused deformation and greenschist facies metamorphism of the sedimentary successions, with coeval deposition of the Sabará syn-orogenic deposits (e.g., Alkmim and Marshak, 1998; Chemale et al., 1994; and references therein).

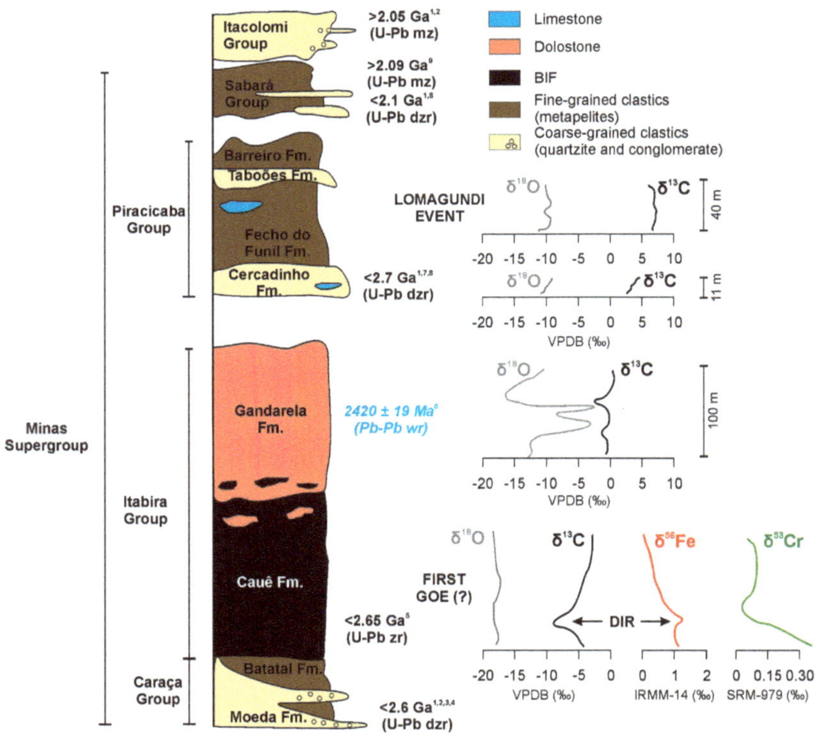

Fig. 4 Simplified composed stratigraphic column of the Minas Supergroup and isotope stratigraphy records (see text for data sources). zr = zircon; dzr = youngest detrital zircon population; wr = whole-rock isochron; g-wr = garnet-whole rock isochron; mz = monazite. Sources for geochronological constraints: 1 Machado et al. (1996), 2 Hartmann et al. (2006), 3 Koglin et al. (2014), 4 Martínez Dopico et al. (2017), 5 Cabral et al. (2012), 6 Babinski et al. (1995), 7 Mendes et al. (2014), 8 Dutra et al. (2019), 9 Brueckner et al. (2000).

Until recently, only scant chemostratigraphic data was available for the chemical units of the Minas Supergroup. As deposition of these units could straddle both the first GOE and the Lomagundi excursion, with BIFs of the Cauê Formation being roughly coeval to the main worldwide peak of Lake Superior-type BIF deposition, those successions are strong candidates to provide important clues to Earth's oxygenation, biosphere evolution, and processes of deposition and enrichment of BIFs.

Sial et al. (2000), Bekker et al. (2003) and Paula, J.R. (Unpublished Data) discuss $\delta^{13}C$ results for the Gandarela Formation, with values ranging from −1.6‰ to 0.5‰; for the overlying Cercadinho Formation, a sandstone-rich succession with thin carbonate layers with $\delta^{13}C$ from +3.3‰ to +5.4‰, and for stromatolitic dolomites of the Fecho do Funil Formation with $\delta^{13}C$ from +5.6‰ to +7.4‰. These data allows to place those carbonates in time, with the Gandarela Formation deposited prior to the Lomagundi $\delta^{13}C$ positive excursion and the Piracicaba Group deposited through the Lomagundi excursion. Recently, Nogueira et al. (2019) presented new geochemical and carbon and oxygen isotope data on carbonate rocks of the Gandarela Formation from two distinct quarries. Most of the studied samples show $\delta^{18}O$ values lower than $-12‰_{VPDB}$, suggesting post-depositional overprinting. The $\delta^{13}C$ values are mostly close to zero and correspond to similar marine carbonates of the same age, but some samples with values around −4‰ are interpreted as the result of decarbonation reactions.

Spier et al. (2007) presented carbon isotope studies of the Cauê BIFs, with $\delta^{13}C$ values of −2.5‰ to −0.8‰ for mineralized dolomitic BIFs from the Águas Claras mine. Those values were then interpreted as reflecting direct precipitation from the seawater column. Recently, Teixeira et al. (2017) presented a systematic petrographic and geochemical study of well-preserved samples of dolomitic/ankeritic BIF from a drill core of the Cauê Formation, including carbon, oxygen, iron and chromium isotope data. This drill core is especially important because it allowed sampling of a well preserved, non-mineralized section, with only mild metamorphism and localized deformation. It also preserves primary features indicating a sedimentary origin for lamination, with a very fine, micrometric size of oxide and carbonate grains, in comparison to the normal coarse hematite-specularite/quartz layering of mineralized and metamorphosed sections of the Cauê BIF. Unlike the results obtained by Spier et al. (2007), values of $\delta^{13}C$ and $\delta^{18}O$ are all negatively fractionated, reaching a minimum of −10.0‰ and −24.6‰, respectively, in the 540–560 m interval. In the same interval, values of $\delta^{56}Fe$ reach a maximum, varying between 0.5‰ and 1.3‰ throughout the section. Dissimilatory Iron Reduction (DIR) of the precursor ferric oxides through the activity of microorganisms in the sedimentary surface was invoked as a likely mechanism for the coupled δ^{56}Fe increase and δ^{13}C decrease. Values of authigenic δ^{53}Cr vary between 0.01‰ and 0.26‰, with an increasing-downward pattern accompanied by a rise in U/Th ratios, suggesting either oxidative continental weathering or local oxygenation in continental microenvironments. The results obtained by the detailed study of Teixeira et al. (2017) add to the growing body of data for BIF sequences of similar age around the world (e.g., Kuruman, South Africa and Hamersley, Australia), suggesting that microbial-mediated DIR was a common and widespread mechanism for the genesis of BIFs at the Archean/Paleoproterozoic transition and illustrating some interesting effects of oxygen and redox stratification in those basins.

Other Fe isotopes results were reported by Mendes et al. (2016) for hydrothermalized BIF from 10 iron deposits of the Cauê Formation, including specularite from veins and iron ores, with varied values of δ^{56}Fe of −0.95‰ to +0.83‰. The authors suggest a combination of anoxic biologically mediated ferrous iron oxidation and abiotic oxidation in oxygen oasis.

3.3 Mesoproterozoic to early Neoproterozoic: Isotope stratigraphy of the Brasília and Araçuaí belts carbonate platforms: Insights into the "Boring Billion"

Exposed in the western São Francisco Craton margin and affected by east-verging deformation of the external thrust fronts of the Ediacaran Brasília Belt, the carbonate-siliciclastic strata of the Paranoá and Vazante groups are interpreted as part of a Meso-Neoproterozoic passive margin facing the large Brasilides ocean to the West (Dardenne, 2000; Valeriano et al., 2004a,b and references therein) (Fig. 5).

The Paranoá Group consists of a succession of sandstone, siltstone, rhythmite, and local carbonate layers, with δ^{13}C values in a narrow interval between +0.6‰ and +3.6‰ and ^{87}Sr/^{86}Sr between 0.7056 and 0.7068 (Alvarenga et al., 2014). Detrital zircon U-Pb data indicate a maximum depositional age of ca. 1.54 Ga, and an approximation of the minimum depositional age is furnished by a U-Pb date of ca. 1.04 Ga obtained from diagenetic xenotime overgrowth on detrital zircon (Matteini et al., 2012). Since the values close to VPDB of the Paranoá carbonates contrast with the strong variations observed for the overlying Neoproterozoic Bambuí Group (see below) and also the ^{87}Sr/^{86}Sr ratios of the two successions are quite different, Alvarenga et al. (2014) suggested that those proxies can be used in the mapping of complex areas where carbonates of the two distinct units are in direct contact.

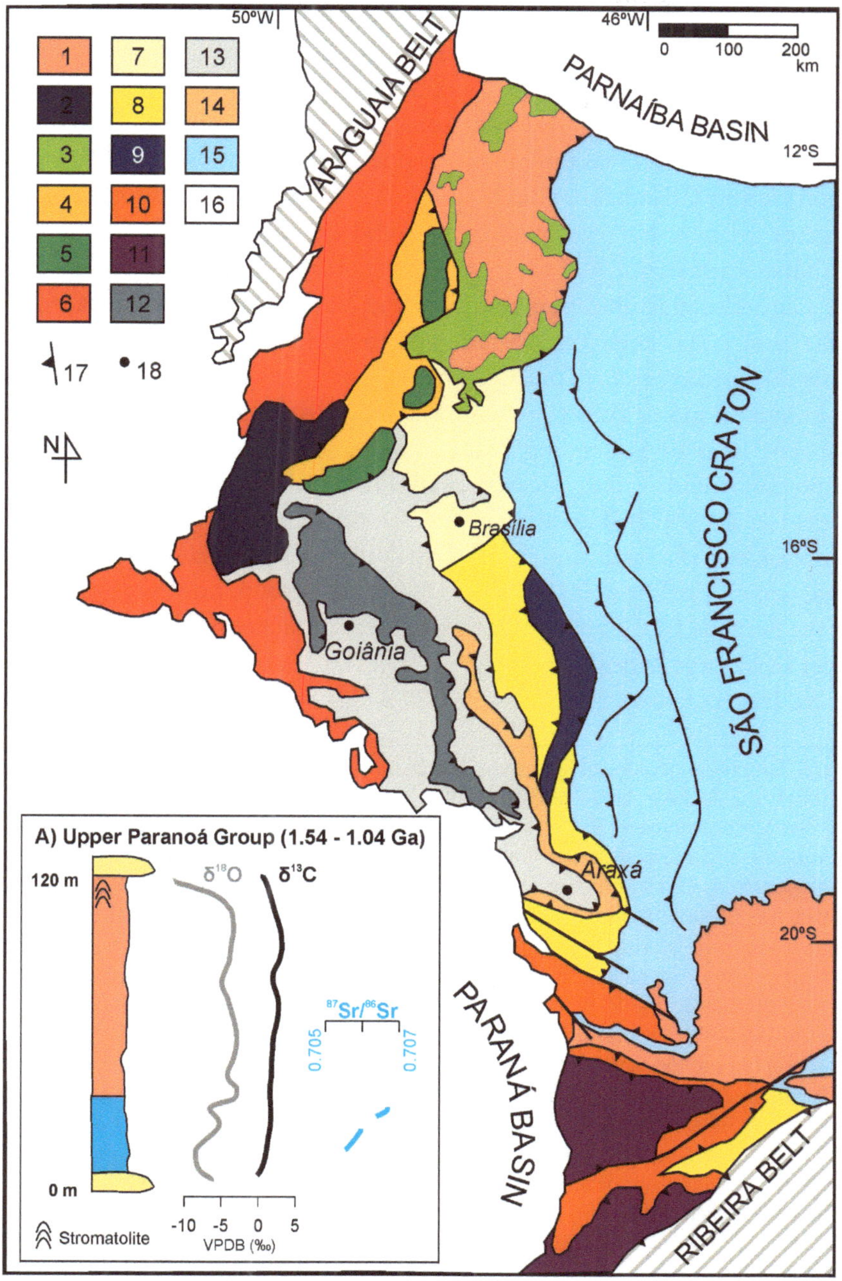

Fig. 5 See legend on next page.

Recently, Viehmann et al. (2019) presented a multi-proxy study of the Paranoá laminated stromatolites, including trace elements and stable carbon, oxygen and cadmium isotopes. Values of $\varepsilon^{112/110}Cd_{dol}$ were found to increase from −3.52 at the mat surface to +3.46 in the interior of the mat. This variation was interpreted as due to the effect of kinetic fractionation during Cd-uptake, which could have happened by adsorption onto organic matter, by precipitation of sulfides, and/or incorporation into carbonate minerals.

Chronostratigraphic positioning of the Vazante Group is still an open issue, and it has been variably assigned to both the Mesoproterozoic and early Neoproterozoic (Tonian). The Vazante Group comprises a widely dolomitized and deformed carbonate-dominated platform that extends in a 300-km long N-S trending belt in the external Brasília Belt, including important sulfide- and silicate-hosted Pb-Zn deposits (Dardenne, 2000; Sotero et al., 2019). For most of the carbonatic intervals, the $\delta^{13}C$ values show only small variations close to the VPDB value, with associated least-radiogenic $^{87}Sr/^{86}Sr$ of 0.7061. At the top of the group, however, carbonates of the Serra da Lapa Formation that overlie a diamictite or carbonatic breccia layer show large perturbations of $\delta^{13}C$, with values as low as −8‰ and associated least-radiogenic $^{87}Sr/^{86}Sr$ of 0.7068 (Azmy et al., 2001, 2006) (Fig. 6).

Fig. 5 Simplified geological features of the Brasília Belt. 1 Archean and Paleoproterozoic basement; 2 Archean Goiás Massif; 3 Paleo/Mesoproterozoic rift systems (Araí Group); 4 Paleo/Mesoproterozoic Serra da Mesa Group; 5 Meso/Neoproterozoic mafic/ultramafic complexes and metavolcanosedimentary units; 6 Neoproterozoic Goiás Magmatic Arc; São Francisco Craton passive margin units: 7 Paranoá Group; 8 Canastra Group; 9 Vazante Group; 10 Andrelândia Megasequence/Group; High-grade Neoproterozoic units: 11 Socorro-Guaxupé high-grade nappe system; 12 High-grade granulite belt; External nappe syn-orogenic and mélange units: 13 Araxá Group; 14 Ibiá Group; 15 Cratonic cover (Bambuí Group); 16 Phanerozoic covers; 17 Thrust faults; 18 Cities. Inset (A) shows the compiled isotope stratigraphy of the Upper Paranoá Group. Key for rock types in the stratigraphic section is the same as in Fig. 4. *Modified from Valeriano, C.M., et al., 2004a. A Evolução Tectônica da Faixa Brasília. In: Mantesso Neto, V., et al. (Eds.) Geologia do Continente Sul-Americano—Evolução da obra de Fernando Flávio Marques de Almeida. São Paulo: Beca;Valeriano, C.M., Dardenne, M.A., Fonseca, M.A., Simões, L.S.A., Seer, H.J., 2004b. A evolução tectônica da Faixa Brasília. In: Mantesso-Neto V., Bartorelli A., Carneiro C. D.R., Brito-Neves B.B (Eds.) Geologia do Continente Sul-Americano: Evolução da Obra de Fernando Flávio Marques De Almeida, São Paulo, pp. 575–592. Beca; Falci, A., Caxito, F.A., Seer, H.J., Valeriano, C.M., Dias, P.H.A., Pedrosa-Soares, A.C., 2018. Provenance shift from a continental margin to a syn-orogenic basin in the Neoproterozoic Araxá nappe system, southern Brasília belt, Brazil. Precambrian Res. 306, 209–219. Inset (A): Data from Alvarenga, C.J.S., et al., 2014. Meso-Neoproterozoic isotope stratigraphy on carbonates platforms in the Brasilia Belt of Brazil. Precambrian Res. Elsevier, 251, 164–180.*

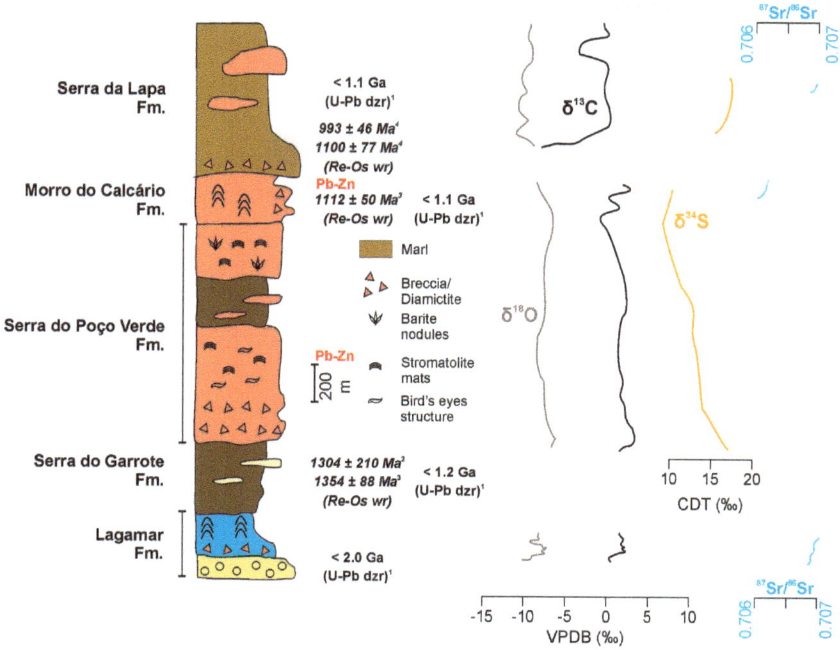

Fig. 6 Simplified stratigraphic column of the Vazante Group with isotope stratigraphy records (see text for data sources). Key for rock types and structures in the stratigraphic section is the same as in Figs. 4 and 5. Sources for geochronological constraints: 1 Rodrigues et al. (2012), 2 Bertoni et al. (2014); 3 Geboy et al. (2013), 4 Azmy et al. (2008).

The authors correlated this isotope signature and the associated diamictite with the Sturtian glacial phase. However, in a later paper, Azmy et al. (2008) presented Re-Os isochron dates of black shales associated with the diamictites of 993 ± 46 and 1100 ± 77 Ma. U-Pb zircon data of the Serra da Lapa Formation, topmost unit of the Vazante Group, indicate a maximum depositional age of ca. 1.08 Ga (Rodrigues et al., 2012). Azmy et al. (2008) then re-interpreted the whole Vazante Group as deposited during the late Mesoproterozoic, in between 1000 and 1100 Ma. Geboy et al. (2013) presented further Mesoproterozoic (ca. 1.3 Ga) Re-Os ages for the Morro do Calcário and Serra do Garrote formations. In addition, a Mesoproterozoic Re-Os date (ca. 1.3 Ga) was obtained on black shales from the Serra do Garrote Formation by Bertoni et al. (2014). Azmy et al. (2001) also presented $\delta^{34}S$ values from sulfates trapped in carbonates, ranging between +10.8‰ and +16.9‰, with an excursion to +21.3‰ in the Serra da Lapa Formation (Fig. 6).

U-Pb detrital zircon ages and Sm-Nd and Lu-Hf model ages presented by Pimentel et al. (2011) and Rodrigues et al. (2012) have depicted a different scenario for the Vazante Group evolution. According to these data, the group seems to record an early Neoproterozoic passive margin basin that was mostly fed by recycled Paleoproterozoic and juvenile Mesoproterozoic cratonic sources. In this scenario, the previously described $\delta^{13}C$ and $^{87}Sr/^{86}Sr$ signatures would mark younger paleoenviromental conditions coeval to the first evolutionary stages of the Brasília Belt evolution (e.g., Reis et al., 2017). Recent chemostratigraphic studies based on stable and radiogenic isotopes and the description of possible vase-shaped microfossils in carbonate rocks of the lower Vazante Group suggest a Tonian age for the whole unit (Alvarenga et al., 2019). Alternative interpretations based both on isotope geochemistry and field data are found in the literature (e.g., Marques et al., 2015; Misi et al., 2014) suggesting slightly different stratigraphic frameworks, stratigraphic positioning and evolution of the Vazante Group.

On the other side of the São Francisco Craton, the Araçuaí Orogen also present units recording Mesoproterozoic/early Neorpoterozoic carbonate platforms that have been studied through isotopic approaches. In the upper portion of the Espinhaço Supergroup, the Conselheiro Mata Group comprises siliciclastic and carbonate marine rocks deposited in a sag basin coeval to the Paranoá passive margin deposits (e.g., Kuchenbecker et al., 2015; Reis et al., 2017; Santos et al., 2015). The group hosts detrital zircon populations dated at c. 1.3 Ga and is intruded by c. 930 Ma mafic dykes, which constrain its deposition between the Ectasian and Tonian periods (e.g., Chemale et al., 2012; Kuchenbecker et al., 2015; Santos et al., 2015). The Rio Pardo Grande Formation is the topmost unit of the Conselheiro Mata Group, and comprises mainly gray massive dolostones with local occurrences of stromatolites (Fraga et al., 2014; Santos et al., 2015). Santos et al. (2004) report a flat isotopic pattern in a section of the unit, with $\delta^{13}C$ ranging from $-0.3‰$ to $+1.9‰$ and $\delta^{18}O$ between $-4.7‰$ and $-2.6‰$. The authors highlight that these values are in good agreement to those found in rocks from the coeval Paranoá Group. Additionally, Fraga (2013) reported $\delta^{13}C$ values from $+1.5‰$ to $+2.2‰$ in dolostones from this unit, as well as $^{87}Sr/^{86}Sr$ ratios ranging between 0.7074 and 0.7079.

Another carbonate platform is recorded within the Araçuaí Orogen in the basal units of the Neoproterozoic Macaúbas Group. The group comprises two different basin cycles: a Tonian aborted rift sequence, which present continental to shallow marine deposits, and a Cryogenian rift to passive margin sequence that include glacially-related rocks, passive margin deposits and oceanic crust remnants (e.g., Kuchenbecker et al., 2015). In the Tonian rift sequence, the Domingas Formation presents stromatolitic dolostones with $\delta^{13}C$ around 0‰ and $^{87}Sr/^{86}Sr$ ratios between 0.7078 and 0.7079 (Fraga, 2013). Another carbonatic unit, called Tijucuçu Farm, is composed by calcilutites and calcarenites that yielded more positive $\delta^{13}C$ values, ranging from +7‰ to +10‰, and $^{87}Sr/^{86}Sr$ ratios ranging from 0.7072 to 0.7076.

3.4 Neoproterozoic: Isotope stratigraphy of cap carbonates and related units: Extreme fluctuations in Earth's chemical cycles at the dusk of the Precambrian

3.4.1 Bambuí Group

The Bambuí Group is a late Neoproterozoic sedimentary cover which outcrops over 350,000 km^2 of the São Francisco basin in east central Brazil (e.g., Alkmim and Martins-Neto, 2012; Caxito et al., 2012 and references therein). The classical lithostratigraphy of the Bambuí Group (Dardenne, 1978) comprises six major units, from base to top (Fig. 7): (1) Jequitaí Formation (diamictites, sandstones and fine-grained siliciclastics); (2) Sete Lagoas (limestones with stromatolites toward the top and subordinated mudstones); (3) Serra de Santa Helena (siltstones and fine-grained sandstones); (4) Lagoa do Jacaré (dark oolithic to muddy limestones); (5) Serra da Saudade (green siltstones, shales and sandstones) and (6) Três Marias (arkoses, sandstones, conglomerates and shales) formations. Stratigraphic positioning of the Jequitaí Formation within the base of the Bambuí Group is sometimes questioned (e.g., Reis et al., 2017). The basal carbonatic units of the Sete Lagoas Formation or, where those are missing, the fine-grained units of the Serra de Santa Helena Formation are underlain by coarse- to fine-grained siliciclastics that have been included either in the Jequitaí Formation or in the Carrancas Formation (e.g., Costa and Branco, 1961; Kuchenbecker et al., 2016a; Reis and Suss, 2016; Uhlein et al., 2016). Despite the different lithostratigraphic interpretations available in the literature for these units (i.e., older successions or part of the basal Bambuí Group), it has been widely accepted that the Jequitaí Formation contains remarkable late Neoproterozoic glacial-related deposits (e.g., Caxito et al., 2012; Cukrov et al., 2005; Karfunkel and Hoppe, 1988; Kuchenbecker et al., 2013; Reis and Suss, 2016), locally sitting atop striated pavements (Isotta et al., 1969). Interpretation of the Carrancas Formation is more contentious (e.g., Caxito et al., 2012; Reis and Suss, 2016; Uhlein et al., 2016; Vieira et al., 2007). Uhlein et al. (2016) describe conglomerates, rhythmites, black shale and dolomite in the Carrancas Formation, and interpret it as representing pre-glacial units deposited under strong influence of gravitational processes in localized grabens carved in the cratonic basement. Other sparse occurrences of coarse-grained siliciclastic rocks related to the basal carbonates of the Sete Lagoas Formation might represent distinct sedimentary processes. Caxito et al. (2012) and Ilambwetsi (2015), for example, interpret one of the classical outcrops of the Carrancas Formation at the MG-424 road, a paraconglomerate with clasts of mainly gneiss and limestone floating on a carbonate-rich sandy matrix, as a meter-thick conglomerate lens within the basal Sete Lagoas Formation. This specific outcrop would then represent a reworked facies of the Sete Lagoas Formation itself, and not a distinct stratigraphic unit.

Other stratigraphic units are progressively being formally included in the Bambuí Group, such as the Samburá (conglomerates, sandstones and pelites) and the Lagoa Formosa (diamictite, sandstone, siltstone, and lenses of limestone and jaspilite) formations (Castro and Dardenne, 2000; de Castro and Dardenne, 1995; Uhlein et al., 2013, 2017). These successions are exposed in the western part of the São Francisco basin and have been considered as lateral equivalents of the Bambuí Group (e.g., Castro and Dardenne, 2000; Fragoso et al., 2011; Reis et al., 2017; Uhlein et al., 2017). The Jaíba Formation or Jaíba Member (Caxito et al., 2016; Chiavegatto et al., 2003; Iglesias and Uhlein, 2008, 2009; Sanchez et al., 2018) of the Serra da Saudade Formation and the Gorutuba Formation (Kuchenbecker et al., 2016b), on the other hand, are exposed exclusively in the upper part of the Bambuí Group in the eastern São Francisco basin. The Jaíba Member encompasses a ca. 40 m-thick limestone unit atop fine-grained siliciclastics of the Serra da Saudade Formation and overlain by coarse-grained deposits of the Três Marias Formation in the Jaíba Range, while the Gorutuba Formation includes a succession of continental to shallow marine breccias, conglomerates, sandstones and pelites in the uppermost Bambuí Group.

The base of the Sete Lagoas Formation has been interpreted as a classical cap carbonate unit composed of a basal thin, patchy, meter-thick pink to gray dolostone with decreasing-upward $\delta^{13}C$ from −3.2‰ down to −6.5‰ and associated $\delta^{18}O$ generally around −5‰, followed by limestones bearing pseudomorphs of aragonite-fans with progressively higher $\delta^{13}C$, which upsection stabilize shortly around 0‰ and then quickly rises to above 10‰ (Alvarenga et al., 2014; Babinski et al., 2007; Caetano-Filho et al., 2019; Caxito et al., 2012, 2018; de Alvarenga et al., 2007; Guacaneme et al., 2017; Hippertt et al., 2019; Iyer et al., 1995; Kaufman et al., 2001;

Kuchenbecker et al., 2016a; Martins and Lemos, 2007; Misi and Veizer, 1998; Paula-Santos et al., 2017; Perrella Júnior et al., 2017; Santos et al., 2000, 2004; Uhlein et al., 2019; Vieira et al., 2007). The impressive positive excursion to values above 10‰, dubbed MIBE (MIddle Bambuí Excursion, Uhlein et al., 2019) stands for the whole upper Bambuí package throughout the Serra de Santa Helena, Lagoa do Jacaré and lower Serra da Saudade formations (Fig. 7). Carbonates of the topmost Bambuí Group (Jaíba Member of the Serra da Saudade Formation) show a decrease to much less fractionated $\delta^{13}C$ values, from +0.9‰ to +3.3‰ (Uhlein et al., 2019).

$^{87}Sr/^{86}Sr$ ratios around 0.7074–0.7076 are characteristic for most of the Bambuí Group, suggesting deposition over a relatively short time span (Alvarenga et al., 2014; Babinski et al., 2007; Caxito et al., 2012, 2018; de Alvarenga et al., 2007; Kaufman et al., 2001; Kuchenbecker et al., 2016a; Misi and Veizer, 1998; Misi et al., 2007). However, a distinct jump to values up to 0.7080 at the middle portion of the cap carbonate succession, followed upsection by a quick return to values around 0.7074, was described by Babinski et al. (2007) and interpreted as a primary signal by Paula-Santos et al. (2017) and Caxito et al. (2018). Caxito et al. (2018) also suggested that distinctly higher $^{87}Sr/^{86}Sr$ and lower $\delta^{18}O$ values in the first 60 m of a carbonate section of the Sete Lagoas Formation in the southern Bahia state can be interpreted as evidence for meteoric alteration of the shallow carbonate platform, which is indeed characterized by evidence of intermittent exposure features such as mudcracks in the area.

Based on the available $\delta^{13}C$ and $^{87}Sr/^{86}Sr$ database for the basal Bambuí Group, Paula-Santos et al. (2017) proposed its subdivision into three major Chemostratigraphic Intervals (CI) which would record large paleoenvironmental changes in the basin. The basal CI-1 comprises the cap carbonates from the Sete Lagoas Formation, which show $\delta^{13}C$ ranging from −3‰ to −5‰ followed by a recover to values around 0‰, and $^{87}Sr/^{86}Sr$ ratios from 0.7074 to 0.7082. In the middle portion of Sete Lagoas Formation, the CI-2 presents $\delta^{13}C$ values around 0‰ and $^{87}Sr/^{86}Sr$ ratios ~0.7084. In turn, the following CI-3 comprises the upper Sete Lagoas, Serra de Santa Helena and Lagoa do Jacaré formations, which show the above mentioned high $\delta^{13}C$ values (+8‰ to +16‰) and $^{87}Sr/^{86}Sr$ ratios ~0.7075. CI-1 would record the initial transgression over the continental area while the CI-2 would correspond to a stage in which the basin was connected to the open ocean (as attested by the presence of Cloudina sp., Warren et al., 2014). On the other hand, CI-3 would represent a stage of basin restriction, in which isolation from the global ocean prevented isotope homogenization, and the high values of $\delta^{13}C$ would result from enhanced burial of organic carbon and/or from methanogenesis in anoxic conditions (Iyer et al., 1995; Paula-Santos et al., 2017). Uhlein et al. (2019) argued that those two processes could not alone explain the unusually high $\delta^{13}C$ values of the middle Bambuí Group (the MIBE excursion), instead suggesting a complex interplay of different causes linked fundamentally to a restricted basin. In this model, a change in total carbon input due to preferential weathering of ancient sedimentary carbonates on the continents and higher burial rates of authigenic carbonate due to methane or organic matter oxidation via sulfate or iron reduction, amplified by poor ocean ventilation, would be the major causes for the MIBE. As reported by Caetano-Filho et al. (2019) the overall chemostratigraphic pattern discussed above is intimately coupled with sequence stratigraphy of the basin, matching its 2nd order stacking patterns. These same authors report a sudden increase in Sr content and Sr/Ca ratios in the first 2nd order highstand system tract at basinal scale, which is also interpreted in the context of the restricted paleoenvironment.

The Bambuí Group has been interpreted as the record of a foreland system, developed during the Brasiliano/Pan-African West Gondwana assembly (Alkmim and Martins Neto, 2001; Martins-Neto, 2009; Reis et al., 2017; Uhlein et al., 2017). In spite of the multiple overloads imposed by the marginal Brasiliano-Pan/African orogens on the borders of the São Francisco-Congo plate, the sedimentary dispersal and evolution of this basin system was mostly controlled by the Brasília Belt load in the western cratonic margin (e.g., Alkmim and Martins-Neto, 2012; Castro and Dardenne, 2000; Martins-Neto, 2009; Reis et al., 2017; Uhlein et al., 2017). This led to sedimentary partitioning of the basin into two major depocenters: (i) a foredeep mostly filled with fine- to coarse-grained siliciclastics and (ii) a forebulge filled with mixed carbonate-siliciclastics, in the western and eastern portions of the São Francisco basin, respectively. This relatively simple evolution could, however, include some nuances and secondary processes within the major foreland system evolution. Some authors, for example, interpret a major unconformity in the middle portion of the Sete Lagoas Formation, separating a lower interval, deposited in an intracratonic basin, from an upper interval deposited in a foreland basin system (Martins and Lemos, 2007; Zalán and Romeiro-Silva, 2007), based mainly on regional seismic data and in a chemostratigraphic break in the $\delta^{13}C$ values of the carbonates.

Uhlein et al. (2017) presented an integrated view of basinal evolution linking the sedimentary record of the Bambuí Group with tectonic pulses of the Brasília Orogen to the West, which generated distinct early to late Ediacaran conglomeratic wedges in the foredeep region and deposition of the mixed carbonatic-siliciclastic units in the forebulge region. According to the authors, the middle Sete Lagoas unconformity was generated by foredeep subsidence and forebulge uplift, separating an early Ediacaran phase in which the foredeep-related Samburá conglomerates (younger zircon crystals around 630 Ma) and the forebulge-related cap carbonate from the basal Sete Lagoas Formation were deposited, from a late Ediacaran phase with deposition of the Lagoa Formosa conglomeratic wedges (younger zircon crystals at 560 Ma). This is coherent with the detrital zircon content recovered from the middle Sete Lagoas Formation, with zircon grains as young as ca. 557 Ma (Paula-Santos et al., 2015) similar to the Lagoa Formosa Formation, and with recent findings of *Cloudina* sp., a terminal Ediacaran index fossil (ca. 550–541 Ma) described by Warren et al. (2014) in the middle Sete Lagoas Formation in the Januária region.

On the other hand, based on recently acquired high resolution seismic and well data and surface information, Reis and Suss (2016) and Reis et al. (2017) interpret the entire Bambuí Group succession (including its basal cap carbonate rocks) as the record of a single late Ediacaran foreland system. Although recognizing multiple lower-rank unconformities within the Bambuí strata, these authors did not describe any stratigraphic gap within the basal to middle Sete Lagoas Formation. The recognized unconformities, however, bound different second-order sequences (e.g., at the uppermost Sete Lagoas Formation) and have been similarly interpreted as the typical record of foredeep-forebulge dynamics. In this view, the Bambuí foreland system would have evolved during the late Ediacaran-Cambrian within an almost completely confined setting, since the Brasiliano orogens surrounding the São Francisco-Congo plate were already uplifted by this time (Paula-Santos et al., 2015; Reis et al., 2017). These authors also consider the glaciogenic Jequitaí Formation (in its type area) as an older and different basin record correlative to the Cryogenian deposits of the Macaúbas Group exposed in the Araçuaí belt to the east (Pedrosa-Soares et al., 2011). A similar interpretation has been recently presented by Caetano-Filho et al. (2019).

The different views of basin evolution presented in the literature have practical implications for the understanding of the isotope record of the Bambuí Group and its basal glaciogenic record. In this sense, they may suggest either a Cryogenian (e.g., Babinski et al., 2007; Vieira et al., 2007), early Ediacaran (Alvarenga et al., 2014; Caxito et al., 2012) or a much younger, late Ediacaran (e.g., Caetano-Filho et al., 2019; Kuchenbecker et al., 2013;

Reis and Suss, 2016) age for the cap carbonate units at the base of the Sete Lagoas Formation. Unfortunately, neither *syn*-sedimentary igneous rocks nor detrital zircons younger than ca. 880 Ma have been reported for both the glacial rocks of the Jequitaí Formation (Rodrigues, 2008 and references therein) and for the basal, cap carbonate unit of the Sete Lagoas Formation. This lack of reliable absolute dating has led to controversies about the absolute age and extent of the glaciation(s) recorded by the diamictites and cap carbonate deposition, which remains unresolved.

Babinski et al. (2007) obtained a Pb-Pb whole-rock isochron array of 740 ± 20 Ma on carbonates containing aragonite fan pseudomorphs at the Sambra Quarry, near Sete Lagoas town (southern São Francisco basin), thus suggesting a middle-Cryogenian (post-Sturtian glaciation) age for the Sete Lagoas cap carbonate (Vieira et al., 2007). Alternatively, Caxito et al. (2012) and Alvarenga et al. (2014) pointed out that the Sete Lagoas cap carbonate bears a set of litho- and chemostratigraphic features that are very similar to post-Marinoan (early Ediacaran) cap carbonates worldwide, such as a basal, meter-thick pale cap dolostone with decreasing-upward $\delta^{13}C$ values, $^{87}Sr/^{86}Sr$ ratios above 0.7074, widespread aragonite crystal fans, and other features such as distinct barite and phosphorite layers (e.g., Crockford et al., 2018; Drummond et al., 2015). Chemostratigraphic correlation with other nearby sequences with more better constrained geochronological bracketing (e.g., the cap carbonates of the Sergipano and West Congo belts overlying diamictites with detrital zircons as young as ca. 650 Ma, probably deposited in the same shallow watermass in different margins of the São Francisco Craton; Caxito et al., 2012 and references therein) was also invoked to suggest a post-Marinoan age for the Sete Lagoas cap carbonate. All of those features, however not necessarily unique to post-Marinoan settings, altogether suggest deposition in a similar setting as the cap dolostone dated at ca. 635 Ma in Namibia and China and the Nuccaleena Formation chosen as the GSSP (Global Stratotype Section and Point) to the formally defined Ediacaran Period in the Flinders Range Park, South Australia (Knoll et al., 2004); and suggested as one of the few globally synchronous depositional events in Earth's history (e.g., Shields, 2005). The recent findings of *Cloudina* shells (Warren et al., 2014), considered as an index fossil for the terminal Ediacaran (ca. 540–510 Ma), and of ca. 557 Ma detrital zircons (Paula-Santos et al., 2015) in the middle Sete Lagoas Formation could imply a potentially even younger Ediacaran age for the Sete Lagoas cap carbonate (e.g., Kuchenbecker et al., 2016a), which would then not be linked neither to the global Sturtian nor Marinoan glaciations. Despite the difficulties in solving this conundrum (at least until absolute dating of not-yet-found interleaved volcanic layers emerges), the amassed dataset seems to suggest that deposition of the major part of the Bambuí Group (at least from the middle Sete Lagoas Formation upward—circa 1000m-thick) occurred during the terminal Ediacaran, probably straddling the Cambrian boundary in its upper reaches (Caxito et al., 2018; Hippertt et al., 2019; Uhlein et al., 2019).

Recently, Crockford et al. (2018) analyzed cm-thin barite layers interleaved in the base of the Sete Lagoas cap carbonate for triple oxygen and sulfur isotopes. Values of $\Delta^{17}O$ are as low as −1.05‰, within range of other negative triple oxygen isotope anomalies found for basal Ediacaran cap carbonates around the world, such as in Svalbard (−1.64‰; Bao et al., 2009) south China (−0.87‰; Peng et al., 2011), northwestern Canada (−0.84‰; Crockford et al., 2016), and Norway (−1.02‰, Crockford et al., 2018). Values of $\Delta^{17}O$ below −0.4‰ are known only from early Ediacaran deposits, and anomalous values are absent from Cryogenian or post-Ediacaran sulfates (Bao et al., 2008). Thus, Crockford et al. (2018) argue that this anomalous horizon could represent a global marker unique to the early Ediacaran, caused by catastrophic changes in the ozone layer following extremely high ρCO_2 due to perturbations in the natural biogeochemical cycles during the pan-glacial state. This stratigraphic marker could be as important as the iridium anomaly marking the Cretaceous-Paleogene boundary, similarly implicating an extreme and unprecedented event in Earth's history.

Barium isotopes were very recently used to investigate the origins of the Sete Lagoas and other post-glacial barite layers (Crockford et al., 2019). Comparing the $^{138}Ba/^{134}Ba$ ratios ($\delta^{138/134}Ba_{NIST}$) of barite samples from various cap carbonate successions around the world, along with almost a hundred samples from distinct modern and ancient environments, those authors reached the conclusion that post-glacial barites represent a well-mixed, effectively limitless barium source, similar to modern marine barium, in contrast with modern cold seep or terrestrial barites which show a wide range of isotopic values related to local predominance of fractionation processes. A model for globally synchronous deposition of post-glacial barite is put forward where a well-mixed Ba-replete reservoir accumulated during the glacial interval, and was then brought upward through the water mass upon deglaciation, encountering continental weathering-derived SO_4 in the shallow waters and promoting widespread barite precipitation during cap carbonate deposition.

A multi-proxy study using C, O, Sr, Nd, Pb and Cr isotopes, as well as trace elements, was conducted in a section of the Sete Lagoas cap carbonate in Bahia state by Caxito et al. (2018). A model for deposition of the cap carbonate based on those proxies is presented by the authors, starting with the glacial aftermath, when the cap dolostone and basal limestones were deposited under the influence of glacier melting causing the sea level to rise and mixing of glacial meltwaters with anoxic bottom waters. This is marked by a decreasing trend of εNd(t) from -6 to -8, indicating continued input of continentally-derived particles to the basin, as also suggested by higher Al and Mn contents. Intermittent subaerial exposure caused meteoric alteration, shifting $^{87}Sr/^{86}Sr$ toward higher ratios (>0.710) and $\delta^{18}O$ toward lower values. After this phase, the late postglacial succession is marked by modern marine-like REE+Y patterns and lower Al and Mn contents, reflecting the diminishing of glacial meltwater input, while a sharp rise of U/Th ratios and of $\delta^{53}Cr$ values suggest oxygenated conditions in postglacial surface waters toward the middle to late Ediacaran. Those results add to a growing body of data for other Ediacaran carbonate intervals around the world which suggest that after ca. 1.0 Ga of widespread anoxia (the Boring Billion) the oceans became increasingly oxygenated following the major Cryogenian glaciations. The higher oxygen levels ultimately could have led or interacted with the explosion of complex macroscopic life following the NOE.

In addition, coupled isotopic C, O, Sr and non-isotopic multiproxy studies conducted in the upper portion of Bambuí strata also yielded important paleoenvironmental and paleoceanographic constraints (Caetano-Filho et al., 2019; de Paula-Santos et al., 2018; Hippertt et al., 2019; Paula-Santos et al., 2017; Uhlein et al., 2019). For instance, extremely positive (up to +16‰) $\delta^{13}C$ values and $^{87}Sr/^{86}Sr$ ratios around 0.7075 observed in the middle to upper Bambuí basin (the MIBE excursion of Uhlein et al., 2019) were associated with a major paleoceanographic shift when the Bambuí watermass probably became physically disconnected from neighboring open marine systems, evolving to a typical restricted scenario, where the lack of homogenization with open marine sources would induce anomalously low $^{87}Sr/^{86}Sr$ ratios around 0.7075 (Paula-Santos et al., 2017). This is supported by $\delta^{13}C$, TOC (Total Organic Carbon) and iron speciation data from Hippertt et al. (2019) which suggests that the upper Bambuí basin developed in a disconnected, restricted environment, which enabled the persistence of anoxic and possibly ferruginous watermass conditions. As proposed by Paula-Santos et al. (2017) and

Uhlein et al. (2019) this anoxic and restricted environment would also favor methanogenesis, evaporation and organic matter preservation, which coupled to preferential weathering of ancient sedimentary carbonates, higher burial rates of authigenic carbonate and loss of oceanic ventilation would explain the anomalously high $\delta^{13}C$ values that characterize the MIBE. Nevertheless, weak correlations between $\delta^{13}C$ values, TOC contents and productivity proxies in the Lagoa do Jacaré Formation samples suggest low bioproductivity, which could have damaged biological evolution and explain the lack of Ediacaran-Cambrian index fossils in the upper Bambuí basin (Caetano-Filho et al., 2019; Hippertt et al., 2019). Those interpretations are in line with the developing picture of a very complex interplay of environmental factors in the Ediacaran-Cambrian transition, where the combination of increasing atmospheric oxygen levels, tectonic activity, watermass connection and nutrient supply would be the trigger of a life supportive environment which enabled the explosion of complex macroscopic life.

Kaufman et al. (2001) presented S isotope data for CAS in the Sete Lagoas cap carbonate with fractionated $\delta^{34}S$ values up to +47.5‰. This extreme enrichment in ^{34}S was interpreted as due to high rates of bacterial sulfate reduction in anoxic waters during glaciation, with post-glacial transgression and photosynthetic production of O_2 likely driving sulfate reducers back into sediments. The values are higher than expected for riverine input (ca. 6–10‰), and thus Kaufman et al. (2001) interpret the S isotope anomaly as a consequence of oceanographic processes rather than intense post-glacial weathering under high ρCO_2. Kaufman et al. (2001) also complemented $\delta^{13}C_{org}$ data previously presented by Iyer et al. (1995), showing constant values of −18‰ to −20‰, considered as anomalously enriched and resulting in lower values of carbon isotope fractionation ($\Delta^{13}C$) between inorganic and organic phases.

Silva-Tamayo et al. (2010a) presented Ca isotope results for a section of the Sete Lagoas cap carbonate, with the lowermost part of the section displaying $\delta^{44}Ca$ values around 0‰, rapidly increasing and stabilizing upsection to near 1‰, and then quickly dropping to ca. 0.5‰ in the top of the 160 m-thick section. The near-zero values of the bottom of the section require a major perturbation in the marine Ca mass balance due to deglaciation, with increased Ca delivery to the post-glacial ocean likely resulting from enhanced continental weathering under high ρCO_2. Ca isotopes then show a quick recovery to normal values, representing a recovery of normal atmospheric and hydrospheric conditions.

Paula-Santos (2018) presented Li and Mg isotope data for carbonates from the basal Bambuí Group. The results show that δ^7Li and δ^{26}Mg vary according to the conditions of restriction and connection with the global ocean, in response to the consequent changes in the silicate weathering influx. According to the authors, local controls play an important role in the variation of Li and Mg isotopes in most of the Sete Lagoas Formation, except for a strong coupled negative δ^7Li (c.10‰ to 3‰) and positive δ^{26}Mg (−3‰ to −1‰) excursions in the very base of the unit, which are similar to other Neoproterozoic cap carbonates successions.

The Una Group in the central São Francisco Craton (Fig. 1) presents similar stratigraphic and isotope stratigraphy record to the basal Bambuí Group, with diamictites of the Bebedouro Formation overlain by cap carbonates of the Salitre Formation (Misi and Veizer, 1998). Those are correlative to the Jequitaí and Sete Lagoas formations, respectively. The Una Group outcropping in the Irecê and Una basins thus probably corresponds to restricted basins deposited at the same time as the basal Bambuí Group in the São Francisco basin, or alternatively to a lateral extension of those deposits (Fig. 7).

3.4.2 Sergipano belt

In the Sergipano Belt bordering the northeastern São Francisco Craton margin, two cap carbonate intervals overlying distinct glaciogenic diamictites occur: (i) the Jacoca Formation carbonates resting atop the Ribeirópolis Formation diamictite, and (ii) the Olhos D'água Formation overlying diamictites of the Palestina Formation (Fig. 8). All of those sedimentary rocks were deformed and metamorphosed during the Brasiliano Orogeny in the area (ca. 600–575 Ma; Oliveira et al., 2010; Sial et al., 2010). Another sequence outcrops to the west, composed of the Acauã Formation dolostones in sharp contact above diamictites of the Juetê Formation. These are interpreted as correlatives to the Jacoca and Ribeirópolis formations, respectively.

Values of δ^{13}C are similar for the Jacoca and Acauã formations, around −4‰ to −5‰, while in the Olhos D'Agua Formation a more complex pattern is observed, with negative values around −5‰ in the basal shallow-marine deposits of the unit superseded by values close to zero and, after a sharp break, by very fractionated values between +8‰ and +10‰ at the top (Sial et al., 2010) (Fig. 8). This is very similar to the Sete Lagoas Formation patterns discussed above, and in fact a chronocorrelation between the Sete Lagoas and Olhos D'água cap carbonates was proposed by

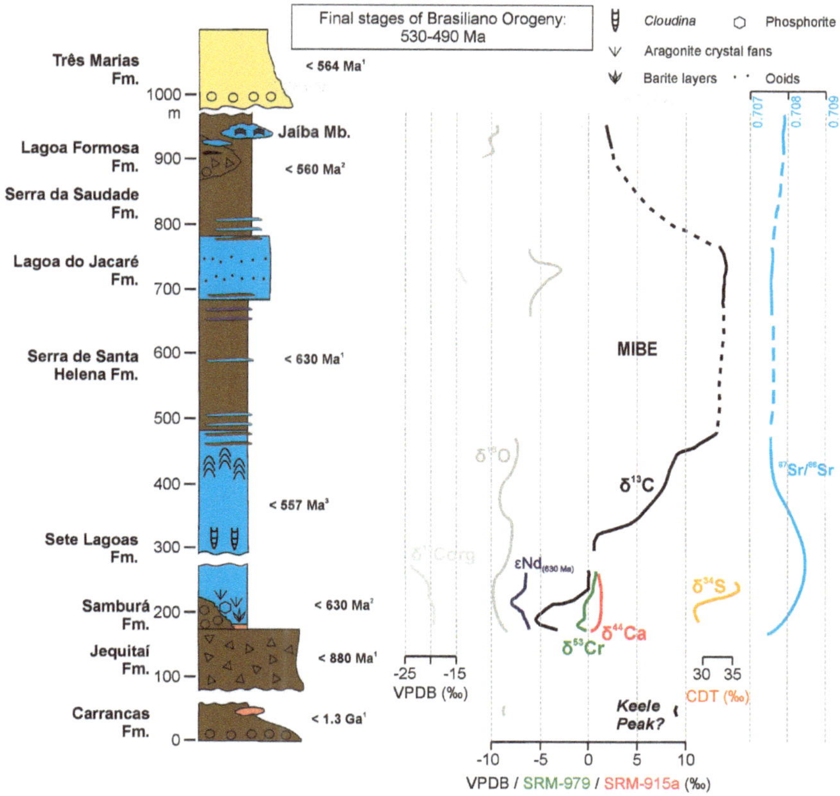

Fig. 7 Simplified stratigraphic column of the Bambuí Group with isotope stratigraphy records (see text for data sources). Key for rock types and structures in the stratigraphic section is the same as in Figs. 4–6. Sources for geochronological constraints: 1 Rodrigues (2008), 2 Uhlein et al. (2017); 3 Paula-Santos et al. (2015).

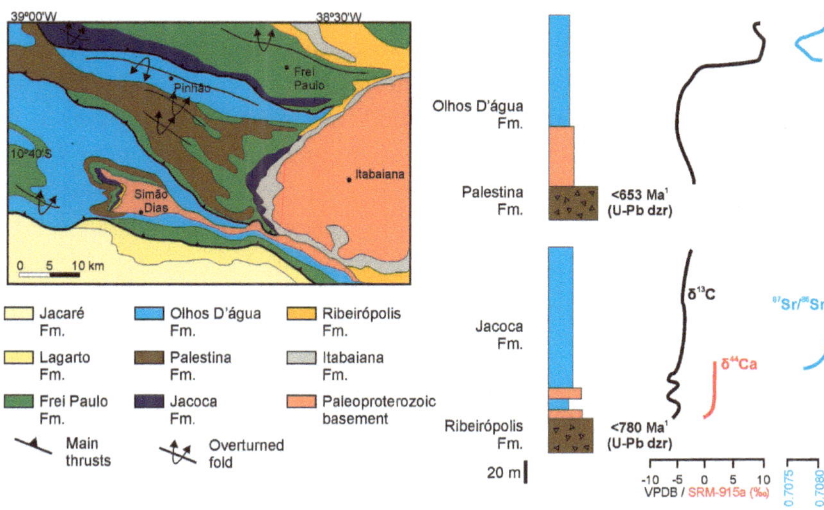

Fig. 8 Simplified geological map and stratigraphic columns of the Sergipano Belt with isotope stratigraphy records (Sial et al., 2010). Sources for geochronological constraints: 1 Oliveira et al. (2015).

Caxito et al. (2012). The Acauã Formation yielded $^{87}Sr/^{86}Sr$ ratios from 0.7072 to 0.7073, while the Jacoca and Olhos D'Agua formations yielded values within the range from 0.7077 to 0.7081 (Sial et al., 2010). These authors suggest that the Jacoca and Acauã formations are probably correlative and correspond to the mid-Cryogenian (Sturtian) cap carbonates, while the Olhos D'Água Formation correspond to the basal Ediacaran (Marinoan) cap carbonate. This interpretation is coherent with the youngest detrital zircon grains recovered from the Palestina diamictite and dated at ca. 653 Ma (Oliveira et al., 2015). Sial et al. (2010) also presented values of Hg concentrations for all of the cap carbonate sequences, with concentrations between 20 and 80 ng/g similar to carbonates deposited under the influence of volcanic activity, thus suggesting that cap carbonate CO_2 is mostly mantle-derived.

Silva-Tamayo et al. (2010a) presented Ca isotope results for the Jacoca Formation, with a pattern of $\delta^{44}Ca$ values similar to those of the Sete Lagoas Formation and interpreted as deposited under similar influence of strong continental weathering disrupting the Ca mass balance in post-glacial seawater.

3.4.3 Seridó belt

The sigmoidal-shaped Seridó Belt of the Borborema Province, northeastern Brazil, represents an area with a former sedimentary basin related to the Brasiliano cycle, bearing detrital zircons as young as ca. 650 Ma (Hollanda et al., 2015; Van Schmus et al., 2003) and deformed and metamorphosed during the Brasiliano Orogeny, at ca. 575 Ma. The Jucurutu Formation at the base of the Seridó Group is composed of marbles overlying restricted occurrences of BIF and iron ores. Locally, diamictites of uncertain stratigraphic positioning also occur. At the Ferro do Bonito mine, the Jucurutu marbles sitting atop BIF show $\delta^{13}C$ values as low as −12‰, followed by values around −5‰ and then superseded by positive values at the top of the section, at +4‰ to +10‰. The $^{87}Sr/^{86}Sr$ ratios are at 0.7074–0.7075, suggesting deposition during the late Cryogenian to early Ediacaran, and a correspondence of the possibly glaciogenic-related diamictites, BIFs and marbles with other worldwide sequences of this age (Nascimento and Sial, 2004; Nascimento et al., 2007; Sial et al., 2015b).

Sial et al. (2015b) also analyzed Cr isotopes in the Jucurutu BIFs, which yielded largely unfractionated values, between −0.42‰ and −0.12‰, within the range of lithogenic reservoirs. These and other chemostratigraphic data led the authors to interpret the Jucurutu BIFs as deposited under the influence of hydrothermal vents, in anoxic and acidic deep waters, in a restricted rift basin.

The overlying marbles, however, yielded slightly positively fractionated $\delta^{53}Cr$, possibly representing post-glacial cap carbonate sequences deposited in anoxic to slightly oxic shallow marine environments.

3.4.4 Paraguay belt

The Paraguay Belt represents a curvilinear Brasiliano/Pan-African W thrust-fold belt exposed in the center of the South American continent, originated by the deformation of Neoproterozoic units in the southeastern border of the Amazon Craton and eastern border of the Rio Apa block during the agglutination of the West Gondwana (Cordani et al., 2009). The Paraguay Belt can be divided into a northern (NW-verging) and a southern (W-verging) sector, with the Corumbá Graben system (Trompette et al., 1998) linking the two arms of the fold belts and the Tucavaca Aulacogen of Bolivia to the west. The sedimentary successions of the fold belt are correlated with a non-metamorphic and undeformed cratonic cover to the west (da Campanha et al., 2011). They constitute important archives of climatic, geochemical and bioevolutive changes occurred during the terminal Neoproterozoic Era, notably the fossiliferous carbonates rocks of the Tamengo Formation (Corumbá Group) and Itapucumi Group (Fig. 9).

The Neoproterozoic record is represented by glaciation (Puga Formation), precipitation of thick BIF and manganese beds (Jacadigo Group) and the deposition of carbonate successions (Araras and Corumbá groups) bearing phosphorites and evidences of the earliest skeletonized animals in Earth history (Alvarenga et al., 2009; Boggiani et al., 2010; Freitas et al., 2011; Gaucher et al., 2003; Nogueira et al., 2003). This geological context also includes the Itapucumi Group in Paraguay, a narrow strip of east-verging folded rocks that belongs to the Vallemí Fold Belt (da Campanha et al., 2010). This mixed siliciclastic-carbonate unit present well-preserved fossils of Ediacaran organisms such as *Cloudina* and *Corumbella* and is considered as coeval to the Corumbá Group in Brazil (Warren et al., 2019).

In general, the sedimentary successions that constitute the Paraguay Belt are characterized by putative glacial diamictites from the Puga Formation covered by carbonate rocks of the Corumbá (South) and Araras (North) groups. Despite the regional occurrence of basal diamictites in the entire extension of the Paraguay Belt, the stratigraphic framework of the southern and northern portions of the belt reflects a distinct sedimentary evolution (Alvarenga et al., 2009, 2011). The Neoproterozoic Puga Formation and Cuiabá Group are interpreted as late Cryogenian/early Ediacaran (representing the diamictite/cap carbonate pair of the Marionan glaciation),

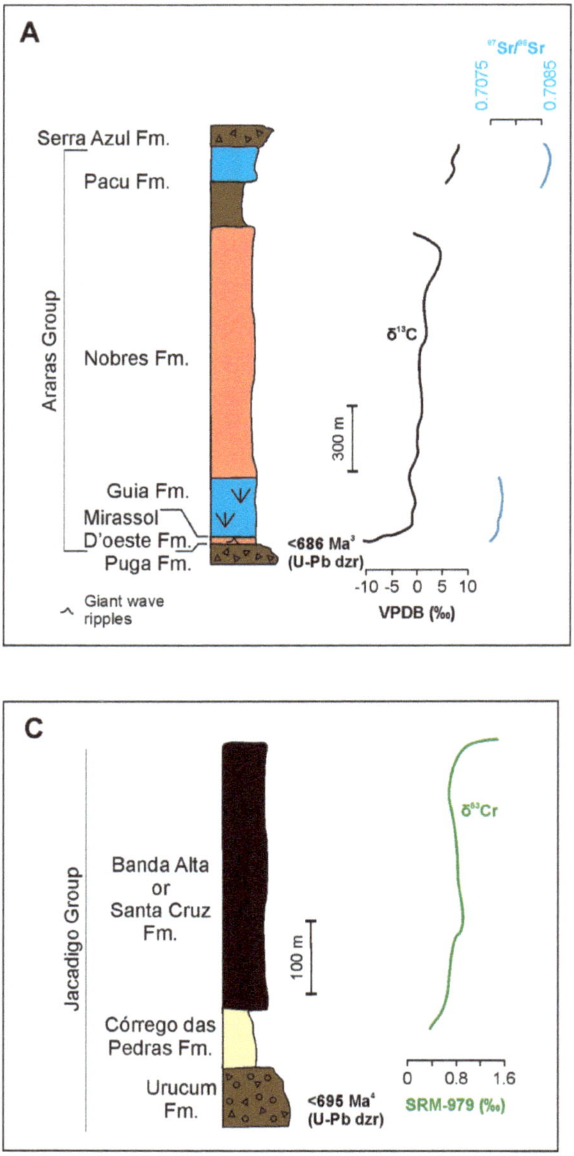

Fig. 9 Simplified geological map and stratigraphic columns of the Paraguay Belt with isotope stratigraphy records for the Northern (A, data sources = Alvarenga et al., 2004, 2008; Font et al., 2006; Nogueira et al., 2003, 2007) and Southern (B, data source = Boggiani et al., 2010) areas and for the Corumbá area (C, data source = Frei et al., 2017). Sources for geochronological constraints: 1 Babinski et al. (2008), 2 Parry et al. (2017), 3 McGee et al. (2018), 4 Frei et al. (2017).

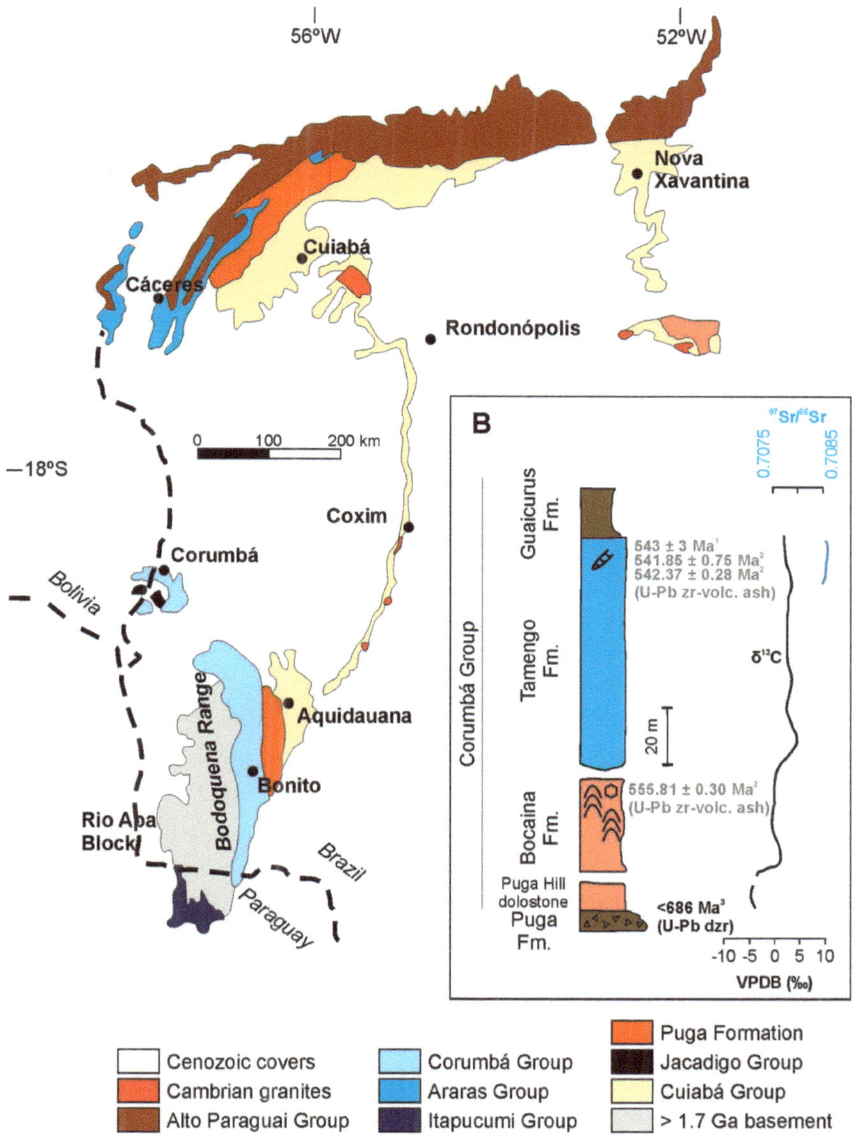
56°W
52°W
Nova Xavantina
Cuiabá
Cáceres
Rondonópolis
0 100 200 km
–18°S
Coxim
Corumbá
Bolivia
Aquidauana
Bodoquena Range
Bonito
Rio Apa Block
Brazil
Paraguay
B
87Sr/86Sr
0.7075
0.7085
Guaicurus Fm.
Tamengo Fm.
Bocaina Fm.
Corumbá Group
Puga Hill dolostone
Puga Fm.
543 ± 3 Ma1
541.85 ± 0.75 Ma2
542.37 ± 0.28 Ma2
(U-Pb zr-volc. ash)
20 m
555.81 ± 0.30 Ma2
(U-Pb zr-volc. ash)
<686 Ma3
(U-Pb dzr)
δ13C
-10 -5 0 5 10
VPDB (‰)
Cenozoic covers
Cambrian granites
Alto Paraguai Group
Corumbá Group
Araras Group
Itapucumi Group
Puga Formation
Jacadigo Group
Cuiabá Group
> 1.7 Ga basement

Fig. 9—Cont'd

due its stratigraphic position below the *Cloudina*-bearing limestones of the Tamengo Formation (Corumbá Group). On the other hand, the maximum depositional age for this unit, obtained by detrital zircons is at ca. 706 Ma (U-Pb SHRIMP), not allowing to completely rule out a correlation to the mid-Cryogenian Sturtian glaciation (Babinski et al., 2013). The possibility of occurrence of more than one glacial record cannot be discarded.

The Puga diamictite present striated and faceted pebbles interpreted as glacially-related features in the northern part of the Paraguay Belt (Alvarenga and Trompette, 1993; Alvarenga et al., 2004, 2008, 2011; Nogueira et al., 2003). In the southern part of the Paraguay Belt, evidences for glacial deposition are not as clear. A ca. 100 m-thick layer of diamictites was described in the Puga Hill and interpreted as glacially-related (Maciel, 1959) in the southern Paraguay Belt, but glacial features such as striated and faceted pebbles were not yet found. The diamictite of the Puga Formation is directly covered by a cap carbonate of the Mirassol d'Oeste Formation (Araras Group) in the northern Paraguay Belt (Alvarenga et al., 2008; Nogueira et al., 2003). In the southern Paraguay Belt similar cap carbonates were described above the diamictites in Puga Hill (Misi et al., 2007), including tubestone-microbialite associations in cap dolomites (Romero et al., 2016).

At North, paleolatitude data obtained from the Puga diamictite indicate deposition under low latitude and a Pb-Pb isochron age of 633 ± 25 Ma on the related cap carbonate suggest deposition of the diamictite-cap carbonate pair related to the Marinoan glaciation (Romero et al., 2013). The Araras Group succession initiate with cap carbonate deposits followed by limestone, stromatolitic dolostone, shale-carbonate and lenses of limestones at the top, followed by terrigenous molassic sediments from the Alto Paraguay Group (Dantas et al., 2009; McGee et al., 2015). The $\delta^{13}C$ values in the Araras Group are negative at the cap carbonate ($\sim -5‰$) and not so negative ($\sim -1‰$) in the lower limestone succession, changing to low positive value in dolostone, around near zero too ($\sim +1‰$), reaching positive values in the last 70 m ($\sim +7‰$) (Alvarenga et al., 2004, 2008, 2011; Font et al., 2006; Nogueira et al., 2003, 2007; Sial et al., 2016). The Araras Group present $^{87}Sr/^{86}Sr$ values between 0.7075 and 0.7079 in the lower formations and $^{87}Sr/^{86}Sr$ of 0.7087 in the uppermost columnar-stromatolite-bearing limestone lenses (Alvarenga et al., 2008). At South, the Morro do Puga cap carbonate that overlays diamictite of the Puga Formation (Boggiani et al., 2003) show consistently negative $\delta^{13}C$ values of about $-5‰$ similar to those obtained from other Marinoan cap carbonates worldwide

(Boggiani et al., 2010). Until now, no metazoan fossils were described in any of these units (Nogueira et al., 2003, 2007).

Silva-Tamayo et al. (2010b) presented Ca isotope results for the Mirassol D'oeste cap carbonate. A basal anomaly with $\delta^{44}Ca = +0.1‰$ is observed, followed by a rise to ca. +2‰ and a decrease to a steady state with $\delta^{44}Ca$ around +0.8‰. This is interpreted as a result of the mass balance between post-glacial Ca input from carbonate and silicate dissolution and Ca precipitation. The oscillating $\delta^{44}Ca$ pattern is predicted as a probable typical feature of Marinoan post-glacial cap carbonate successions.

The diamictites of the Puga Formation are locally associated with ferriferous diamictite and possibly present a stratigraphic correlation to the huge manganese and iron deposits of Jacadigo Group in the Corumbá Graben System, interpreted as glaciogenic (Trompette et al., 1998). Authigenic chromium isotope signatures (average $\delta^{53}Cr = 1.10 \pm 0.4‰$; $n = 16$; 2σ) were identified in the entire stratigraphic section of iron formation (Jacadigo Group), compatible with those from other iron formations deposited during the late Neoproterozoic and Precambrian-Cambrian transition worldwide. This isotope record indicates prevailing high atmospheric O_2 levels following the late Neoproterozoic glaciations that accompanied the evolution of macroscopic multicellular organisms (Frei et al., 2017).

All siliciclastic units (e.g., Puga and Jacadigo in Brazil and Vallemi Formation in Paraguay) are interpreted as deposited in a series of small rift basins during the last stages of the fragmentation of the Rodinia supercontinent (Warren et al., 2019). In a post-rift stage, the continuity of the thermal subsidence led to formation of several carbonate ramps surrounding the margins of the Amazon Craton and Rio Apa block, allowing the deposition of stromatolitic dolostones and phosphorites of the Bocaina Formation. After this initial stage, a new uplift cycle resulted in an expressive and regional drop in sea level causing the erosion of these recently deposited carbonates (Boggiani et al., 2010). This tectono-sedimentary event is recorded by a breccia deposit positioned at the base of the Tamengo Formation, which is marked by negative $\delta^{13}C$ values ($\sim -3‰$) followed by a positive excursion (+ 5‰) in an expressive transgressive event. The metazoan fossil assemblage of the Tamengo Formation include the index fossil *Cloudina* and *Corumbella* remnants and occur in carbonates with consistent positive $\delta^{13}C$ values around +3‰ to +5‰ (Boggiani et al., 2010). In Paraguay, a similar fossil assemblage also occur (Warren et al., 2017) and is associated with peritidal carbonates of the Tagatiya Guazu Formation, also presenting $\delta^{13}C$ positive values of about +2‰ (Warren et al., 2019).

The presence of the *Cloudina* index fossil and similarities in the C isotope curves allow to assign a terminal Ediacaran to Cambrian age for both correlative units. Zircons crystals recovered from of volcanic ash layer yielded a 541.2 Ma (U-Pb) date consistent with 0.7084–0.7085 $^{87}Sr/^{86}Sr$ ratios reinforcing the deposition of these units close to the Precambrian-Cambrian transition. Another tuff layer at the top of the Bocaina Formation yielded a U-Pb zircon age of ca. 555 Ma constraining the age of the Tamengo Formation to the 555–542 Ma interval (Parry et al., 2017).

The final cycle of sedimentation in the Corumbá Group is represented by 200–300 m thick siltstones of Guaicurus Formation that, in theory, can comprise the Precambrian-Cambrian boundary. These successions were deposited during a regional transgression that flooded the carbonate-dominated strata of the Corumbá (Brazil) and Itapucumi (Paraguay) groups.

Deformation of the Corumbá and Itapucumi groups initiated at ca. 528 Ma following the last compressional events of the Brasiliano Cycle. The compressional reactivation of listric faults from the rift stage related to the compression caused by the collage of the Pampia, Amazon and Rio Apa lithospheric blocks led to the formation of the Vallemí and Paraguay fold-and-thrust belts with opposite vergence (Warren et al., 2019). The end of this tectono-thermal event probably occurred around 484 Ma, possibly in the Ordovician period.

3.5 Isotopes as proxies for sedimentary provenance: Sm-Nd mapping of stratigraphic sequences in Brazilian thrust-and-fold belts

An interesting application of Nd isotopes coupled to stratigraphic sections is the study of variations in the sedimentary provenance of Precambrian sequences, especially in places where original stratigraphic relations have been modified by subsequent deformation and metamorphism. This is the case for the Brasiliano (ca. 630–530 Ma) thrust-and-fold belts that border the major cratonic blocks of the Brazilian territory.

Dantas et al. (2009) applied this tool in the Paraguay Belt bordering the southeastern margin of the Amazon Craton, plotting T_{DM} model ages and $\varepsilon Nd(t)$ against the stratigraphic stacking to reveal changes in the source areas as functions of the erosion/deposition ratios and uplift during orogenic processes. The results show that the lower metasedimentary successions of the Cuiabá Group bear distinctly older T_{DM} of ca. 2.3 Ga and negative $\varepsilon Nd(t)$ values around -11, changing abruptly in the upper succession of the Alto Paraguay Group to T_{DM} as young as 1.5 Ga and $\varepsilon Nd(t)$

up to −6. Those changes are interpreted by the authors as the result of a provenance shift from a continental margin (coming from the erosion of the Amazon Craton to the north) to a foreland basin developed in response to orogenic loading.

Caxito et al. (2014) presented a similar study using Nd isotope changes versus stratigraphy in the Rio Preto Fold Belt of the northwestern São Francisco Craton margin, coupled to detrital zircon U-Pb data. Similar differences in εNd(t) and T_{DM} have been observed between sedimentary successions interpreted as deposited in distinct settings. Interestingly, within the rift-related deposits of the Canabravinha Formation (younger detrital zircon grains at ca. 900 Ma), the T_{DM} values present a trend with older values toward the top, from ca. 1.9 to ca. 2.7 Ga, with accompanying εNd(t) from −8 to −20. This trend was interpreted as reflecting the progressive exposure of older basement with continued erosion of the source areas located mainly in the São Francisco Craton to the south, a hypothesis also supported by variations in detrital zircon ages throughout the stratigraphic section.

4. Conclusions and future directions

Isotope stratigraphy studies have been extensively applied in Brazilian Precambrian sedimentary successions throughout the last decades. Due to the availability of suitable equipment and laboratories, the state of understanding of natural cycles and fractionation processes, the easiness of collection and preparation of samples and treatment of large datasets, C and O isotope studies surpass by far studies that use other isotope proxies, followed by radiogenic Sr isotopes, and, to a minor extent, stable S isotopes. However, this is a field in full expansion, and some *avant-garde* studies are being conducted in the last years in Brazilian Precambrian rocks using non-traditional isotope systems whose development is under way in the most advanced laboratories of the world (e.g., Ca, Ba, Cr, Fe, Cd, triple oxygen isotopes, etc.).

Brazil is privileged in containing expressive sedimentary intervals of Precambrian age, spanning the most important extreme and unidirectional events in Earth's history, such as the major pulses of atmospheric oxygenation (GOE and NOE), metazoan diversification during the Ediacaran/Cambrian transition, and the wide climatic variations caused by the Cryogenian greenhouse/icehouse conditions. Chemical sedimentary rocks such as carbonates, BIFs and evaporites, which are prime recorders of ancient seawater geochemical and biochemical conditions, are present in successions throughout the stratigraphic record in Brazil (e.g., the Archean Carajás province; the Paleoproterozoic Minas Supergroup bearing large Lake Superior-type BIFs and carbonates deposited before, during and after the first GOE and spanning the global Lomagundi event; Mesoproterozoic to early Neoproterozoic carbonates spanning the "Boring Billion"; Neoproterozoic strata containing widespread records of global glaciations, glacial-related BIFs, cap carbonate successions and the first metazoans of the late Ediacaran). Brazil is thus a varied and promising natural laboratory for the application of isotope stratigraphy tools in unraveling ancient biogeochemical, climate and tectonic patterns, and will surely provide important pages in the book of Earth's history for the years to come.

Acknowledgments

The database discussed here was acquired through scientific work of various groups during the course of the last decades, mainly supported by the Brazilian federal and state governments and fully or co-developed within the infrastructure of Brazilian's public universities and Higher Education Institutions. The authors acknowledge the support of state and federal research funding agencies such as CNPq, FAPEMIG and FAPESP, of state-owned companies such as CODEMGE and PETROBRAS, and of private companies such as VALE S/A. Many of the authors are Fellows of the Brazilian Research Council (CNPq) and acknowledge for the support received.

References

Abouchami, W., et al., 2014. Biogeochemical cycling of cadmium isotopes in the Southern Ocean along the Zero Meridian. Geochim. Cosmochim. Acta 127, 348–367. https://doi.org/10.1016/j.gca.2013.10.022.

Alkmim, F.F., Marshak, S., 1998. Transamazonian orogeny in the Southern São Francisco Craton region, Minas Gerais, Brazil: evidence for Paleoproterozoic collision and collapse in the Quadrilátero Ferrı´fero. Precambrian Res. 90 (1), 29–58. https://doi.org/10.1016/S0301-9268(98)00032-1.

Alkmim, F.F., Martins Neto, M.M., 2001. A Bacia intracratônica do São Francisco: Arcabouço estrutural e cenários evolutivos. In: Pinto, C.P., Martins Neto, M.M. (Eds.), Bacia do São Francisco—Geologia e Recursos Naturais. SBG/MG, pp. 9–30.

Alkmim, F.F., Martins-Neto, M.A., 2012. Proterozoic first-order sedimentary sequences of the São Francisco craton, eastern Brazil. Mar. Pet. Geol. 33 (1), 127–139. https://doi.org/10.1016/j.marpetgeo.2011.08.011.

Alvarenga, C.J.S., Trompette, R., 1993. Evolução tectônica brasiliana da faixa paraguai: a estruturação da região de cuiabá. Rev. Bras. Geosci. 23 (1), 18–30. https://doi.org/10.25249/0375-7536.1993231183O.

Alvarenga, C.J.S., Santos, R.V., Dantas, E.L., 2004. C–O–Sr isotopic stratigraphy of cap carbonates overlying Marinoan-age glacial diamictites in the Paraguay Belt, Brazil. Precambrian Res. 131 (1–2), 1–21. https://doi.org/10.1016/J.PRECAMRES.2003.12.006. Elsevier.

Alvarenga, C.J.S., et al., 2008. Isotope stratigraphy of Neoproterozoic cap carbonates in the Araras Group, Brazil. Gondw. Res. 13 (4), 469–479. https://doi.org/10.1016/J.GR.2007.05.004. Elsevier.

Alvarenga, C.J.S., et al., 2009. Chapter 2 The Amazonian Palaeocontinent. In: Developments in Precambrian Geology. Elsevier, pp. 15–28. https://doi.org/10.1016/S0166-2635(09)01602-8.

Alvarenga, C.J.S., et al., 2011. Chapter 45 Glacially influenced sedimentation of the Puga Formation, Cuiabá Group and Jacadigo Group, and associated carbonates of the Araras and Corumbá groups, Paraguay Belt, Brazil. Geol. Soc. Lond. Mem. 36 (1), 487–497. https://doi.org/10.1144/M36.45.

Alvarenga, C.J.S., et al., 2014. Meso-Neoproterozoic isotope stratigraphy on carbonates platforms in the Brasilia Belt of Brazil. Precambrian Res. 251, 164–180. Elsevier.

Alvarenga, C.J.S., et al., 2019. Carbonate chemostratigraphy of the Vazante Group, Brazil: a probable Tonian age. Precambrian Res. 331, 105378. https://doi.org/10.1016/j.precamres.2019.105378.

Azmy, K., et al., 2001. Dolomitization and isotope stratigraphy of the Vazante Formation, São Francisco Basin, Brazil. Precambrian Res. 112 (3–4), 303–329. https://doi.org/10.1016/S0301-9268(01)00194-2. Elsevier.

Azmy, K., et al., 2006. Isotope stratigraphy of the Lapa Formation, São Francisco Basin, Brazil: implications for Late Neoproterozoic glacial events in South America. Precambrian Res. 149 (3–4), 231–248. https://doi.org/10.1016/J.PRECAMRES.2006.07.001. Elsevier.

Azmy, K., et al., 2008. Global correlation of the Vazante Group, São Francisco Basin, Brazil: Re–Os and U–Pb radiometric age constraints. Precambrian Res. 164 (3–4), 160–172. https://doi.org/10.1016/J.PRECAMRES.2008.05.001. Elsevier.

Babinski, M., Chemale, F., Van Schmus, W.R., 1995. The Pb/Pb age of the Minas Supergroup carbonate rocks, Quadrilátero Ferrífero, Brazil. Precambrian Res. 72 (3–4), 235–245. https://doi.org/10.1016/0301-9268(94)00091-5.

Babinski, M., Vieira, L.C., Trindade, R.I.F., 2007. Direct dating of the Sete Lagoas cap carbonate (Bambuí Group, Brazil) and implications for the Neoproterozoic glacial events. Terra Nova 19 (6), 401–406. https://doi.org/10.1111/j.1365-3121.2007.00764.x.

Babinski, M., Boggiani, P.C., Fanning, C.M., Fairchild, T.R., Simon, C.M., Sial, A.N., 2008. U-Pb SHRIMP geochronology and isotope chemostratigraphy (C, O, Sr) of the Tamengo Formation, Southern Paraguay Belt, Brazil. In: VI South American Symposium on Isotope Geology, Book of Abstracts, San Carlos de Bariloche, p. 160.

Babinski, M., et al., 2013. Detrital zircon ages and geochronological constraints on the Neoproterozoic Puga diamictites and associated BIFs in the southern Paraguay Belt, Brazil. Gondw. Res. 23 (3), 988–997. Elsevier.

Bao, H., Lyons, J.R., Zhou, C., 2008. Triple oxygen isotope evidence for elevated CO2 levels after a Neoproterozoic glaciation. Nature 453 (7194), 504–506. https://doi.org/10.1038/nature06959.

Bao, H., et al., 2009. Stretching the envelope of past surface environments: Neoproterozoic glacial lakes from Svalbard. Science 323 (5910), 119–122. https://doi.org/10.1126/science.1165373.

Barovich, K.M., Foden, J., 2000. A Neoproterozoic flood basalt province in southern-central Australia: geochemical and Nd isotope evidence from basin fill. Precambrian Res. 100 (1), 213–234. https://doi.org/10.1016/S0301-9268(99)00075-3.

Barovich, K., Hand, M., 2008. Tectonic setting and provenance of the Paleoproterozoic Willyama Supergroup, Curnamona Province, Australia: geochemical and Nd isotopic constraints on contrasting source terrain components. Precambrian Res. 166 (1), 318–337. https://doi.org/10.1016/j.precamres.2007.06.024.

Beard, B.L., Johnson, C.M., 2004. Fe isotope variations in the modern and ancient Earth and other planetary bodies. Rev. Mineral. Geochem. 55 (1), 319–357. https://doi.org/10.2138/gsrmg.55.1.319. GeoScienceWorld.

Bekker, A., et al., 2003. Chemostratigraphy of carbonates from the Minas Supergroup, Quadrilatero Ferryifero (Iron Quadrangle), Brazil: a stratigraphic record of Early Proterozoic atmospheric, biogeochemical and climactic Change. Am. J. Sci. 303 (10), 865–904. https://doi.org/10.2475/ajs.303.10.865.
Bekker, A., Karhu, J.A., Kaufman, A.J., 2006. Carbon isotope record for the onset of the Lomagundi carbon isotope excursion in the Great Lakes area, North America. Precambrian Res. 148 (1–2), 145–180. https://doi.org/10.1016/J.PRECAMRES.2006.03.008. Elsevier.
Bertoni, M.E., et al., 2014. Neoproterozoic Re–Os systematics of organic-rich rocks in the São Francisco Basin, Brazil and implications for hydrocarbon exploration. Precambrian Res. 255, 355–366. Elsevier.
Boggiani, P.C., et al., 2003. The cap carbonate of the Puga Hill (central south America) in the context of the post-varanger glaciation. In: South American Symposium on Isotope Geology. CBPM, Salvador, pp. 324–327.
Boggiani, P.C., et al., 2010. Chemostratigraphy of the Tamengo Formation (Corumbá Group, Brazil): a contribution to the calibration of the Ediacaran carbon-isotope curve. Precambrian Res. 182 (4), 382–401. https://doi.org/10.1016/j.precamres.2010.06.003.
Boyle, E.A., Sclater, F., Edmond, J.M., 1976. On the marine geochemistry of cadmium. Nature 263 (5572), 42–44. https://doi.org/10.1038/263042a0.
Brueckner, H.K., Cunningham, D., Alkmim, F.F., Marshak, S., 2000. Tectonic implications of Precambrian Sm-Nd dates from the southern Sao Francisco craton and adjacent Aracuai and Ribeira belts, Brazil. Precambrian Res. 99 (3–4), 255–269. https://doi.org/10.1016/S0301-9268(99)00065-0.
Bullen, T.D., et al., 2001. Demonstration of significant abiotic iron isotope fractionation in nature. Geology 29 (8), 699. https://doi.org/10.1130/0091-7613(2001)029<0699:DOSAII>2.0.CO;2. GeoScienceWorld.
Cabral, A.R., Zeh, A., Koglin, N., Gomes Jr., A.A.S., Viana, D.J., Lehmann, B., 2012. Dating the Itabira iron formation, Quadrilátero Ferrífero of Minas Gerais, Brazil, at 2.65 Ga: depositional U–Pb age of zircon from a metavolcanic layer. Precambrian Res 204–205, 40–45.
Cabral, A.R., et al., 2013. Trace-element and multi-isotope geochemistry of Late-Archean black shales in the Carajás iron-ore district, Brazil. Chem. Geol. 362, 91–104. https://doi.org/10.1016/J.CHEMGEO.2013.08.041. Elsevier.
Caetano-Filho, S., et al., 2019. Sequence stratigraphy and chemostratigraphy of an Ediacaran-Cambrian foreland-related carbonate ramp (Bambuí Group, Brazil). Precambrian Res. 331, 105365. https://doi.org/10.1016/J.PRECAMRES.2019.105365. Elsevier.
Calver, C.R., et al., 2013. Globally synchronous Marinoan deglaciation indicated by U-Pb geochronology of the Cottons Breccia, Tasmania, Australia. Geology 41 (10), 1127–1130. https://doi.org/10.1130/G34568.1. GeoScienceWorld.
Canfield, D.E., 1998. A new model for Proterozoic ocean chemistry. Nature 396 (6710), 450–453. https://doi.org/10.1038/24839.
Canfield, D.E., Teske, A., 1996. Late Proterozoic rise in atmospheric oxygen concentration inferred from phylogenetic and sulphur-isotope studies. Nature 382 (6587), 127–132. https://doi.org/10.1038/382127a0.
Castro, P.T.A., Dardenne, M.A., 2000. The sedimentology stratigraphy and Tectonic context of the São Francisco supergroup at the southwestern domain of the São Francisco Craton, Brazil. Rev. Bras. Geosci. 30 (3), 439–441. Sociedade Brasileira de Geologia.
Caxito, F.A., Halverson, G.P., Uhlein, A., Stevenson, R., Dias, T.G., Uhlein, G.J., 2012. Marinoan glaciation in east central Brazil. Precambrian Res. 200–203, 38–58. https://doi.org/10.1016/j.precamres.2012.01.005.

Caxito, F.A., et al., 2014. Detrital zircon (U–Pb) and Sm–Nd isotope studies of the provenance and tectonic setting of basins related to collisional orogens: the case of the Rio Preto fold belt on the northwest São Francisco Craton margin, NE Brazil. Gondw. Res. 26 (2), 741–754. https://doi.org/10.1016/j.gr.2013.07.007.

Caxito, F.A., et al., 2016. ESTRATIGRAFIA ISOTÓPICA DE CARBONO E OXIGÊNIO DA FORMAÇÃO JAÍBA, GRUPO BAMBUÍ SUPERIOR, EDIACARANO DA BACIA DO SÃO FRANCISCO. In: 48 Congresso Brasileiro de Geologia, Porto Alegre, RS.

Caxito, F.A., et al., 2018. Multiproxy geochemical and isotope stratigraphy records of a Neoproterozoic Oxygenation Event in the Ediacaran Sete Lagoas cap carbonate, Bambuí Group, Brazil. Chem. Geol. 481, 119–132. https://doi.org/10.1016/j.chemgeo.2018.02.007.

Chemale, F., Rosière, C.A., Endo, I., 1994. The tectonic evolution of the Quadrilátero Ferrífero, Minas Gerais, Brazil. Precambrian Res. 65 (1–4), 25–54. https://doi.org/10.1016/0301-9268(94)90098-1.

Chemale, F., et al., 2012. Unravelling a Proterozoic basin history through detrital zircon geochronology: the case of the Espinhaço Supergroup, Minas Gerais, Brazil. Gondw. Res. 22 (1), 200–206. https://doi.org/10.1016/j.gr.2011.08.016.

Chiavegatto, J.R.S., et al., 2003. Estratigrafia do Grupo Bambuí nas regiões do Norte de Minas Gerais: Uma nova unidade estratigráfica em um contexto de inversão de bacia. In: Simpósio de Geología de Minas Gerais. vol. 12. SBG, Ouro Preto.

Condon, D., et al., 2005. U-Pb ages from the neoproterozoic Doushantuo Formation, China. Science (New York, N.Y.) 308 (5718), 95–98. https://doi.org/10.1126/science.1107765. American Association for the Advancement of Science.

Conway, T.M., John, S.G., 2014. Quantification of dissolved iron sources to the North Atlantic Ocean. Nature 511 (7508), 212–215. https://doi.org/10.1038/nature13482. Nature Publishing Group.

Cordani, U.G., et al., 2009. The position of the Amazonian Craton in supercontinents. Gondw. Res. 15 (3–4), 396–407. https://doi.org/10.1016/J.GR.2008.12.005. Elsevier.

Costa, M.T., Branco, J.J.R., 1961. Roteiro para excursão Belo Horizonte-Brasília. SBG, Congresso Brasileiro de Geologia, 14, Belo Horizonte.

Cox, G.M., Halverson, G.P., Poirier, A., et al., 2016a. A model for Cryogenian iron formation. Earth Planet. Sci. Lett. 433, 280–292. https://doi.org/10.1016/j.epsl.2015.11.003.

Cox, G.M., Halverson, G.P., Stevenson, R.K., et al., 2016b. Continental flood basalt weathering as a trigger for Neoproterozoic Snowball Earth. Earth Planet. Sci. Lett. 446, 89–99. https://doi.org/10.1016/j.epsl.2016.04.016.

Cox, G.M., et al., 2018. South Australian U-Pb zircon (CA-ID-TIMS) age supports globally synchronous Sturtian deglaciation. Precambrian Res. 315, 257–263. https://doi.org/10.1016/J.PRECAMRES.2018.07.007. Elsevier.

Cozzi, A., Allen, P.A., Grotzinger, J.P., 2004. Understanding carbonate ramp dynamics using δ13C profiles: examples from the Neoproterozoic Buah Formation of Oman. Terra Nova 16 (2), 62–67. https://doi.org/10.1111/j.1365-3121.2004.00528.x.

Crockford, P.W., et al., 2016. Triple oxygen and multiple sulfur isotope constraints on the evolution of the post-Marinoan sulfur cycle. Earth Planet. Sci. Lett. 435, 74–83. https://doi.org/10.1016/j.epsl.2015.12.017.

Crockford, P.W., et al., 2018. Linking paleocontinents through triple oxygen isotope anomalies. Geology 46 (2), 179–182. https://doi.org/10.1130/G39470.1.

Crockford, P.W., et al., 2019. Barium-isotopic constraints on the origin of post-Marinoan barites. Earth Planet. Sci. Lett. 519, 234–244. https://doi.org/10.1016/J.EPSL.2019.05.018. Elsevier.

Cukrov, N., de Alvarenga, C.J.S., Uhlein, A., 2005. Litofácies da glaciação neoproterozóica nas porções sul do Cráton do São Francisco: exemplos de Jequitaí (MG) e Cristalina (GO). Rev. Bras. Geosci. 35 (1), 69–76. https://doi.org/10.25249/0375-7536.20053516976.

da Campanha, G.A.C., et al., 2010. Structural analysis of the Itapucumí Group in the Vallemí region, northern Paraguay: evidence of a new Brasiliano/Pan-African mobile belt. J. South Am. Earth Sci. 30 (1), 1–11. https://doi.org/10.1016/J.JSAMES.2010.04.001Pergamon.

da Campanha, G.A.C., et al., 2011. A faixa de dobramento Paraguai na Serra da Bodoquena e depressão do Rio Miranda, Mato Grosso do Sul. Geologia USP. Série Científica 11 (3), 79–96. https://doi.org/10.5327/Z1519-874X2011000300005.

Dantas, E.L., et al., 2009. Using Nd isotopes to understand the provenance of sedimentary rocks from a continental margin to a foreland basin in the Neoproterozoic Paraguay Belt, Central Brazil. Precambrian Res. 170 (1–2), 1–12. https://doi.org/10.1016/J.PRECAMRES.2008.11.005Elsevier.

Dardenne, M.A., 1978. Síntese sobre a estratigrafia do Grupo Bambuí no Brasil Central. In: Annals, Congresso Brasileiro de Geologia. vol. 30. SBG, Recife.

Dardenne, M.A., 2000. The Brasília Fold Belt. In: Cordani, U.G. et al., (Ed.), Tectonic Evolution of South America. FINEP, Rio de Janerio, p. 854.

de Alvarenga, C.J.S., et al., 2007. Variações dos isótopos de C e Sr em carbonatos pré e pós-glaciação Jequitaí (Esturtiano) na região de Bezerra-Formosa, Goiás. Rev. Bras. Geosci. 37, 147–155. https://doi.org/10.25249/0375-7536.200737s4147155.

de Castro, P.T.A., Dardenne, M.A., 1995. O Conglomerado Samburá (Grupo Bambuí, Neoproterozóico) e rochas sedimentares associadas no flanco leste da Serra da Pimenta, SW de Minas Gerais: um sistema de fan delta. Geonomos 3, 35–41. https://doi.org/10.18285/geonomos.v3i2.207.

de Paula-Santos, G.M., et al., 2018. Rare earth elements of carbonate rocks from the Bambuí Group, southern São Francisco Basin, Brazil, and their significance as paleoenvironmental proxies. Precambrian Res. 305, 327–340. https://doi.org/10.1016/j.precamres.2017.12.023.

Derry, L.A., Kaufman, A.J., Jacobsen, S.B., 1992. Sedimentary cycling and environmental change in the Late Proterozoic: evidence from stable and radiogenic isotopes. Geochim. Cosmochim. Acta 56 (3), 1317–1329. https://doi.org/10.1016/0016-7037(92)90064-P.

Dessert, C., et al., 2003. Basalt weathering laws and the impact of basalt weathering on the global carbon cycle. Chem. Geol. 202 (3), 257–273. https://doi.org/10.1016/j.chemgeo.2002.10.001.

Detmers, J., et al., 2001. Diversity of sulfur isotope fractionations by sulfate-reducing prokaryotes. Appl. Environ. Microbiol. 67 (2), 888–894. https://doi.org/10.1128/AEM.67.2.888-894.2001.

Dorr, J.V.N., 1969. Physiographic, stratigraphic, and structural development of the Quadrilatero Ferrifero, Minas Gerais, Brazil. Professional Paper. https://doi.org/10.3133/PP641A.

Drummond, J.B.R., Pufahl, P.K., Porto, C.G., Carvalho, M., 2015. Neoproterozoic peritidal phosphorite from the Sete Lagoas Formation (Brazil) and the Precambrian phosphorus cycle. Sedimentology 62, 1978–2008.

Dutra, L.F., Martins, M., Lana, C., 2019. Sedimentary and U-Pb detrital zircons provenance of the Paleoproterozoic Piracicaba and Sabará groups, Quadrilátero Ferrífero, Southern São Francisco craton, Brazil. Braz. J. Geol. 49, e20180095.

Fabre, S., et al., 2011. Iron and sulphur isotopes from the Carajás mining province (Pará, Brazil): implications for the oxidation of the ocean and the atmosphere across the Archaean–Proterozoic transition. Chem. Geol. 289 (1–2), 124–139. https://doi.org/10.1016/J.CHEMGEO.2011.07.019. Elsevier.

Fairchild, I.J., et al., 1989. Late Proterozoic glacial carbonates in northeast Spitsbergen: new insights into the carbonate–tillite association. Geol. Mag. 126 (05), 469. https://doi.org/10.1017/S0016756800022809.

Farquhar, J., Wing, B.A., 2003. Multiple sulfur isotopes and the evolution of the atmosphere. Earth Planet. Sci. Lett. 213 (1–2), 1–13. https://doi.org/10.1016/S0012-821X(03)00296-6.

Faure, G., Mensing, T.M., 2005. Isotopes: Principles and Applications. Wiley.

Font, E., et al., 2006. Chemostratigraphy of the Neoproterozoic Mirassol d'Oeste cap dolostones (Mato Grosso, Brazil): an alternative model for Marinoan cap dolostone formation. Earth Planet. Sci. Lett. 250 (1–2), 89–103. https://doi.org/10.1016/J.EPSL.2006.06.047. Elsevier.

Fraga, L.M.S., 2013. Análise estratigráfica do grupo Macaúbas no domínio meridional da Serra do Espinhaço. UFMG, Minas Gerais.

Fraga, L.M.S., et al., 2014. C-, Sr-isotope stratigraphy of carbonate rocks from the Southern Espinhaço Ridge, Minas Gerais, southeastern Brazil. An. Acad. Bras. Cienc. 86 (2), 633–648. https://doi.org/10.1590/0001-3765201420130152.

Fragoso, D.G.C., et al., 2011. Geologia dos Grupos Bambuí, Areado e Mata da Corda na Folha Presidente Olegário (1:100.000), MG: Registro deposicional do Neoproterozóico ao Neocretáceo da bacia do São Francisco. Geonomos 19, 28–38. https://doi.org/10.18285/geonomos.v19i1.60.

Frei, R., et al., 2009. Fluctuations in Precambrian atmospheric oxygenation recorded by chromium isotopes. Nature 461 (7261), 250–253. https://doi.org/10.1038/nature08266. Nature Publishing Group.

Frei, R., et al., 2011. Chromium isotopes in carbonates—A tracer for climate change and for reconstructing the redox state of ancient seawater. Earth Planet. Sci. Lett. 312, 114–125. https://doi.org/10.1016/j.epsl.2011.10.009.

Frei, R., et al., 2017. Extensive oxidative weathering in the aftermath of a late Neoproterozoic glaciation—evidence from trace element and chromium isotope records in the Urucum district (Jacadigo Group) and Puga iron formations (Mato Grosso do Sul, Brazil). Gondw. Res. 49, 1–20. https://doi.org/10.1016/J.GR.2017.05.003. Elsevier.

Freitas, B.T., Warren, L.V., Boggiani, P.C., Almeida, R.P., Piacentini, T., 2011. Tectono-sedimentary evolution of the Neoproterozoic BIF-bearing Jacadigo Group, SW-Brazil. Sediment. Geol. 238, 48–70.

Gaucher, C., Boggiani, P.C., Sprechmann, P., Sial, A.N., Fairchild, T.R., 2003. Integrated correlation of Vendian to Cambrian Arroyo del Soldado and Corumbá Groups (Uruguay and Brazil): palaeogeographic, palaeoclimatic and palaeobiologic implications. Precambrian Res. 120, 241–278.

Geboy, N.J., et al., 2013. Re–Os age constraints and new observations of Proterozoic glacial deposits in the Vazante Group, Brazil. Precambrian Res. 238, 199–213. Elsevier.

Georgiev, S.V., et al., 2015. Cadmium-isotopic evidence for increasing primary productivity during the Late Permian anoxic event. Earth Planet. Sci. Lett. 410, 84–96. https://doi.org/10.1016/j.epsl.2014.11.010.

Goddéris, Y., et al., 2003. The Sturtian "snowball" glaciation: fire and ice. Earth Planet. Sci. Lett. 211 (1–2), 1–12. https://doi.org/10.1016/S0012-821X(03)00197-3. Elsevier.

Guacaneme, C., et al., 2017. C, O, and Sr isotopic variations in Neoproterozoic-Cambrian carbonate rocks from Sete Lagoas Formation (Bambuí Group), in the Southern São Francisco Basin, Brazil. Braz. J. Geol. 47 (3), 521–543. https://doi.org/10.1590/2317-4889201720160126.

Halverson, G.P., et al., 2002. A major perturbation of the carbon cycle before the Ghaub glaciation (Neoproterozoic) in Namibia: prelude to snowball Earth? Geochem. Geophys. Geosyst. 3 (6), 1–24. https://doi.org/10.1029/2001GC000244.

Halverson, G.P., Maloof, A.C., Hoffman, P.F., 2004. The Marinoan glaciation (Neoproterozoic) in northeast Svalbard. Basin Res. 16 (3), 297–324. https://doi.org/10.1111/j.1365-2117.2004.00234.x.

Halverson, G.P., et al., 2005. Toward a Neoproterozoic composite carbon-isotope record. Bull. Geol. Soc. Am. 117, 9–10. https://doi.org/10.1130/B25630.1.

Halverson, G.P., et al., 2007. Evolution of the 87Sr/86Sr composition of Neoproterozoic seawater. Palaeogeogr. Palaeoclimatol. Palaeoecol. 256 (3), 103–129. https://doi.org/10.1016/j.palaeo.2007.02.028.

Halverson, G.P., et al., 2010. Neoproterozoic chemostratigraphy. Precambrian Res. 182 (4), 337–350. https://doi.org/10.1016/J.PRECAMRES.2010.04.007. Elsevier.

Halverson, G.P., et al., 2011. Fe isotope and trace element geochemistry of the Neoproterozoic syn-glacial Rapitan iron formation. Earth Planet. Sci. Lett. 309 (1–2), 100–112. https://doi.org/10.1016/J.EPSL.2011.06.021. Elsevier.

Hartmann, L.A., et al., 2006. Provenance and age delimitation of Quadrilátero Ferrífero sandstones based on zircon U–Pb isotopes. J. South Am. Earth Sci. 20 (4), 273–285. https://doi.org/10.1016/j.jsames.2005.07.015.

Hayes, J.M., Strauss, H., Kaufman, A.J., 1999. The abundance of 13C in marine organic matter and isotopic fractionation in the global biogeochemical cycle of carbon during the past 800 Ma. Chem. Geol. 161, 103–125. https://doi.org/10.1016/S0009-2541(99)00083-2. Elsevier.

Higgins, J.A., Schrag, D.P., 2003. Aftermath of a snowball Earth. Geochem. Geophys. Geosyst. 4 (3). https://doi.org/10.1029/2002GC000403.

Hippertt, J.P., et al., 2019. The fate of a Neoproterozoic intracratonic marine basin: trace elements, TOC and IRON speciation geochemistry of the Bambuí Basin, Brazil. Precambrian Res. 330, 101–120. https://doi.org/10.1016/J.PRECAMRES.2019.05.001. Elsevier.

Hoefs, J., 2009. Stable Isotope Geochemistry. Springer Berlin Heidelberg, Berlin, Heidelberg. https://doi.org/10.1007/978-3-540-70708-0.

Hoffman, P.F., Schrag, D.P., 2002. The snowball Earth hypothesis: testing the limits of global change. Terra Nova 14 (3), 129–155. https://doi.org/10.1046/j.1365-3121.2002.00408.x.

Hoffman, P.F., et al., 1998. A Neoproterozoic Snowball Earth. Science 281 (5381), 1342–1346. https://doi.org/10.1126/science.281.5381.1342.

Hoffman, P.F., et al., 2007. Are basal Ediacaran (635 Ma) post-glacial "cap dolostones" diachronous? Earth Planet. Sci. Lett. 258 (1–2), 114–131. https://doi.org/10.1016/j.epsl.2007.03.032.

Hoffman, P.F., et al., 2017. Snowball Earth climate dynamics and Cryogenian geology-geobiology. Sci. Adv. 3 (11), e1600983. https://doi.org/10.1126/sciadv.1600983. American Association for the Advancement of Science.

Hoffmann, K.-H., et al., 2004. U-Pb zircon date from the Neoproterozoic Ghaub Formation, Namibia: constraints on Marinoan glaciation. Geology 32 (9), 817. https://doi.org/10.1130/G20519.1.

Hohl, S.V., et al., 2017. Cadmium isotope variations in Neoproterozoic carbonates—a tracer of biologic production? Geochem. Persp. Lett. 3, 32–44. https://doi.org/10.7185/geochemlet.1704.

Hollanda, M.H.B.M., et al., 2015. Detrital zircon ages and Nd isotope compositions of the Seridó and Lavras da Mangabeira basins (Borborema Province, NE Brazil): evidence for exhumation and recycling associated with a major shift in sedimentary provenance. Precambrian Res. 258, 186–207. https://doi.org/10.1016/j.precamres.2014.12.009.

Hurtgen, M.T., Arthur, M.A., Halverson, G.P., 2005. Neoproterozoic sulfur isotopes, the evolution of microbial sulfur species, and the burial efficiency of sulfide as sedimentary pyrite. Geology 33 (1), 41. https://doi.org/10.1130/G20923.1.

Iglesias, M., Uhlein, A., 2008. Estratigrafia do Grupo Bambuí na serra do Jaíba, norte de Minas Gerais. Geonomos 16, 21–27. https://doi.org/10.18285/geonomos.v16i1.90.

Iglesias, M., Uhlein, A., 2009. Estratigrafia do Grupo Bambuí e coberturas fanerozóicas no vale do rio São Francisco, norte de Minas Gerais. Rev. Bras. Geosci. 39 (2), 256–266.

Ilambwetsi, A.M., 2015. Geologia das unidades Carrancas, Pedro Leopoldo e Lagoa Santa da porção basal do Grupo Bambuí nas proximidades de Belo Horizonte—Minas Gerais. Trabalho de Conclusão de Curso de Geologia. Instituto de Geociências, Universidade Federal de Minas Gerais, Belo Horizonte-MG, Brazil. 100 pp.

Isotta, C.A.L., Rocha-campos, A.C., Yoshida, R., 1969. Striated Pavement of the Upper Pre-Cambrian Glaciation in Brazil. Nature 222 (5192), 466–468. https://doi.org/10.1038/222466a0.

Iyer, S.S., et al., 1995. Highly 13C-enriched carbonate and organic matter in the Neoproterozoic sediments of the BambuíGroup, Brazil. Precambrian Res. 73, 271–282. https://doi.org/10.1016/0301-9268(94)00082-3. Elsevier.

Jacobsen, S.B., Kaufman, A.J., 1999. The Sr, C and O isotopic evolution of Neoproterozoic seawater. Chem. Geol. 161, 37–57. https://doi.org/10.1016/S0009-2541(99)00080-7. Elsevier.

John, S.G., et al., 2017. Zinc and cadmium stable isotopes in the geological record: a case study from the post-snowball Earth Nuccaleena cap dolostone. Palaeogeogr. Palaeoclimatol. Palaeoecol. 466, 202–208. https://doi.org/10.1016/j.palaeo.2016.11.003.

Johnson, C.M., Beard, B.L., 2006. Fe isotopes: an emerging technique for understanding modern and ancient biogeochemical cycles. GSA Today 16, 11. https://doi.org/10.1130/GSAT01611A.1.

Johnson, C.M., Beard, B.L., Beukes, N.J., Klein, C., O'Leary, J.M., 2003. Ancient geochemical cycling in the Earth as inferred from Fe isotope studies of banded iron formations from the Transvaal Craton. Contrib. Mineral. Petrol. 144, 523–547.

Kah, L.C., Lyons, T.W., Chesley, J.T., 2001. Geochemistry of a 1.2 Ga carbonate-evaporite succession, northern Baffin and Bylot Islands: implications for Mesoproterozoic marine evolution. Precambrian Res. 111 (1–4), 203–234. https://doi.org/10.1016/S0301-9268(01)00161-9. Elsevier.

Karfunkel, J., Hoppe, A., 1988. Late Proterozoic glaciation in central-eastern Brazil: synthesis and model. Palaeogeogr. Palaeoclimatol. Palaeoecol. 65 (1), 1–21. https://doi.org/10.1016/0031-0182(88)90108-3.

Kasemann, S.A., et al., 2005. Boron and calcium isotope composition in Neoproterozoic carbonate rocks from Namibia: evidence for extreme environmental change. Earth Planet. Sci. Lett. 231 (1–2), 73–86. https://doi.org/10.1016/J.EPSL.2004.12.006. Elsevier.

Kaufman, A.J., Knoll, A.H., Narbonne, G.M., 1997. Isotopes, ice ages, and terminal Proterozoic earth history. Proc. Natl. Acad. Sci. U.S.A. 94 (13), 6600–6605. https://doi.org/10.1073/pnas.94.13.6600.

Kaufman, A.J., et al., 2001. Anomalous d34S signatures in trace sulfate from a potential cap carbonate in the Neoproterozoic Bambuí Group, Brazil. In: Misi, A., Teixeira, J.B.G. (Eds.), Proterozoic Sediment-Hosted Base Metal Deposits of Western Gondwana, I Field Workshop, Belo Horizonte and Paracatu (Minas Gerais), Brazil, pp. 62–65.

Kennedy, M.J., Christie-Blick, N., Sohl, L.E., 2001. Are Proterozoic cap carbonates and isotopic excursions a record of gas hydrate destabilization following Earth's coldest intervals? Geology 29 (5), 443. https://doi.org/10.1130/0091-7613(2001)029<0443:APCCAI>2.0.CO;2.

Keto, L.S., Jacobsen, S.B., 1988. Nd isotopic variations of Phanerozoic paleoceans. Earth Planet. Sci. Lett. 90 (4), 395–410. https://doi.org/10.1016/0012-821X(88)90138-0. Elsevier.

Knoll, A.H., Walter, M.R., 1992. Latest Proterozoic stratigraphy and Earth history. Nature 356, 673–678. https://doi.org/10.1038/356673a0.

Knoll, A.H., et al., 2004. A new period for the geologic time scale. Science 305 (5684), 621–622.

Koglin, N., et al., 2014. Depositional age and sediment source of the auriferous Moeda Formation, Quadrilátero Ferrífero of Minas Gerais, Brazil: new constraints from U–Pb–Hf isotopes in zircon and xenotime. Precambrian Res. 255, 96–108. https://doi.org/10.1016/j.precamres.2014.09.010.

Konhauser, K.O., Newman, D.K., Kappler, A., 2005. The potential significance of microbial Fe(III) reduction during deposition of Precambrian banded iron formations. Geobiology 3 (3), 167–177. https://doi.org/10.1111/j.1472-4669.2005.00055.x. John Wiley & Sons, Ltd (10.1111).

Kuchenbecker, M., et al., 2013. Proveniência e análise sedimentar da porção basal do Grupo Bambuí em Arcos (MG). Geologia USP. Série Científica 13 (4), 49–61. https://doi.org/10.5327/Z1519-874X201300040003.

Kuchenbecker, M., Reis, H.L.S., da Silva, L.C., da Costa, R.D., Fragoso, D.G.C., Knauer, L.G., Dussin, I.A., Soares, A.C.P., 2015. Age constraints for deposition and sedimentary provenance of espinhaço supergroup and bambuí group in eastern são francisco craton. Geonomos 23, 14–28. https://doi.org/10.18285/geonomos.v23i2.708.

Kuchenbecker, M., Atman, D., da Costa, R.D., Pedrosa-Soares, A.C., Babinski, M., 2016a. A Formação Gorutuba: sedimentação litorânea a continental na margem leste da Bacia Bambuí (MG). Geologia USP. Série Científica 16 (2), 67. https://doi.org/10.11606//issn.2316-9095.v16i2p67-81.

Kuchenbecker, M., Babinski, M., Pedrosa-Soares, A.C., Lopes-Silva, L., Pimenta, F., et al., 2016b. Chemostratigraphy of the lower Bambuí Group, southwestern São Francisco Craton, Brazil: insights on Gondwana paleoenvironments. Braz. J. Geol. 46 (Suppl. 1), 145–162. https://doi.org/10.1590/2317-488920160030285. Sociedade Brasileira de Geologia.

Kump, L., 1991. Interpreting carbon-isotope excursions: Stranglove oceans. Geology 29, 443–446.

Macdonald, F.A., et al., 2010. Calibrating the Cryogenian. Science 327 (5970), 1241–1243. https://doi.org/10.1126/science.1183325.

Machado, N., et al., 1991. U-Pb geochronology of Archean magmatism and basement reactivation in the Carajás area, Amazon shield, Brazil. Precambrian Res. 49 (3–4), 329–354. https://doi.org/10.1016/0301-9268(91)90040-H. Elsevier.

Machado, N., et al., 1996. Ages of detrital zircon from Archean-Paleoproterozoic sequences: implications for Greenstone Belt setting and evolution of a Transamazonian foreland basin in Quadrilátero Ferrífero, southeast Brazil'. Earth Planet. Sci. Lett. 141 (1–4), 259–276. https://doi.org/10.1016/0012-821X(96)00054-4.

Maciel, P., 1959. Tilito Cambriano (?) no Estado de Mato Grosso. Bol. Soc. Brasil. Geol. 8 (1), 31–39. Sociedade Brasileira de Geologia.

MacLennan, S., et al., 2018. The arc of the Snowball: U-Pb dates constrain the Islay anomaly and the initiation of the Sturtian glaciation. Geology 46 (6), 539–542. https://doi.org/10.1130/G40171.1. GeoScienceWorld.

Maheshwari, A., et al., 2010. Global nature of the Paleoproterozoic Lomagundi carbon isotope excursion: a review of occurrences in Brazil, India, and Uruguay. Precambrian Res. 182 (4), 274–299. https://doi.org/10.1016/j.precamres.2010.06.017.

Marques, C.S.S., et al., 2015. Geologia e quimioestratigrafia isotópica do Grupo Vazante em Lagamar (mg). Geonomos 23, 26–41. https://doi.org/10.18285/geonomos.v23i1.659.

Martínez Dopico, C.I., et al., 2017. U–Pb ages and Hf-isotope data of detrital zircons from the late Neoarchean-Paleoproterozoic Minas Basin, SE Brazil. Precambrian Res. 291, 143–161. https://doi.org/10.1016/j.precamres.2017.01.026.

Martins, M., Lemos, V.B., 2007. Análise estratigráfica das seqüências neoproterozóicas da Bacia do São Francisco. Rev. Bras. Geosci. 37 (Suppl. 4), 156–167.

Martins-Neto, M.A., 2009. Sequence stratigraphic framework of Proterozoic successions in eastern Brazil. Mar. Pet. Geol. 26 (2), 163–176. https://doi.org/10.1016/j.marpetgeo.2007.10.001.

Matteini, M., et al., 2012. U–Pb and Hf isotope study on detrital zircons from the Paranoá Group, Brasília Belt Brazil: constraints on depositional age at Mesoproterozoic—Neoproterozoic transition and tectono-magmatic events in the São Francisco craton. Precambrian Res. 206–207, 168–181. https://doi.org/10.1016/j.precamres.2012.03.007.

McArthur, J.M., Howarth, R.J., Bailey, T.R., 2001. Strontium isotope stratigraphy: LOWESS Version 3: best fit to the Marine Sr-Isotope Curve for 0–509 Ma and accompanying look-up table for deriving numerical age. J. Geol. 109 (2), 155–170. https://doi.org/10.1086/319243.

McGee, B., et al., 2015. Age and provenance of the Cryogenian to Cambrian passive margin to foreland basin sequence of the northern Paraguay Belt, Brazil. Geol. Soc. Am. Bull. 127 (1–2), 76–86. https://doi.org/10.1130/B30842.1.

McGee, B., et al., 2018. Tracing final Gondwana assembly: age and provenance of key stratigraphic units in the southern Paraguay Belt, Brazil. Precambrian Res. 307, 1–33.

McLennan, S.M., et al., 1993. Geochemical approaches to sedimentation, provenance, and tectonics. In: Processes Controlling the Composition of Clastic Sediments. Geological Society of America, pp. 21–40. https://doi.org/10.1130/SPE284-p21.

Melezhik, V.A., et al., 1999. Extreme 13Ccarb enrichment in ca. 2.0 Ga magnesite–stromatolite–dolomite–'red beds' association in a global context: a case for the world-wide signal enhanced by a local environment. Earth Sci. Rev. 48 (1–2), 71–120. https://doi.org/10.1016/S0012-8252(99)00044-6. Elsevier.

Melezhik, V.A., et al., 2001. Chemostratigraphy of Neoproterozoic carbonates: implications for "blind dating" Terra Nova 13, 1–11. https://doi.org/10.1046/j.1365-3121.2001.00318.x.

Melezhik, V.A., et al., 2007. Temporal constraints on the Paleoproterozoic Lomagundi-Jatuli carbon isotopic event. Geology 35 (7), 655. https://doi.org/10.1130/G23764A.1.

Mendes, M., Lobato, L.M., Suckau, V., Lana, C., 2014. In situ LAICPMS U-Pb dating of detrital zircons from the Cercadinho Formation, Minas Supergroup. Geologia USP Série Científica 14 (1), 55–68. https://doi.org/10.5327/Z1519-874X201400010004.

Mendes, M., et al., 2016. Iron isotope and REE + Y composition of the Cauê banded iron formation and related iron ores of the Quadrilátero Ferrífero, Brazil. Miner. Deposita 52, 159–180. https://doi.org/10.1007/s00126-016-0649-9.

Middag, R., et al., 2018. The relationship between cadmium and phosphate in the Atlantic Ocean unravelled. Earth Planet. Sci. Lett. 492, 79–88. https://doi.org/10.1016/j.epsl.2018.03.046.

Misi, A., Veizer, J., 1998. Neoproterozoic carbonate sequences of the Una Group, Irecê Basin, Brazil: chemostratigraphy, age and correlations. Precambrian Res. 89 (1), 87–100. https://doi.org/10.1016/S0301-9268(97)00073-9.

Misi, A., et al., 2007. Chemostratigraphic correlation of Neoproterozoic successions in South America. Chem. Geol. 237, 143–167. Elsevier.

Misi, A., et al., 2014. Review of the geological and geochronological framework of the Vazante sequence, Minas Gerais, Brazil: implications to metallogenic and phosphogenic models. Ore Geol. Rev. 63, 76–90. https://doi.org/10.1016/J.OREGEOREV.2014.05.002. Elsevier.

Nascimento, R., Sial, A.N., 2004. Carbon Isotope Fluctuations in Precambrian Marbles of the Seridó Belt, Borborema Province, Northeast Brazil. Rev. Bras. Geosci. 30, 446–449. https://doi.org/10.25249/0375-7536.2000303446449.

Nascimento, R.S.C., Sial, A.N., Pimentel, M.M., 2007. C- and Sr-isotope systematics applied to Neoproterozoic marbles of the Seridó belt, northeastern Brazil. Chem. Geol. 237 (1–2), 191–210. https://doi.org/10.1016/j.chemgeo.2006.06.017.

Nogueira, A.C., et al., 2003. Soft-sediment deformation at the base of the Neoproterozoic Puga cap carbonate (southwestern Amazon craton, Brazil): confirmation of rapid icehouse to greenhouse transition in snowball Earth. Geology 31 (7), 613. https://doi.org/10.1130/0091-7613(2003)031<0613:SDATBO>2.0.CO;2. GeoScienceWorld.

Nogueira, A.C.R., et al., 2007. Carbon and strontium isotope fluctuations and paleoceanographic changes in the late Neoproterozoic Araras carbonate platform, southern Amazon craton, Brazil. Chem. Geol. 237, 168–190. Elsevier.

Nogueira, L.B., et al., 2019. Geochemistry and C and O isotope composition of carbonate rocks from Bemil and Lagoa Seca quarries, Gandarela Formation, Quadrilátero Ferrífero—Brazil. J. South Am. Earth Sci. 92, 609–630. https://doi.org/10.1016/j.jsames.2019.04.001.

Oliveira, E.P., Windley, B.F., Araújo, M.N.C., 2010. The Neoproterozoic Sergipano orogenic belt, NE Brazil: a complete plate tectonic cycle in western Gondwana. Precambrian Res. 181 (1), 64–84. https://doi.org/10.1016/j.precamres.2010.05.014.

Oliveira, E.P., et al., 2015. Detrital zircon U–Pb geochronology and whole-rock Nd-isotope constraints on sediment provenance in the Neoproterozoic Sergipano orogen, Brazil: from early passive margins to late foreland basins. Tectonophysics 662, 183–194. https://doi.org/10.1016/j.tecto.2015.02.017.

Parry, L.A., et al., 2017. Ichnological evidence for meiofaunal bilaterians from the terminal Ediacaran and earliest Cambrian of Brazil. Nat. Ecol. Evol. 1 (10), 1455–1464. https://doi.org/10.1038/s41559-017-0301-9. Nature Publishing Group.

Paula-Santos, G.M., 2018. Quimioestratigrafia isotópica (C, O, Sr, Li, Mg) e proveniência sedimentar (U-Pb, Hf, Sm-Nd) do grupo Bambuí no sul da bacia do São Francisco. Biblioteca Digital de Teses e Dissertações da Universidade de São Paulohttps://doi.org/10.11606/T.44.2018.tde-26042018-092122.

Paula-Santos, G.M., et al., 2015. New evidence of an Ediacaran age for the Bambuí Group in southern São Francisco craton (eastern Brazil) from zircon U–Pb data and isotope chemostratigraphy. Gondw. Res. 28 (2), 702–720. https://doi.org/10.1016/j.gr.2014.07.012.

Paula-Santos, G.M., et al., 2017. Tracking connection and restriction of West Gondwana São Francisco Basin through isotope chemostratigraphy. Gondw. Res. 42, 280–305. Elsevier.

Paytan, A., et al., 1993. Marine barite as a monitor of seawater strontium isotope composition. Nature 366 (6454), 445–449. https://doi.org/10.1038/366445a0.

Pedrosa-Soares, A.C., et al., 2011. Chapter 49 The Neoproterozoic Macaúbas Group, Araçuaí orogen, SE Brazil. Geol. Soc. Lond. Mem. 36 (1), 523–534. https://doi.org/10.1144/M36.49.

Peng, Y., Bao, H., Zhou, C., Yuan, X., 2011. 17O-depleted barite from two Marinoan cap dolostone sections, south China. Earth Planet. Sci. Lett. 305, 21–31.

Perrella Júnior, P., et al., 2017. Facies analysis, sequence stratigraphy and chemostratigraphy of the Sete Lagoas Formation (Bambui Group), northern Minas Gerais State, Brazil: evidence of a cap carbonate deposited on the Januária basement high. Braz. J. Geol. 47 (1), 59–77. https://doi.org/10.1590/2317-4889201720160112.

Pimentel, M.M., et al., 2011. The tectonic evolution of the Neoproterozoic Brasília Belt, central Brazil, based on SHRIMP and LA-ICPMS U–Pb sedimentary provenance data: a review. J. South Am. Earth Sci. 31 (4), 345–357. https://doi.org/10.1016/j.jsames.2011.02.011.

Poulton, S.W., Canfield, D.E., 2005. Development of a sequential extraction procedure for iron: implications for iron partitioning in continentally derived particulates. Chem. Geol. 214 (3), 209–221. https://doi.org/10.1016/j.chemgeo.2004.09.003.

Poulton, S.W., Fralick, P.W., Canfield, D.E., 2004. The transition to a sulphidic ocean ~1.84 billion years ago. Nature 431 (7005), 173–177. https://doi.org/10.1038/nature02912.

Prave, A.R., et al., 2016. Duration and nature of the end-Cryogenian (Marinoan) glaciation. Geology 44 (8), 631–634. https://doi.org/10.1130/G38089.1. GeoScienceWorld.

Reis, H.L.S., Suss, J.F., 2016. Mixed carbonate–siliciclastic sedimentation in forebulge grabens: an example from the Ediacaran Bambuí Group, São Francisco Basin, Brazil. Sediment. Geol. 339, 83–103. Elsevier.

Reis, H.L.S., et al., 2017. Ediacaran forebulge grabens of the southern São Francisco basin, SE Brazil: Craton interior dynamics during West Gondwana assembly. Precambrian Res. 302, 150–170. https://doi.org/10.1016/j.precamres.2017.09.023.

Renger, F.E., et al., 1994. Evolução Sedimentar do Supergrupo Minas: 500 Ma. de Registro Geológico no Quadrilátero Ferrífero, Minas Gerais, Brasil. Geonomos 2, 1–11. https://doi.org/10.18285/geonomos.v2i1.227.

Rodrigues, J.B., 2008. Proveniência de sedimentos dos grupos Canastra, Ibiá, Vazante e Bambuí–Um estudo de zircões detríticos e Idades Modelo Sm-Nd. Universidade de Brasília.

Rodrigues, J.B., et al., 2012. Provenance of the Vazante Group: new U–Pb, Sm–Nd, Lu–Hf isotopic data and implications for the tectonic evolution of the Neoproterozoic Brasília Belt. Gondw. Res. 21 (2–3), 439–450. https://doi.org/10.1016/J.GR.2011.07.017. Elsevier.

Romero, J.A.S., et al., 2013. Sr isotope geochemistry and Pb–Pb geochronology of the Neoproterozoic cap carbonates, Tangará da Serra, Brazil. Int. Geol. Rev. 55 (2), 185–203. https://doi.org/10.1080/00206814.2012.692517.

Romero, G.R., et al., 2016. Tubestone microbialite association in the Ediacaran cap carbonates in the southern Paraguay Fold Belt (SW Brazil): geobiological and stratigraphic implications for a Marinoan cap carbonate. J. South Am. Earth Sci. 71, 172–181. https://doi.org/10.1016/j.jsames.2016.06.014.

Rooney, A.D., et al., 2015. A Cryogenian chronology: two long-lasting synchronous Neoproterozoic glaciations. Geology 43 (5), 459–462. https://doi.org/10.1130/G36511.1. GeoScienceWorld.

Sanchez, E.A.M., et al., 2018. Microbialitos fósseis da Formação Jaíba, Grupo Bambuí, Minas Gerais, Brasil. Revista Brasileira de Paleontologia 21 (2), 175–186. https://doi.org/10.4072/rbp.2018.2.07.

Sansjofre, P., et al., 2011. A carbon isotope challenge to the snowball Earth. Nature 478 (7367), 93–96. https://doi.org/10.1038/nature10499. Nature Publishing Group.

Santos, R.V., et al., 2000. Carbon and oxygen isotope profiles across Meso-Neoproterozoic limestones from central Brazil: Bambuı´ and Paranoá groups. Precambrian Res. 104 (3), 107–122. https://doi.org/10.1016/S0301-9268(00)00082-6.

Santos, R.V., et al., 2004. Carbon isotopes of Mesoproterozoic–Neoproterozoic sequences from Southern São Francisco craton and Araçuaí Belt, Brazil: paleographic implications. J. South Am. Earth Sci. 18 (1), 27–39. https://doi.org/10.1016/j.jsames.2004.08.009.

Santos, M.N., et al., 2015. Provenance and paleogeographic reconstruction of a mesoproterozoic intracratonic sag basin (Upper Espinhaço Basin, Brazil). Sediment. Geol. 318, 40–57. https://doi.org/10.1016/j.sedgeo.2014.12.006.

Schidlowski, M., Eichmann, R., Junge, C.E., 1976. Carbon isotope geochemistry of the Precambrian Lomagundi carbonate province, Rhodesia. Geochim. Cosmochim. Acta 40 (4), 449–455. https://doi.org/10.1016/0016-7037(76)90010-7. Pergamon.

Schoenberg, R., Zink, S., Staubwasser, M., von Blanckenburg, F., 2008. The stable Cr isotope inventory of solid Earth reservoirs determined by double spike MC-ICP-MS. Chem. Geol. 249, 294–306.

Schrag, D.P., et al., 2002. On the initiation of a snowball Earth. Geochem. Geophys. Geosyst. 3 (6), 1–21. https://doi.org/10.1029/2001GC000219.

Shields, G.A., 2005. Neoproterozoic cap carbonates: a critical appraisal of existing models and the plumeworld hypothesis. Terra Nova 17 (4), 299–310. https://doi.org/10.1111/j.1365-3121.2005.00638.x.

Shields, G.A., 2007. A normalised seawater strontium isotope curve: possible implications for Neoproterozoic-Cambrian weathering rates and the further oxygenation of the Earth. eEarth 2 (2), 35–42. https://doi.org/10.5194/ee-2-35-2007.

Shields, G., Veizer, J., 2002. Precambrian marine carbonate isotope database: version 1.1. Geochem. Geophys. Geosyst. 3 (6), 1–12. https://doi.org/10.1029/2001GC000266.

Shields-Zhou, G., Och, L., 2011. The case for a Neoproterozoic oxygenation event: geochemical evidence and biological consequences. GSA Today 21 (3), 4–11. https://doi.org/10.1130/GSATG102A.1.

Sial, A.N., et al., 2000. Carbon isotope fluctuations in Precambrian carbonate sequences of several localities in Brazil. An. Acad. Bras. Cienc 72 (4), 539–558. https://doi.org/10.1590/S0001-37652000000400006. Academia Brasileira de Ciências.

Sial, A.N., et al., 2010. C-, Sr-isotope and Hg chemostratigraphy of Neoproterozoic cap carbonates of the Sergipano Belt, Northeastern Brazil. Precambrian Res. 182 (4), 351–372. https://doi.org/10.1016/j.precamres.2010.05.008.

Sial, A.N., et al., 2015a. Algoma-type Neoproterozoic BIFs and related marbles in the Seridó Belt (NE Brazil): REE, C, O, Cr and Sr isotope evidence. J. South Am. Earth Sci. 61, 33–52. https://doi.org/10.1016/j.jsames.2015.04.001.

Sial, A.N., et al., 2015b. Isotope and elemental chemostratigraphy. In: Chemostratigraphy. Elsevier, pp. 23–64. https://doi.org/10.1016/B978-0-12-419968-2.00002-9.

Sial, A.N., et al., 2016. Correlations of some Neoproterozoic carbonate-dominated successions in South America based on high-resolution chemostratigraphy. Braz. J. Geol. 46 (3), 439–488. https://doi.org/10.1590/2317-4889201620160079.

Sial, A.N., et al., 2018. Chemostratigraphy as a Formal Stratigraphic Method. American Geophysical Union (AGU), pp. 1–25. https://doi.org/10.1002/9781119382508.ch1.

Silva-Tamayo, J.C., et al., 2010a. Global Ca isotope variations in c. 0.7 Ga old post-glacial carbonate successions. Terra Nova 22 (3), 188–194. https://doi.org/10.1111/j.1365-3121.2010.00933.x.

Silva-Tamayo, J.C., et al., 2010b. 'Global perturbation of the marine Ca isotopic composition in the aftermath of the Marinoan global glaciation. Precambrian Res. 182 (4), 373–381. https://doi.org/10.1016/j.precamres.2010.06.015.

Sotero, M.P., et al., 2019. Análise de fácies e ambientes deposicionais do Grupo Vazante na região da mina de Morro Agudo, Paracatu, noroeste do estado de Minas Gerais. Geologia USP. Série Científica 19 (2), 195–214. https://doi.org/10.11606/issn.2316-9095.v19-144761.

Spier, C.A., et al., 2007. Geochemistry and genesis of the banded iron formations of the Cauê Formation, Quadrilátero Ferrífero, Minas Gerais, Brazil. Precambrian Res. 152 (3–4), 170–206. https://doi.org/10.1016/J.PRECAMRES.2006.10.003. Elsevier.

Teixeira, N.L., et al., 2017. Trace elements and isotope geochemistry (C, O, Fe, Cr) of the Cauê iron formation, Quadrilátero Ferrífero, Brazil: evidence for widespread microbial dissimilatory iron reduction at the Archean/Paleoproterozoic transition. Precambrian Res. 298, 39–55. https://doi.org/10.1016/j.precamres.2017.05.009.

Teng, F.-Z., Dauphas, N., Watkins, J.M., 2017. Non-traditional stable isotopes: retrospective and prospective. Rev. Mineral. Geochem 82 (1), 1–26. https://doi.org/10.2138/rmg.2017.82.1. GeoScienceWorld.

Thomas, C., et al., 2004. $^{87}Sr/^{86}Sr$ chemostratigraphy of Neoproterozoic Dalradian limestones of Scotland and Ireland: constraints on depositional ages and time scales. J. Geol. Soc. London 161 (2), 229–242. https://doi.org/10.1144/0016-764903-001.

Trendall, A.F., et al., 1998. SHRIMP zircon U–Pb constraints on the age of the Carajás formation, Grão ParáGroup, Amazon Craton. J. South Am. Earth Sci. 11 (3), 265–277. https://doi.org/10.1016/S0895-9811(98)00015-7.

Trompette, R., de Alvarenga, C.J.S., Walde, D., 1998. Geological evolution of the Neoproterozoic Corumbágraben system (Brazil). Depositional context of the stratified Fe and Mn ores of the Jacadigo Group. J. South Am. Earth Sci. 11 (6), 587–597. https://doi.org/10.1016/S0895-9811(98)00036-4.

Uhlein, A., et al., 2013. A formação lagoa formosa, grupo bambuí (MG): sistema deposicional de leque submarino em bacia de ante-país. Geonomos 19, 163–172. https://doi.org/10.18285/geonomos.v19i2.51.

Uhlein, G.J., et al., 2016. The Carrancas Formation, Bambuí Group: a record of pre-Marinoan sedimentation on the southern São Francisco craton, Brazil. J. South Am. Earth Sci. 71, 1–16. https://doi.org/10.1016/j.jsames.2016.06.009.

Uhlein, G.J., et al., 2017. Early to late Ediacaran conglomeratic wedges from a complete foreland basin cycle in the southwest São Francisco Craton, Bambuí Group, Brazil. Precambrian Res. 299, 101–116. https://doi.org/10.1016/j.precamres.2017.07.020.

Uhlein, G.J., et al., 2019. Ediacaran paleoenvironmental changes recorded in the mixed carbonate-siliciclastic Bambuí Basin, Brazil. Palaeogeogr. Palaeoclimatol. Palaeoecol. 517, 39–51. Elsevier.

Valeriano, C.M., et al., 2004a. A Evolução Tectônica da Faixa Brasília. In: Mantesso Neto, V. et al., (Ed.), Geologia do Continente Sul-Americano—Evolução da obra de Fernando Flávio Marques de Almeida. Beca, São Paulo.

Valeriano, C.M., Dardenne, M.A., Fonseca, M.A., Simões, L.S.A., Seer, H.J., 2004b. A evolução tectônica da Faixa Brasília. In: Mantesso-Neto, V., Bartorelli, A., Carneiro, C.D.R., Brito-Neves, B.B. (Eds.), Geologia do Continente Sul-Americano: Evolução da Obra de Fernando Flávio Marques De Almeida. Beca, São Paulo, pp. 575–592.

Van Schmus, W., et al., 2003. The Seridó Group of NE Brazil, a late Neoproterozoic pre- to syn-collisional basin in West Gondwana: insights from SHRIMP U–Pb detrital zircon ages and Sm–Nd crustal residence (TDM) ages. Precambrian Res. 127 (4), 287–327. https://doi.org/10.1016/S0301-9268(03)00197-9.

Veizer, J., et al., 1999. 87Sr/86Sr, δ13C and δ18O evolution of Phanerozoic seawater. Chem. Geol. 161 (1–3), 59–88. https://doi.org/10.1016/S0009-2541(99)00081-9. Elsevier.

Viehmann, S., et al., 2019. Metal cycling in Mesoproterozoic microbial habitats: insights from trace elements and stable Cd isotopes in stromatolites. Gondw. Res. 67, 101–114. https://doi.org/10.1016/J.GR.2018.10.014. Elsevier.

Vieira, L.C., et al., 2007. Identification of a Sturtian cap carbonate in the Neoproterozoic Sete Lagoas carbonate platform, Bambuí Group, Brazil'. C. R. Geosci. 339 (3), 240–258. https://doi.org/10.1016/j.crte.2007.02.003.

Wade, B.P., Hand, M., Barovich, K.M., 2005. Nd isotopic and geochemical constraints on provenance of sedimentary rocks in the eastern Officer Basin, Australia: implications for the duration of the intracratonic Petermann Orogeny. J. Geol. Soc. London 162 (3), 513–530. https://doi.org/10.1144/0016-764904-001.

Warren, L.V., et al., 2014. The puzzle assembled: Ediacaran guide fossil Cloudina reveals an old proto-Gondwana seaway. Geology 42 (5), 391–394. https://doi.org/10.1130/G35304.1.

Warren, L.V., et al., 2017. Cloudina-Corumbella-Namacalathus association from the Itapucumi Group, Paraguay: increasing ecosystem complexity and tiering at the end of the Ediacaran. Precambrian Res. 298, 79–87. Elsevier.

Warren, L.V., et al., 2019. Sedimentary evolution and tectonic setting of the Itapucumi Group, Ediacaran, northern Paraguay: from Rodinia break-up to West Gondwana amalgamation. Precambrian Res. 322, 99–121. Elsevier.

Wen, J., Thiemens, M.H., 1993. Multi-isotope study of the $O(^1D) + CO_2$ exchange and stratospheric consequences. J. Geophys. Res. 98 (D7), 12801. https://doi.org/10.1029/93JD00565.

Zalán, P.V., Romeiro-Silva, P.C., 2007. Proposta de mudança significativa na coluna estratigráfica da Bacia do São Francisco. In: Simpósio de Geologia do Sudeste. vol. 10. SBG-MG, Programação e Livro de Resumos, Diamantina, p. 96.

Further reading

Arnold, G.L., et al., 2004. Molybdenum isotope evidence for widespread anoxia in mid-Proterozoic oceans. Science 304 (5667), 87–90. https://doi.org/10.1126/science.1091785.

Babinski, M., et al., 2018. Comparing two arms of an orogenic belt during Gondwana amalgamation: age and provenance of the Cuiabá Group, northern Paraguay Belt, Brazil. J. South Am. Earth Sci. 85, 6–42.

Caxito, F.A., Gonçalves Dias, T., 2018. Ferro. In: Recursos Minerais de Minas Gerais On Line: síntese do conhecimento sobre as riquezas minerais, história geológica, e meio ambiente e mineração em Minas Gerais; coord: Pedrosa-Soares, A.C., Voll, E., Campos, E. CODEMGE, Belo Horizonte. http://recursomineralmg.codemge.com.br/.

Dardenne, M.A., Schobbenhaus, C., 2001. Metalogênese do Brasil. UNB; CPRM, Brasília, ISBN: 8523006478, p. 392.

Falci, A., Caxito, F.A., Seer, H.J., Valeriano, C.M., Dias, P.H.A., Pedrosa-Soares, A.C., 2018. Provenance shift from a continental margin to a syn-orogenic basin in the Neoproterozoic Araxá nappe system, southern Brasília belt, Brazil. Precambrian Res. 306, 209–219.

Martins-Neto, M.A., Hercos, C.M., 2002. Sedimentation and tectonic Setting of Early Neoproterozoic glacial deposits in South-Eastern Brazil. In: Precambrian Sedimentary Environments. Blackwell Publishing Ltd, Oxford, UK, pp. 383–403. https://doi.org/10.1002/9781444304312.ch18.

Scott, C., et al., 2008. Tracing the stepwise oxygenation of the Proterozoic ocean. Nature 452 (7186), 456–459. https://doi.org/10.1038/nature06811.

Uhlein, A., Trompette, R.R., Alvarenga, C.J., 1999. Neoproterozoic glacial and gravitational sedimentation on a continental rifted margin: the Jequitaı´–Macaúbas sequence (Minas Gerais, Brazil). J. South Am. Earth Sci. 12 (5), 435–451. https://doi.org/10.1016/S0895-9811(99)00032-2.

Uhlein, A., et al., 2011. Chapter 51 The glaciogenic Jequitaí Formation, southeastern Brazil. Geol. Soc. Lond. Mem. 36 (1), 541–546. https://doi.org/10.1144/M36.51.

Veizer, J., Prokoph, A., 2015. Temperatures and oxygen isotopic composition of Phanerozoic oceans. Earth Sci. Rev. 146, 92–104. https://doi.org/10.1016/J.EARSCIREV.2015.03.008. Elsevier.

CHAPTER FOUR

Stable isotopes as a tool for stratigraphic studies: Insights from the Brazilian sedimentary record

René Rodrigues[a,*], Egberto Pereira[a], Sérgio Bergamaschi[a], Lucas Pinto Heckert Bastos[b]

[a]Department of Stratigraphy and Paleontology—DEPA, Universidade do Estado do Rio de Janeiro—UERJ, Rio de Janeiro, Brazil

[b]Basin Analysis and Mobile Belts Graduate Program, Universidade do Estado do Rio de Janeiro—UERJ, Rio de Janeiro, Brazil

*Corresponding author: e-mail address: rene@uerj.br

Contents

Abstract

The application of stable isotopes in stratigraphic studies is here demonstrated with selected sedimentary Brazilian sections from the Paleozoic, Mesozoic and Cenozoic. In those different age-case we present the application of $\delta^{13}C_{org\ and\ carb}$, $\delta^{15}N_{total}$ and $\delta^{18}O_{carb}$ stable isotopes showing how past global events and environmental conditions variation can imprint characteristic isotopic signatures in sedimentary rocks

Stratigraphy & Timescales, Volume 4
ISSN 2468-5178
https://doi.org/10.1016/bs.sats.2019.08.007

profiles and how they can be useful in stratigraphic studies. Also, key points for the interpretation of stable isotope data derived from sedimentary organic matter and carbonates are discussed demonstrating how bulk-rocks analyses can be really useful for stratigraphic correlations and paleoenvironmental interpretations when a high-resolution interpretation is conducted.

1. Introduction

In earth sciences, stable isotopes measurements have been widely applicated as a tool to establish interpretations and solve scientific problems. Specifically, in stratigraphic studies, the application of organic and inorganic stable isotopes data to characterize and correlate strata consists of the fundamental principle of isotope stratigraphy. In this sense, the sedimentary record can be sub-divided into stratigraphic units with diagnostic isotopic signatures that, in some cases, can be globally correlated expressing global abrupt climatic and oceanic changes. Such type of study is especially important in areas where traditional methods of correlation, as biostratigraphy and wire log signatures, have proven to be inadequate for precise correlations. In this sense, Weissert et al. (2008) introduce the Chemostratigraphy or chemical stratigraphic unit, which uses the chemical fingerprints in sedimentary sequences for stratigraphic correlation. By the authors, it is a very powerful stratigraphic tool if it is applied in combination with bio- and magnetostratigraphy. Geochemical proxies (stable isotope included), applied in stratigraphy, were originally developed and utilized as proxies for paleoenvironmental change. However, stable isotope signatures fixed in sedimentary inorganic and organic matter are among the most powerful proxies used in chemostratigraphy. Stable isotopes are extremely useful particularly in cases of poorly or non-fossiliferous sequences; areas with very high sedimentation rates or in condensate zone.

Further, many regional and local variables are also involved in isotopic fractionation and they should be evaluated considering which type of material is analyzed; organic matter or carbonates. Changes in the constituents of organic matter (lipids, proteins, and lignin), diagenetic process, variations in water-chemistry and thermal effect, for example, are potentially process that can affect the isotopic composition of rocks and should be evaluated before the establishment of correlations.

In this context, we present here key points for the interpretation of stable isotope data derived from sedimentary organic matter and carbonates, as well some insights from the application of isotope stratigraphy in the Brazilian Paleozoic, Mesozoic, Cenozoic, and recent sedimentary record. In those different age-cases, we explore the application of $\delta^{13}C_{org\ and\ carb}$, $\delta^{15}N_{total}$ and $\delta^{18}O_{carb}$ stable isotopes showing how past global events and environmental conditions variation can imprint characteristic isotopic signatures in sedimentary rocks profiles and how they can be useful in stratigraphic studies.

2. Isotopic studies of organic matter

Sedimentary organic matter provides a variety of indicators that can be used to reconstruct ancient marine and continental environments. The Organic matter concentration in rocks is regulated by the biological and geological process that control your production, preservation, and incorporation in the geosphere (Tyson, 2012).

In aquatic environments, the major source of organic matter buried in sediments is the organic-walled phytoplankton that thrived in the photic zone and through the photosynthesis, is capable to synthesize new organic carbon on earth. However, in terrestrial environments, higher plants dominate the production of new organic matter by photosynthetic fixation of CO_2 and, in this sense, can also be important additional contributors to the organic matter preserved in the sediments.

The preservation of the Organic matter requires an oxygen-depleted condition. Given your constitution of reduced form of carbon, Organic matter is unstable under normal oxygenated water conditions (over 2.0 mL oxygen/L water) and can suffer relatively rapid degradation by respiratory process by aerobic bacteria and other organisms during its transport from the water surface to the bottom (Peters et al., 2005). Also, in the layer sediment degradation process can continue to occur accelerated by bioturbation of benthonic organisms that create spaces that facilitate penetration of oxygen and sulfate in the layer sediment.

On the other hand, under anoxic conditions (less than about 0.1 mL oxygen/L water) the degradation of organic matter suffer significant decrease due to the absence of aerobic organisms that require higher levels of oxygen (Peters et al., 2005) being the degradation due to (anaerobic) sulfate reduction by anaerobic bacteria more relevant in this condition. During times of strong stratification of water column and/or of high primary production, dissolved oxygen is used before it can be replenished, and anoxic bottom waters are developed, enhancing Organic matter preservation.

2.1 Organic carbon stable isotopes

The carbon isotope composition ($^{13}C/^{12}C$) of any synthesized organic compound is related to the Carbon source, isotopic effects related to the carbon assimilation during photosynthesis as well fractionation due to metabolic process and biosynthesis of the organism (Hayes, 1993). Also, preferential degradation of components of organic matter can affect your isotopic composition.

First, the inorganic carbon source available for photosynthesis process could suffer significant variations due to global or local aspects. Input of volcanic CO_2, and thermogenic release of CH_4, related to sill intrusions (^{13}C depleted carbon), for example, can affect the pCO_2 and impact in carbon isotope fractionation in marine and terrestrial organisms. Several interesting examples are observed during Early Aptian Anoxic Event and in the Paleocene–Eocene Thermal Maximum (Méhay et al., 2009; Svensen et al., 2004). Methane hydrate release related to plate tectonic movements (Jahren et al., 2001) and effects related with a meteorite impact (Schulte et al., 2010 are some of the other examples of global events that can affect the carbon cycle and be responsible for significant isotopic excursions.

In terms of local aspects, significant variations in the inorganic carbon source can occur especially in shallow restrict water bodies. Variations in the pressure of aqueous CO_2 could occur due to variations in water temperature, salinity and photosynthesis activity and, in this sense, can strongly affect the composition of phytoplankton and cyanobacteria living in those environments (Freeman and Hayes, 1992; Rau et al., 1992; Schidlowski et al., 1992; Stiller et al., 1985). Evaporation process, for example, discriminates against ^{13}C making available a pool of dissolved $^{13}CO_2$ that is incorporated to the biomass leading to an organic matter enriched in the heavy isotope (Stiller et al., 1985).

As important as the source is the metabolic process involved in carbon assimilation. The majority of the plants, for example, incorporate organic matter carbon using C3, Calvin pathway, which can produce new carbon with a shift of −20‰, when comparing to the atmospheric source. However, plants that use the C4 Hatch-Slack pathway fractionate around −7‰, related to the inorganic carbon source (O'Leary, 1988).

Although, as discussed before, it is particularly important to evaluate the parameters that can affect the carbon source and not just work with absolute values in a direct correlation with organisms. For example, modern marine organic matter presents a $\delta^{13}C$ composition between −22‰ and −20‰ (Meyers, 1997). Although, it's not correct to apply the same association between marine organic matter and this isotopic composition along the geological time, given that the concentration of CO_2 in the atmosphere was not constant. In the Middle Cretaceous, for example, concentrations of CO_2 were higher, and this more available CO_2 may be responsible for the ^{12}C enriched values of marine organic matter when compared to modern marine OM (Dean et al., 1986). The same is observed in Silurian, Devonian and Permian marine siliciclastic sequences in Brazilian record (Rodrigues et al., 2010; Pereira et al., 2010)

2.2 Nitrogen isotopes

Nitrogen is an essential element for the synthesis of proteins and nucleic acids in living organisms being the diatomic atmospheric Nitrogen (N_2) the major source for the organisms. Due to the high molecular stability, as consequence of the triple bond between the two atoms, metabolic processes capable of transferring it to the biotic world (bioavailable nitrogen species) are specific and well defined. In this sense, the conversion of N_2 to reactive Nitrogen (Nr) is restrict to some groups of bacteria and archaebacteria. Moreover, transformations between different nitrogen chemical species in aquatic environments are result of microbiological processes and also linked with redox conditions of the environment (Canfield et al., 2010). Biological process largely regulates the nitrogen cycle with nitrate as the main bioavailable nitrogen species.

Nitrogen fixation is responsible to reduce N_2 to ammonia which is transformed in ammonium in the process of organic matter remineralization. Fractionation involved in microbial nitrogen fixation do not exceed 3‰ and, because a large amount of energy is required to break apart the triply bonded N_2 molecule, this process is mostly restricted to environments that are deficient in biologically available fixed nitrogen.

In oxic conditions, a process of oxidation transforms ammonium in nitrite (NO_2) and after in nitrate (NO_3^-) being this process denominated nitrification. Otherwise, in anoxic conditions, anaerobic reduction of NO_3^- to $N_{2(gas)}$ (denitrification) and/or anaerobic ammonium oxidation could be responsible for significant bioavailable nitrogen losses and large isotopic fractionation (~20‰ in the water column and <3‰ in sediment pore waters).

3. Isotopic studies of carbonates

3.1 Oxygen isotopes

The primary variables which determine the oxygen isotope signal in marine carbonates are water temperature and the $^{18}O/^{16}O$ composition of the sea water, where 99.8% of all the oxygen is in the form of ^{16}O, (Weissert et al., 2008).

The isotopic composition of oxygen in a solid phase is not the same as that in water when the two are in equilibrium. There is a differential partition in the solid and liquid phases, with ^{18}O being enriched in the solid phase. Therefore, the oxygen isotope composition of calcium carbonate is dependent on the isotopic composition of the ambient fluid and on the temperature of precipitation (Weissert et al., 2008). The enrichment in ^{18}O during carbonate formation is temperature-dependent: at higher temperatures, more carbonate ions with ^{18}O atoms could remain in solution, and therefore the carbonates precipitated under these temperature conditions should be less enriched in ^{18}O than those formed under lower temperatures (e.g., Emiliani, 1955). For a constant oxygen isotopic composition of the sea water, the $^{18}O/^{16}O$ ratio in carbonates deposited in isotopic equilibrium will decrease by 0.23‰ with each degree centigrade increase of temperature (e.g., Emiliani, 1955; Epstein et al., 1953). In an ocean with constant isotopic composition of the water, an increase in temperature of little more than 4 °C corresponds to a decrease in $\delta^{18}O$ of about 1‰. Thus, with these findings it was possible to estimate at which water temperature the carbonate precipitation took place (e.g. Epstein et al., 1953).

The most important process that can produces modification in the oxygen isotope composition of the sea water is evaporation. This process concentrates the heavy ^{18}O isotope in the liquid phase, which so becomes relatively enriched in ^{18}O as compared to the vapor phase that becomes enriched in the light ^{16}O isotope. During glaciations, water vapor in the cloud's precipitates ^{16}O-rich water on ice caps in the form of snow, leaving the ocean water enriched in ^{18}O. This means that changes in oxygen isotope ratios in carbonates during ice ages are primarily an ice volume signal, with a minor effect due to temperature (e.g., Shackleton, 1967).

The above relationships have provided the basis for numerous studies of past climate changes, continental glaciations and changes in marine current systems. Furthermore, by studying the isotopic compositional changes in both benthic and planktonic foraminifera it is possible to investigate the structure between surface and deep water and their evolution in the geological history.

3.2 Carbon isotopes

Primary $\delta^{13}C$ absolute values in marine carbonates are a function of the $^{13}C/^{12}C$ of dissolved inorganic carbon (DIC) in surface water, since a small fractionation is associated with carbonate precipitation. Significant changes in carbon isotope ratios reflect changes in carbon cycling of the ocean, particularly between reservoirs of very different isotopic compositions.

Metabolic processes can contribute to significant change in the carbon pool of an environment. As discussed for the organic matter (OM), phytoplankton photosynthesis strongly discriminates against ^{13}C and preferentially incorporate ^{12}C in OM formed by this process. Thus, phytoplankton productivity or bacterial oxidation of organic matter in the water column will significantly affect the $\delta^{13}C$ values of the CO_2 in the surface water. As organic matter is enriched in the light isotope ^{12}C, its removal by burial from oxidative recycling makes the ocean richer in ^{13}C, and therefore increases the $\delta^{13}C$ values of carbonates precipitated under these conditions (Scholle and Arthur, 1980). This water conditions normally characterize dysoxic–anoxic events as will be discussed later.

Of the two stable carbon isotopes, the abundance of ^{12}C in nature is around 98.89%, while ^{13}C forms are responsible for the remaining 1.11% (Craig, 1953). According to Weissert et al., (2008), mass differences of the two isotope species lead to strong fractionation during photosynthetic incorporation of carbon into organic matter, while inorganic carbonate precipitates formed in the aquatic environment are less affected by fractionation processes as discussed and reviewed in a number of articles.

It is also well-known fact that the isotope composition of planktonic foraminifera tests is enriched in ^{13}C (heavy) and in ^{16}O (light) compared to that in benthic foraminifera, which is lighter in carbon isotopes and heavier in oxygen isotopes. Thus, by utilizing isotopic compositional changes in both benthic and planktonic foraminiferal tests it is possible to investigate how the structure between surface and deep water responded to the influence of changing sources of deep waters and the process of their formation, as well as the cycling of nutrients during their transit around the ocean basins (Kennett and Stott, 1990).

During extinction phenomena, as that of the Cretaceous–Tertiary (K–T) boundary, the ocean would have no gradient of carbon isotope from top to bottom, because the water surface would no longer be depleted in ^{12}C due to decreasing photosynthesis (reduction of primary productivity). Carbonates deposited at that time should have a more negative $\delta^{13}C$ signature as compared to values from the underlying and overlying carbonates, which would explain the negative $\delta^{13}C$ excursion at the K–T boundary (e.g., Hsü and Mckenzie, 1985).

During diagenesis carbon isotopes are also fractionated by bacterial reactions, forming CO_2 rich in ^{12}C through organic matter oxidation or sulfate reduction, and CO_2 enriched in ^{13}C due to methanogenesis. If both oxygen and carbon isotopes become simultaneously negative, this could mean fresh-water diagenesis, and therefore a possible sequence boundary (see case studies of the Cenozoic).

4. Brazilian case studies

In this chapter is presented examples where the stable isotopes can be successfully applied and allows the establishment of good stratigraphic correlations in association with other tools of paleoenvironmental analysis. In this sense, case studies from Brazilian sedimentary basins, comprising specific intervals ranging in age from the Devonian to Holocene are separated and explained here in three sections: Paleozoic, Mesozoic, Cenozoic and recent (Fig. 1).

In each case, bulk analyses of organic matter and/or carbonates are presented and discussed evaluating primary signal and potential diagenetic overprint. However, several studies had demonstrated that, even for carbonates or organic matter, the carbon and oxygen isotope signals resemble, in general, those derived from careful analyses of individual foraminifera species (Azevedo et al., 1997).

Moreover, the general trend is not masked when evaluated within the framework of stratigraphy succession and strong diagenetic overprints may be related to sea-level low stands that can be correlated with sequence stratigraphic boundaries. Based on these findings, we demonstrate here the applicability of whole-rock sample analyses to stratigraphic studies.

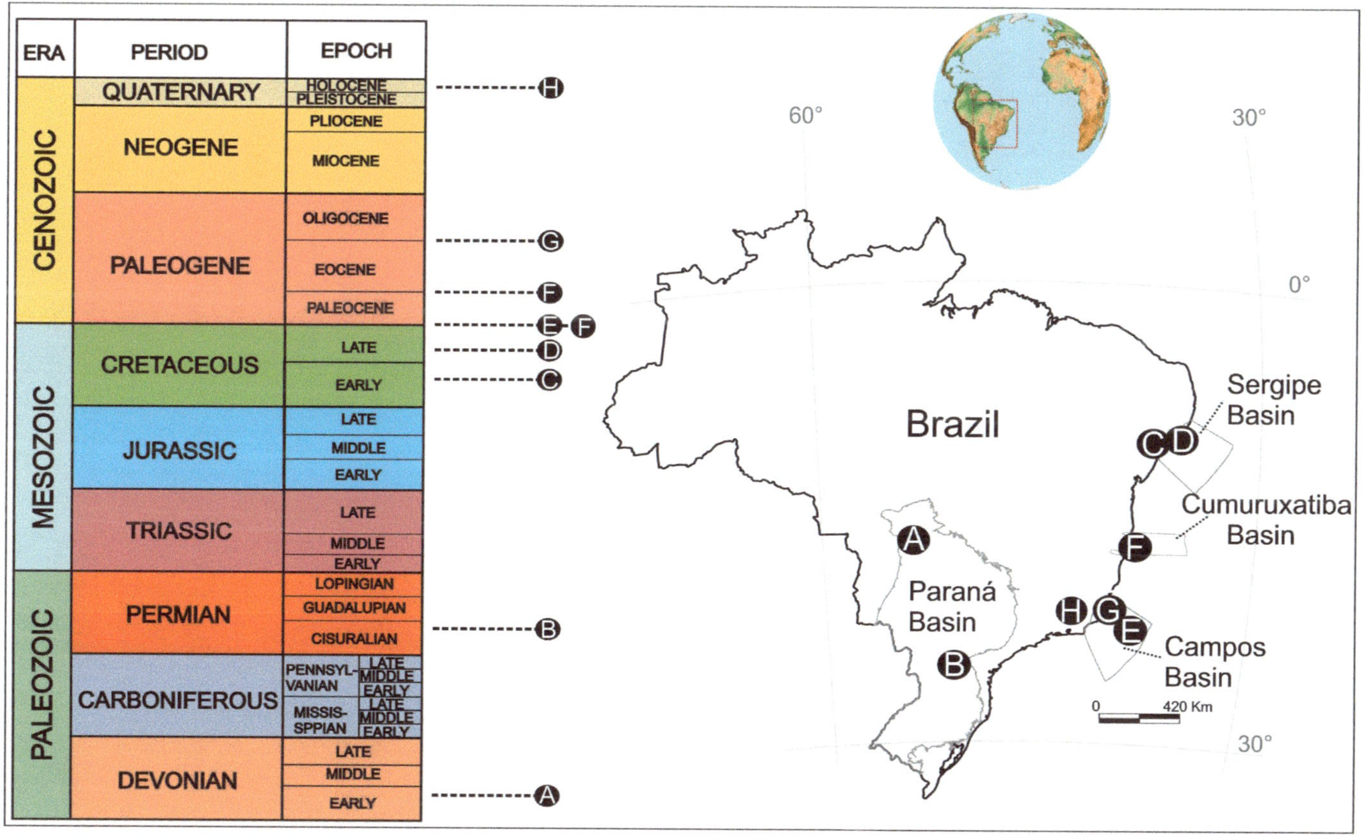

Fig. 1 Geologic timescale with the age of the different Brazilian examples that are presented in the discussion in this chapter. The location of those sections in the Brazilian territory are presented in the map.

4.1 Methods

In the discussions about Brazilian examples geochemical data from Organic matter and for carbonates is here presented. For all analyses, rock powder samples were used being the different analytical procedures described below.

For the TOC analysis, inorganic carbon was removed from samples with 50% HCl acid treatment. The quantification of the organic carbon was determined in an elemental analyzer LECO SC-632 (CO_2 infrared detector combustion system) with result expressed in % relative to the original weight of the sample. LECO-Soil Reference material standard (Part No = 502-308/Lot No1017) was used to calibrate the equipment and assure the quality control of the obtained results.

Rock-Eval pyrolysis analysis followed the procedures described by Espitalie et al. (1977). These analyses were performed on the Rock-Eval VI equipment, using the IFP160000 (Part No 5-081940) certified standard for the control and calibration of the equipment. The parameters S1, S2 and Tmax (°C) were determined and the calculation of the Hydrogen Index (IH) was made using parameter S_2 and TOC content of each sample using the formula $HI = (S2/TOC) \times 100$ in mg HC/g TOC.

Stable total organic carbon and total Nitrogen isotope analyses were performed in a Delta V Advantage IRMS (Thermo Fisher) coupled to an EA-1112 Flash Element Analyzer (Thermo Fisher). For Carbon isotopes analysis 1 mg of acid treated (50% HCl) powder sample was weighed and transferred to a tin capsule that was introduced into the elemental analyzer where the combustion occurs. A flow of Helium carries the CO_2 that was generated in the combustion process to the IRMS where the $^{13}C/^{12}C$ ratio is measured being the result (δ) expressed in part per mil (‰) in relation to Pee Dee Belemnite–VPDB. For Nitrogen isotopes measurement, 15–20 mg of pre-acidified sample (50% HCl) was weighed and transferred to a tin capsule which was introduced into the elemental analyzer for the measurement of the $^{15}N/^{14}N$ ratio. An international standard (Low Organic Content Soil Standard OAS/certificate N° 201613) was measured every 10 samples and the reproducibility of samples and standard was better than ±0.15 (‰) for $\delta^{13}C_{org}$ and ±0.20 (‰) for $\delta^{15}N$.

For biomarker analysis, the bitumen extraction procedure was carried out in a Soxhlet (250 mL) type extractor using dichloromethane P.A. as the solvent. The extracted bitumen was fractionated in saturated compounds, aromatics, resins, and asphaltenes in a glass column (10 mm diameter/25 cm extension) with a solid phase composed of silica gel (1/3) and alumina (2/3) and mobile phase for extraction of the saturated compounds, hexane and, for aromatic compounds, a mixture of dichloromethane P.A. and hexane. Gas chromatography was performed on an Agilent brand chromatograph (6890 series) coupled to an Agilent 5973 Network mass spectrometer (GC/MS) using a HP-5MS capillary column from J&W Scientific (30 m long × 0.25 mm internal diameter × 0.25 mm film thickness) and helium as the carrier gas (temperature programming: 70–290 °C, 6 °C/min).

Carbon and oxygen isotope analyses in carbonates were conducted in a Kiel IV Carbonate Device coupled to the Mass Spectrometer Thermo Delta V Advantage. Powder samples reacts with phosphoric acid in the Kiel Carbonate producing CO_2 and H_2O being the last one retained in cryogenic system. A microvolume of CO_2 is then transferred to the Mass Spectrometer Thermo Delta V Advantage where occurs the isotope ratio measurement. The $\delta^{13}C$ and $\delta^{18}O$ values are reported in relation to the PDB standard (IAEA-CO-1-Marble). Analyses were carried out at Chemostratigraphy and Organic Geochemistry Laboratory from Rio de Janeiro University.

4.2 The Paleozoic

4.2.1 Devonian

Maximum flooding surfaces may be an important marker in the history of a sedimentary basin. The Devonian of the Paraná Basin (Ponta Grossa Formation) records a succession of shales related to a maximum flooding that would have reached the basin during Praguian-Emsian age (Bergamaschi, 1999). This event was responsible for the increase of anoxia and allowed the deposition of organic-rich levels with TOC contents reaching values up to 2%.

The section below shows the geochemical record of this Praguian-Emsian event (Fig. 2). It is possible to observe that a progressively enrichment in organic matter occurs in direction to the top of the section. The enrichment in TOC levels, with values reaching 1,9%, shows good correspondence an enrichment in Hydrogen Index (HI) with a maximum of 600 mg HC/gTOC. Higher values of TOC and HI are possible reflecting the progressive improvement of the conditions of preservation of organic matter as well as enhanced bioproduction that could have been stimulated by fertilization with the marine waters.

A good correlation between negative isotopic values (^{12}C enriched) with TOC and IH results can be observed in the top of the section where the maximum flooding surface is positioned. This association is very well

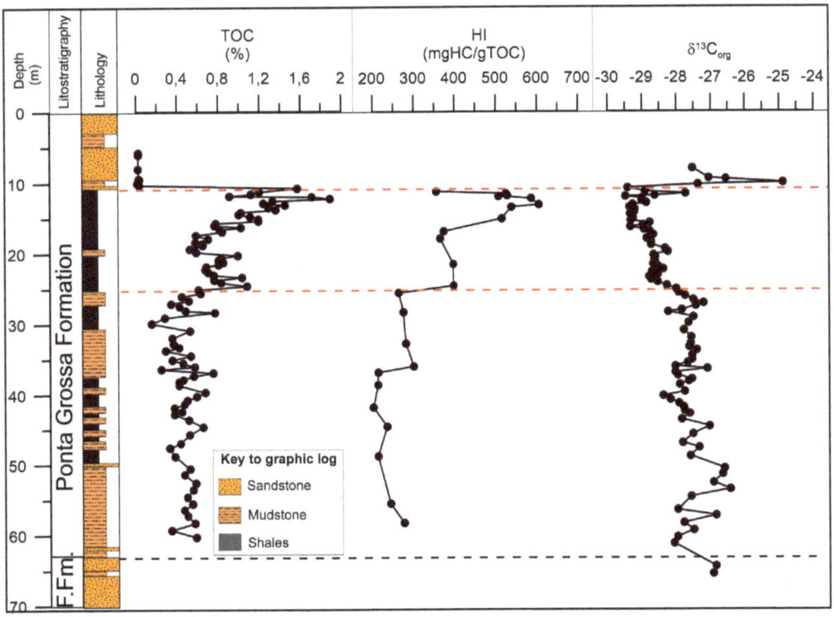

Fig. 2 Total Organic Carbon content (TOC), Hydrogen Index (HI) and organic carbon stable isotopes ($\delta^{13}C$) for a section of Ponta Grossa Formation. Red lines show maximum values of TOC and HI that are correspondent with the most negative $\delta^{13}C$ results for the section. F.Fm = Furnas Formation. *Modified from Dias, A.E.S. 2004. Estratigrafia química e identificação de intervalos potencialmente geradores no devoniano inferior da Formação Ponta Grossa. 62 f. Monograph. (Faculdade de Geologia) - Universidade do Estado do Rio de Janeiro.61 pp.*

defined by distinct proxies as demonstrated by Pereira et al. (2007, 2010). Theses authors show a progressive decrease of the biotubartion index into the negative isotope excursion and a progressive deepening of the basin. The deepening is record by the progressive ichnofacies changing. The skolithos ichnofacies predominant in the Furnas Formation is progressive replace by Cruziana, and finally, in the flooding surface, the Zoophycus ichnofacies is observed.

Although an improvement in the preservation condition of the organic matter could be raised as a cause of lighter isotopic values, petrographic studies in this section observed a good correlation between ^{12}C enriched values and changes in the composition of organic matter (Dias, 2004). An association between negative peaks with alginite accumulations suggests that isotopic shifts can be related with the preservation of a more lipidic organic matter that is characteristic of algae which usually presents an enrichment in the light isotope (Meyers, 1997).

4.2.2 Permian

The late Paleozoic records complex climatic changes that have affected ecosystems around the globe. Sedimentary records of the transition from Icehouse conditions to extreme greenhouse conditions can be found today, for example, in South America, African and Falkland Islands. During this climatic context of progressively warming, in the South of the Gondwana a tectonic convergent scenario was responsible for the entrapment of a huge water-mas that lead to the deposition of almost ~5 million km^2 of organic rich rocks mainly in South-Africa (Whitehill Formation) and in Brazil (Irati Formation) (Faure and Cole, 1999).

The Brazilian record of this extreme event is preserved at Paraná Basin, being composed of organic-rich rocks with an age of 278.4±2.2 Ma (Santos et al., 2006). Previously studies suggested that the deposition of Irati rocks occurred in a restrict marine environment with saline and anoxia fluctuations (Araujo, 2001; Rodrigues et al., 2010; Alferes et al., 2011).

Further, the Formation is subdivided in two members, Taquaral and Assistência (Milani and Zalán, 1999). The lowermost Taquaral Member is composed, in general, of dark-gray siltstones and shales where the occurrence of marine acritarchs suggest a marine epicontinental to restricted environment with normal salinity. The boundary between Taquaral and Assistência Members can be easily identified by the occurrence of the carbonates that, in some cases, can present dolomitic composition. Also, carbonates are interbedded with shales and organic-rich shales deposited restrict marine environment. In terms of thermal evolution, Irati Rocks are immature with exception of specific places where Jurassic intrusions can be found affecting the sedimentary layers (Araujo, 2001).

In the section of Irati Formation (Fig. 3) is presented bulk geochemistry parameters, molecular indicators usually associated with hypersaline conditions and organic carbon and Nitrogen stable isotopes. The Uppermost Member Assistência presents significant higher values of TOC, S and hydrogen index when comparing to the Taquaral.

This indicates that in the carbonate system bioproduction and preservation conditions was probably better than the siliciclastic. Also, two well-marked intervals of increase in carbonate content and tow levels of progressively organic enrichment can be identified and mapped in major part of Paraná Basin reflecting significant events of salinity and anoxia variation, respectively. As noted in Fig. 3, the increase in carbonate content is associated with the increase in the biomarkers iC_{25}, iC_{30} and Gammacerane that have your sources associated with specific organisms that are able to tolerate high salinity rates (Grice et al., 1998; Sinninghe Damste et al., 1995; Wang, 1998).

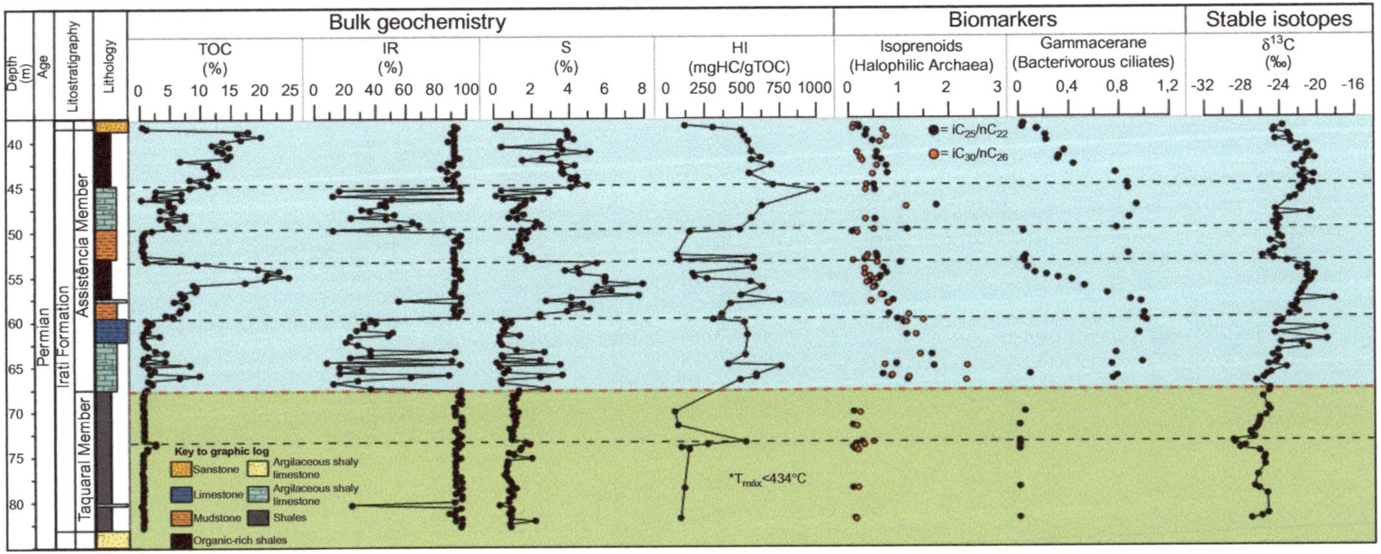

Fig. 3 Total Organic Carbon (TOC), Insoluble Residue (IR), total Sulfur (S), Hydrogen Index (HI) for Irati samples from Paleosul-05 borehole (São Mateus, PR). Relative proportion of the isoprenoids (*i*C25/nC_{22} and *i*C30/nC_{26}) whose origin is possibly derived from ether-bound membrane lipids of halophilic archaea, is also presented as well the Gammacerane index (G/C30hopane) that can be associated with bacterivorous ciliates organisms that habits hypersaline environments.

In the $\delta^{13}C$ profile, it is possible to observe a slightly isotopic negative excursion at Taquaral Member where the values around −28‰ occur associated with an organic enrichment and higher HI. This excursion can be identified in several profiles (distances reaching 800 km) being associated with a compositional change in the organic matter that is associated with a marine incursion that reached an extensive area of the basin. In this level, higher proportion of biomarkers derived of green and/or red algae are found indicating lipidic sources to the organic matter, material that in general is enriched ^{12}C (Fig. 4). After this negative excursion, a large isotopic positive excursion begins within the Taquaral, going to the Member assistance. The transition between these two members reflects a process of environmental restriction and entrapment of an aqueous body.

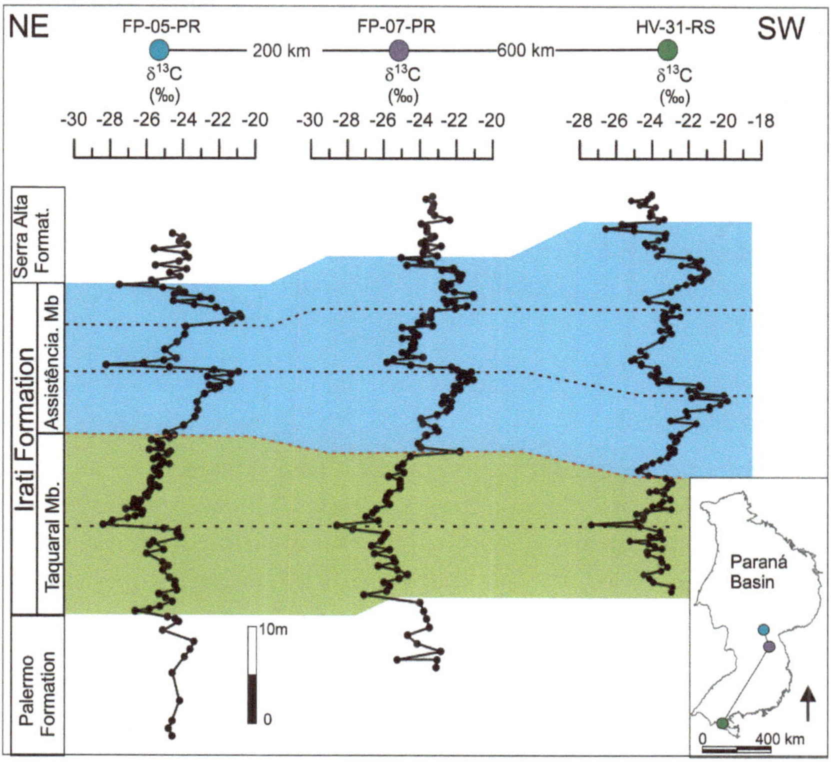

Fig. 4 Correlation of three different sections of Irati Formation based on carbon isotopic composition of preserved organic matter. In the siliciclastic Member (Taquaral) a negative excursion reflects a compositional change caused by a marine incursion while isotopic fluctuations in the carbonate system (Assistência Member) was mainly caused by salinity and anoxia variations.

As a restrict evaporative and hypersaline environment (carbonate system), the carbon sources are more susceptible to suffer isotopic modifications due to physical/chemical process (Stiller et al., 1985). Evaporation processes probably discriminated against the heavy isotope leading an enrichment of ^{13}C in the Assistência environment that was consequently assimilated by the biomass. Additionally, the increase in salinity and temperature may also have affected the dissolved CO_2 pressure discriminating the ^{13}C contributing to the positive $\delta^{13}C$ excursion that occurs associated with low values of IR and high proportions of isoprenoids and gammacerane.

However, as observed at 60 m (Fig. 3), the expressive increase in TOC concentration and higher values of insoluble residue indicate the input of water into the paleoenvironment. This process probably induced salinity-driven water stratification allowing excellent preservation of the organic matter and an event of expansion of the ciliate's organisms, as evidenced by gammacerane index, as noted in Fig. 3. In this context of vertical separation, low pressures of dissolved CO_2 could have been reached by intense bioproduction leading to a decrease in isotope fractionation between dissolved CO_2 and the microbial biomass, increasing its ^{13}C concentration (Hsü and McKenzie, 1985; Rau et al., 1992; Schidlowski et al., 1992). As noted above (Fig. 4), major events of ^{12}C depletion could be traced in distinct sections of Irati Formation being this signature also described in your equivalent formation in Karoo Basin (Whitehill Formation) (Faure and Cole, 1999).

4.3 The Mesozoic

The sedimentary record of Brazilian marginal basins is directly related to the rifting of the African and South American plates. Thick sequences of lacustrine and fluvial deposits were accumulated during the Early Cretaceous rifting phase while the first intermittent marine transgressions occurred in the lower Aptian, when tectonic activity has been restricted to subsidence (Estrella et al., 1984). The geochemical record of these marine transgressions is characterized by the presence of chemical fossils derived from marine algae and carbonates with higher $\delta^{13}C$ values, as compared to the underlying sequences that present biomarkers and carbon isotopic signatures consistent with freshwater to saline lacustrine environments (e.g. Takaki and Rodrigues, 1984; Mello et al., 1988; Rodrigues, 1995; Bastos et al., 2014).

As opposed to what is observed in the Middle Permian record of the Paraná Basin (Irati Formation) where the convergent scenario of the south

Gondwana entrapped a huge water mass, the Aptian of the Sergipe-Alagoas Basin records the first steps of the development of the South Atlantic Ocean. The transition from a restricted environment where different types of salt deposits can be found (Muribeca Formation) to an open marine setting with the development of an extensive carbonate platform represents a key moment in the Brazilian geological history.

Such transition in the studied section is evidenced by the occurrence between 268 and 180 m of marine dinoflagellates of the *Subtilisphaera* community (Carvalho et al., 2016) that is associated with a marine incursion that took place during Early Cretaceous and can be identified by irrefutable marine evidence (dinoflagellates, echinoids, foraminifera, molluscs and fishes) (Arai, 2014). The $\delta^{13}C_{carb}$ record of the saline lacustrine phase of Sergipe Basin presents significant ^{12}C enrichment with values reaching −10‰ while in the marine phase (Riachuelo Formation) values fluctuated between −2‰ and 2‰ (Fig. 5).

A sharp negative $\delta^{15}N$ (~4‰) marks the boundary between the lacustrine saline and marine phase (Fig. 5). Strong enrichment in ^{15}N is usually associated with denitrification and/or annamox (anaerobic oxidation of ammonium) which makes senses in the restrict lacustrine context of the Muribeca Formation that probably received inputs of NO_3 an NH_4^+. Although, in saline

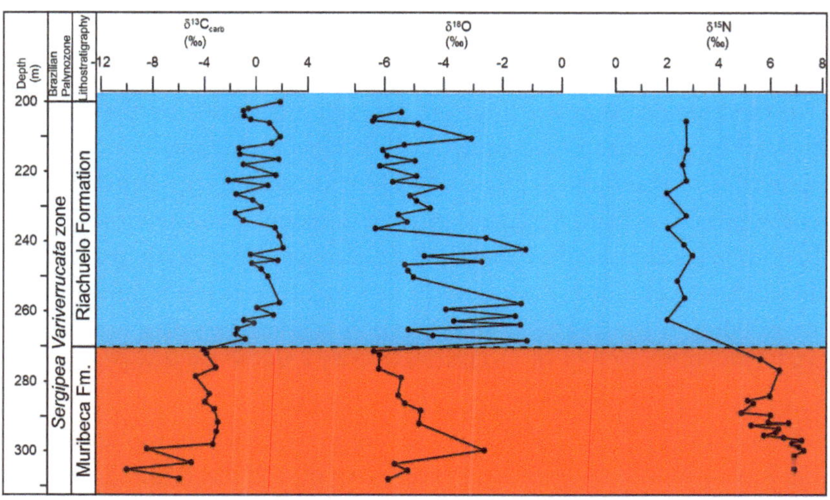

Fig. 5 Total nitrogen stable isotopes and Carbon and oxygen isotopes measures in the inorganic fraction of Aptian rocks from Sergipe basin where the transition from a restrict environment to an open marine is recorded.

environments (alkaline conditions) volatilization of NH_3 can also play a role in nitrogen cycle. Experimental analysis show that with increasing in pH and temperature, ammonium tends to dissociate (with preferential loss of $^{14}NH_4^+$), being the produced ammonia quickly released by degasification process (Li et al., 2012). This process can lead to an extreme ^{15}N in the nitrogen species available for microbial biomass.

On the other hand, in the marine phase a decrease in $\delta^{15}N$ values could reflect a switch in the nitrogen source (Fig. 5). With a reduction in the Nitrate/ammonium, N_2 fixation by diazotrophs may have supplied the biological deficiency in nitrogen in the environment through Microbial nitrogen fixation that generally yields only a minor negative isotope fractionation (<3‰) relative to the atmospheric N_2 source (0‰) (Brandes and Devol, 2002; Gruber and Galloway, 2008) (Junium and Arthur, 2007; Kuypers et al., 2004).

Another interesting record in the Sergipe-Alagoas Basin occurs at the Cenomanian-Turonian boundary when, due to the advanced rift process of the South America and Africa plates, strictly marine conditions were already reached with a well-preserved ammonite fauna. Worldwide this interval is marked by the deposition of organic rich shales that would be the result of a period of increase of anoxia and due to the global character of these horizons, such events are called global anoxic events. An accelerated hydrological cycle, intensified silicate weathering, and intense large igneous province (LIP) volcanism may have served as direct or indirect nutrient sources to enhanced bioproduction leading the intensification of global anoxia.

In the Sergipe basin the limit can be well established by the ammonia fauna and by the isotopic signature itself that can be identified in drill holes or outcrops. Takaki and Rodrigues (1993) identified this dysoxic-anoxic event in boreholes from Sergipe Basin and suggested that the more negative $\delta^{18}O$ values indicates that the dysoxic–anoxic event took place under mild climatic conditions, associated with higher relative temperatures of seawater. Fig. 6 shows the isotopic carbon and oxygen data for the limit in the Sergipe basin and the isotope from East Kent (England) where the event was characterized by Jenkyns et al. (1994).

In outcrops of Sergipe Basin, the event is well marked in carbon inorganic record with a positive $\delta^{13}C$ excursion of ~2,5‰ (Fig. 7), record in the Cenomanian—Turonian boundary indicate by the foraminiferous biozone *Whiteinella archeocretacea—Heterohelix reussi* (Dias et al., 2004).

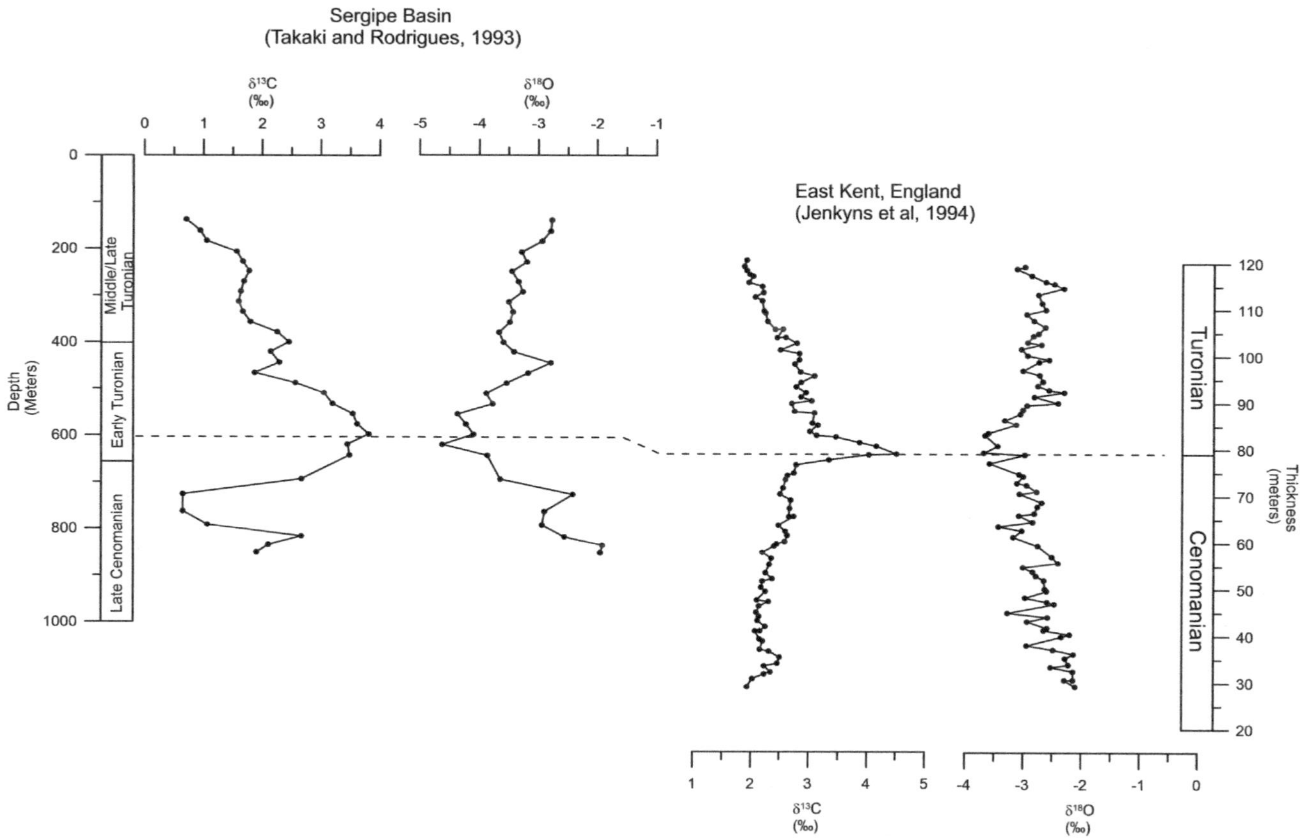

Fig. 6 Correlation of the isotopic $\delta^{13}C$ and $\delta^{18}O$ data for the Cenomanian-Turonian boundary in Sergipe Basin identified by Takaki and Rodrigues (1993) with the isotopic record of the Cenomanian-Turonian $\delta^{13}C$ record from East Kent, England (Jenkyns et al., 1994).

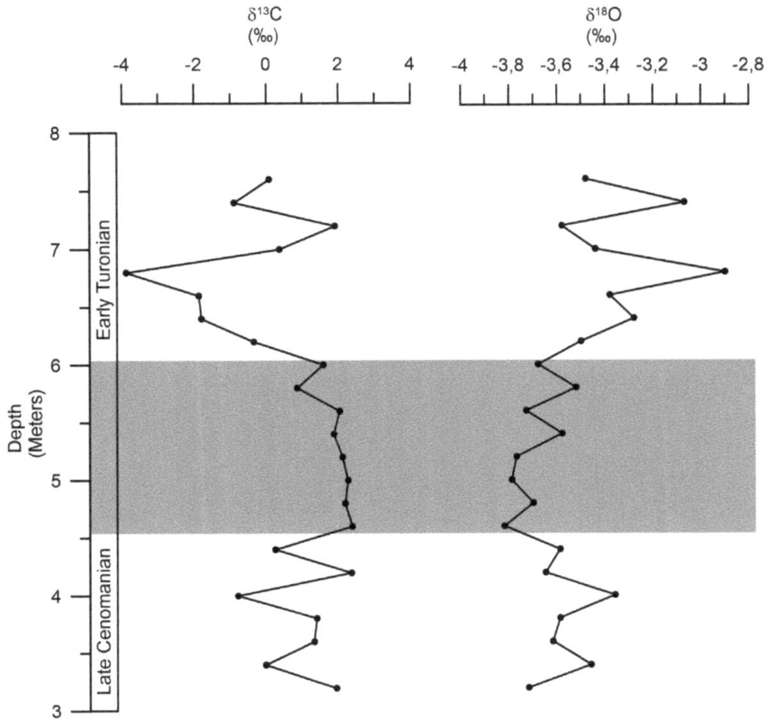

Fig. 7 Carbon and oxygen isotopes measures in the inorganic fraction of Cenomanian-Turonian rocks from Segipe Basin (Dias et al., 2003) with a well marker positive $\delta^{13}C$ excursion that is characteristic of the event and allows correlation with other sections like East Kent in England (Jenkyns et al., 1994).

4.4 The Cenozoic

4.4.1 Cretaceous–Paleogene boundary

The Cretaceous–Paleogene boundary represents one of the most impressive geological events of the earth history. For this time, substantial evidences support an association of changes in ocean thermal conditions (greenhouse effect) and the result of rapid reduction in the rates of primary productivity and nutrient accumulation (mass mortality of oceanic plankton) as a consequence of the effects of a possible meteorite impact Chicxulub, Mexico (e.g., Boersma et al., 1979; Hsü and Mckenzie, 1985; Schmitz et al., 1992; Schulte et al., 2010)

The isotopic record of this event is characterized by a sharp negative $\delta^{13}C$ suggesting a major disruption in the marine carbon vertical gradient and productivity. Further, the observed $\delta^{13}C$ anomaly for the event suggests that the release of methane, input of soot, or the dependency of the isotopic signal on the metabolism of different species are possible explanation for the $\delta^{13}C$ signature (Thomas, 2007). In the Cumuruxatiba Basin (Brazil), Rodrigues (2005) identified negative oxygen and carbon isotopic excursions (1.0–2.0‰) in the K-Pg boundary.

According to the author they are probably related to changes in ocean thermal conditions (greenhouse effect) and the result of rapid reduction in the rates of primary productivity and nutrient accumulation (mass mortality of oceanic plankton) also as an effect of a possible meteorite impact (e.g., Boersma et al., 1979; Hsü and Mckenzie, 1985; Schmitz et al., 1992).

In the peculiar example below (Fig. 8), Rodrigues et al. (1998) documented anomalous Carbon and Oxygen isotopic signature in Calcilutites deposited in the Cretaceous–Paleogene boundary (*Micula prinsii*/*Markalius* nannofossil zone) of Campos Basin, Brazil. As noted, an extremely ^{12}C enrichment with values around −40‰ indicates that the carbonate was precipitated by CO_2 derived from methane oxidation. The possibility of an isotopic imprint from diagenesis of bioturbated pelagic sediments under methane venting conditions was raised as a possible cause for the observed ^{12}C enrichment.

Although, detailed petrographic studies together with SEM images suggested that the precipitation o those carbonates was primarily by induced by methanotrophic bacteria. (Rodrigues et al., 1998). Also, regarding the non-diagenetic isotopic overprint, less negative $\delta^{18}O$ values suggests the formation of carbonates was under cooler bottom waters.

4.4.2 *Paleocene–Eocene Thermal Maximum and the glacial Eocene–Oligocene event*

Significant rises in the sea-surface temperature was experimented during the Paleocene–Eocene Thermal Maximum reflecting a global warming which causes have been associated mainly with a large release of thermogenic methane from contact metamorphism of sill intrusions with organic material, resulting in the production of $^{14}CH_4$ enriched gas ($\delta^{13}C = -35$–50‰) (Svensen et al., 2004). Therefore, the warming strongly affected physical chemistry and biological life of terrestrial and marine ecosystem (Gingerich, 2003; Livsey et al., 2019).

As shown by the $\delta^{18}O$ isotope data from Cumuruxatiba Basin, the rapid climatic change of the latest Paleocene was one of the most dramatic warming events in the geological record (Fig. 9). Oxygen isotope values decrease during late Paleocene to reach a minimum in the earliest Eocene, thus suggesting maximum sea surface temperatures at that time, as pointed out by Shackleton and Kennett (1975).

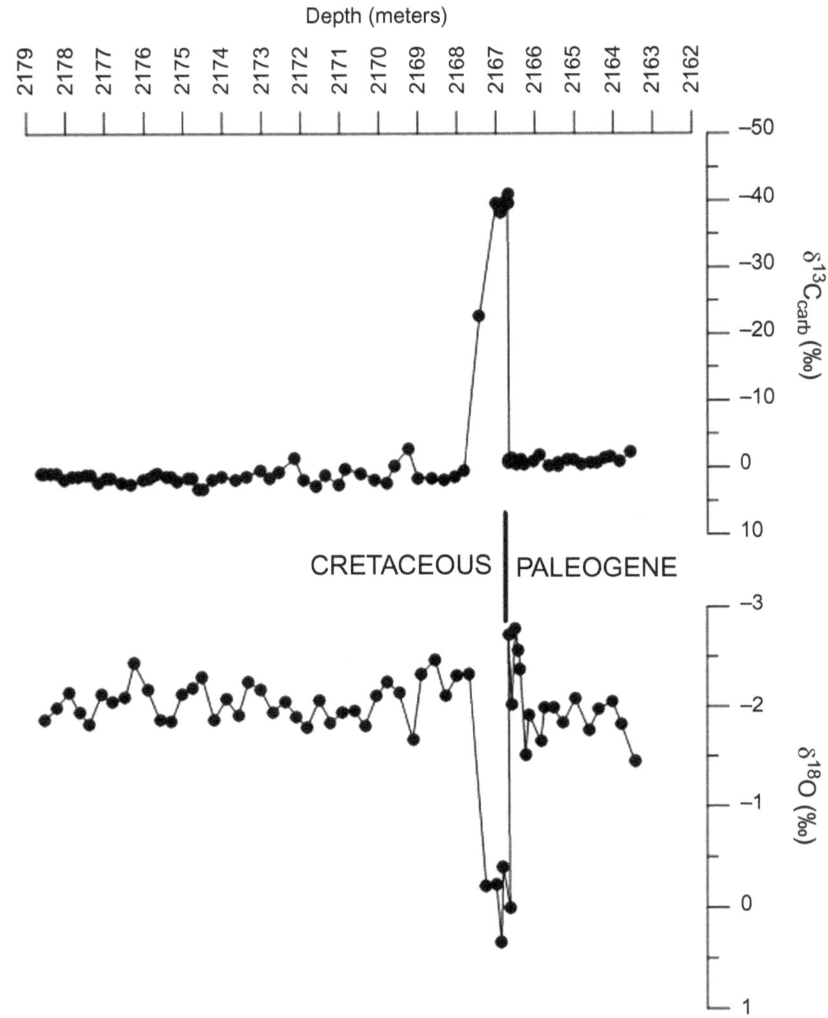

Fig. 8 Anomalous Carbon and Oxygen isotopic signature in Calcilutites deposited in the Cretaceous–Paleogene boundary (Micula prinsii/Markalius nannofossil zone) of Campos Basin, Brazil (Rodrigues et al., 1998).

Cooling periods in Earth history can also be recorded in stable isotopes. A significative global climatic change can be observed and mapped with isotopic proxies during the transition from the Eocene greenhouse world to the icehouse conditions of the early Oligocene.

The classic work of Miller et al. (1991) presents the $\delta^{18}O$ isotopic signature of benthonic foraminiferal for the deglaciation-glaciation events in Antarctica during the Eocene to Miocene where a set of positive $\delta^{18}O$ excursions denominated Oi1, Oi2, Oi 2a and Mi1 events can be identified.

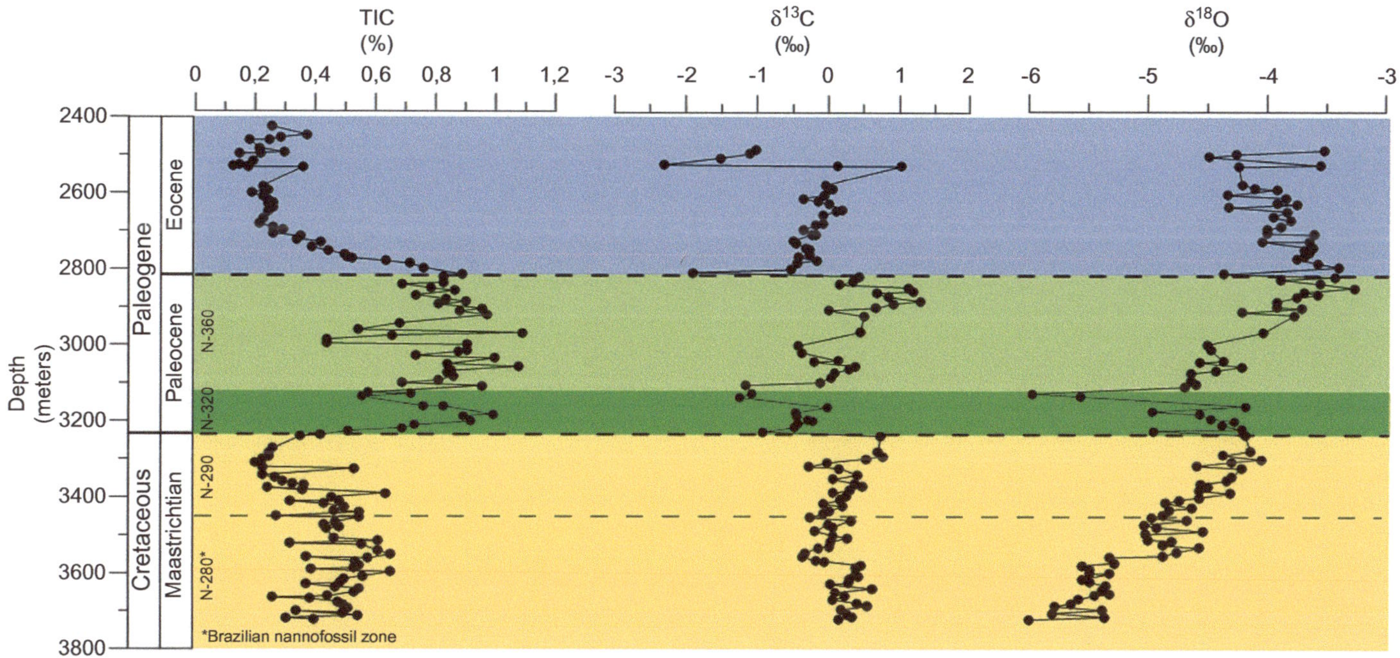

Fig. 9 Total Organic Carbon (TOC) and inorganic Carbon and Oxygen stable isotopes for the Cretaceous–Paleogene and Paleocene–Eocene boundaries, Cumuruxatiba Basin, northeast Brazil (Rodrigues, 2005).

In Brazil, at Campo Basin, the isotopic signature of those episodes was identified in bulk carbonates rocks samples by Rodrigues (2005) being possible to note a set of isotopic events in the lower Oligocene that are associated with specific deglaciation-glaciation events, which are globally identified being probably related to Milankovitch cycles of 400 Kyr. The specific events Oi1 and Mi 1 are markers of the Eocene/Oligocene and Oligocene/Miocene boundaries, respectively.

Previously studies suggest different explanations for the Antarctic glaciation during the Eocene/Oligocene transition. Changes in ocean circulation through tectonic opening of Southern Ocean gateways or changes in the global carbon cycle are raised as the two main categories for the triggers of the glaciation event (Merico et al., 2008 and references therein). However, changes in the global carbon cycle have been proposed as the most likely root for the Antarctic glaciation Further, hypothesis of an increase in organic carbon burial; weathering of silicate rocks; an increase in global siliceous (vs calcareous) plankton export production and a shift of global $CaCO_3$ sedimentation from shelf to deep ocean basins are discussed as possibilities to the carbo cycle perturbation across the Eocene/Oligocene transition.

The total organic carbon data are another interesting point to be considered. In fact, in the studied section at Campos Basin, the coeval variation between the oxygen isotope data and the total organic carbon content, associated with organic matter predominantly derived from higher plants, suggest an increase of terrestrial organic matter transported to the sea during the sea-level fall. Probably, this process could have been responsible for the supply of nutrients (mainly bioavailable species of Nitrogen and phosphorous) enhancing bioproduction.

The hypothesis of micronutrients supply provided by glacial dust and debris is discussed by Salamy et al. (1999) for a Southern Ocean record. The authors identified a significant increase in opal production associated with higher terrigenous sediments accumulation during the Oi1 and Oi1a glaciation events, supporting the thesis of ocean fertilization. In addition, according to Villa et al. (2014) a turnover of oligotrophic to eutrophic conditions can be traced by nannofossil assemblage shift toward the Eocene-Oligocene boundary and may also be a consequence of ocean fertilization. This fertilization process could have been responsible to promote the bioproductivity increase not only in the lower Oligocene (Oi events), as also observed by Salamy et al. (1999), but also in the lower Miocene (Mi1 event) explaining this positive $\delta^{13}C$ isotopic event (Fig. 10).

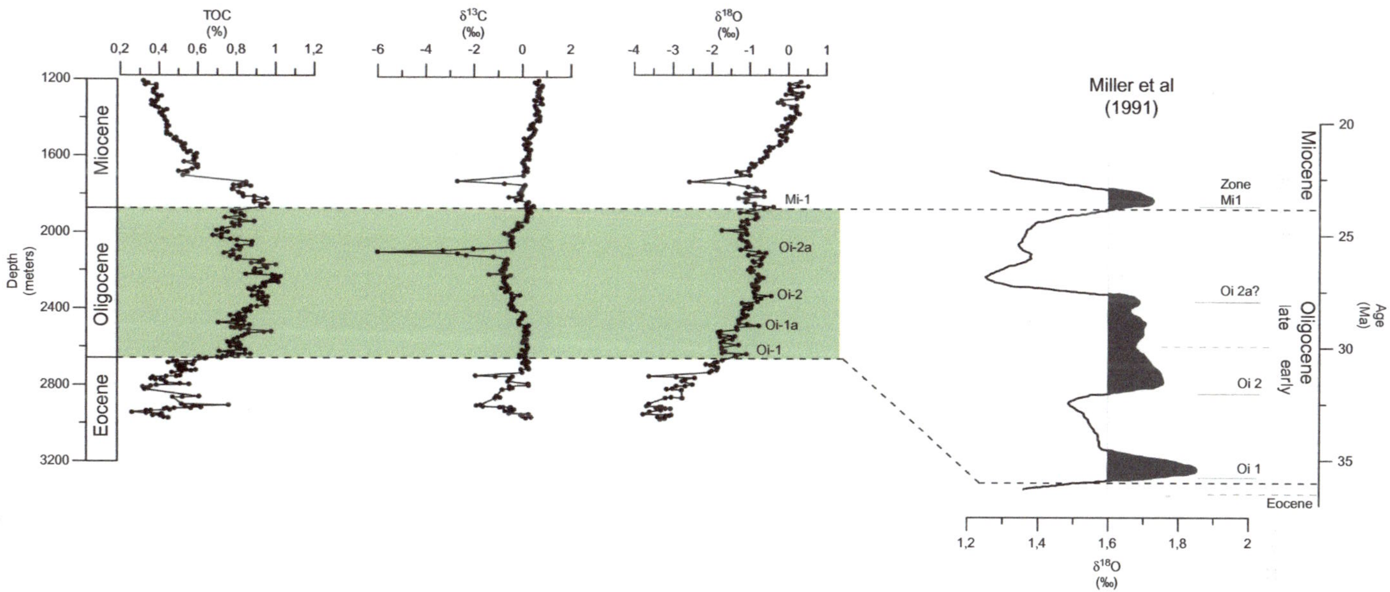

Fig. 10 Total Organic Carbon (TOC), Carbon and Oxygen stable isotopes for a Eocene-Miocene record of Campos Basin (Rodrigues, 2005). Note that, in green, higher values of TOC can be associated with Oxygen isotopic variations and could reflect increase in ocean fertilization. The main isotopic positive shifts were identified by the author and can be correlated with the data presented by Miller et al. (1991).

4.5 A recent example

High saline lagoons are potential analogs environments that can help to understand extreme ancient environments. The coastal lagoon Lagoa Vermelha (Fig. 11), situated in Rio de Janeiro State, is a good Brazilian example of modern hypersaline environment where biogeochemical conditions apparently indicate that salinity variation and bacterial activity regulates mineral precipitation (Vasconcelos and Mackenzie, 1997, Laut et al., 2017). This lagoon is located in the protected area of Massambaba, between the cities of Araruama and Saquarema (Brazil) and displays two quite uncommon particularities: it exhibits carbonate sedimentation and displays the development of Holocene stromatolites (Laut et al., 2017).

Here is presented geochemical and isotopic data from a sediment core from Lagoa Vermelha where, in stratigraphic terms, decomposed material of microbial mats, in addition to thin layers of sand and silt and levels where shells and mollusks can be found. The geochemical and sedimentological data showed in the high-resolution scale, two cycles of Organic Carbon and Sulfur enrichment in the bottom with a decreasing toward the top of each cycle. These cycles are highlighted by the darkest basal levels (black), associated with degraded material of microbial mats, followed ascendingly by light gray material, with millimetric interpolations of carbonate slides. Occasionally, these light gray ranges are succeeded by sandy and silica levels, culminating at the top by levels of shells of mollusks (Fig. 12).

Fig. 11 Photo of Lagoa Vermelha, a Coastal isolated shallow water lagoon located in Rio de Janeiro State (Brazil). *Photo source: https:/guialagos.com/lagoa-vermelha-regiao-dos-lagos/.*

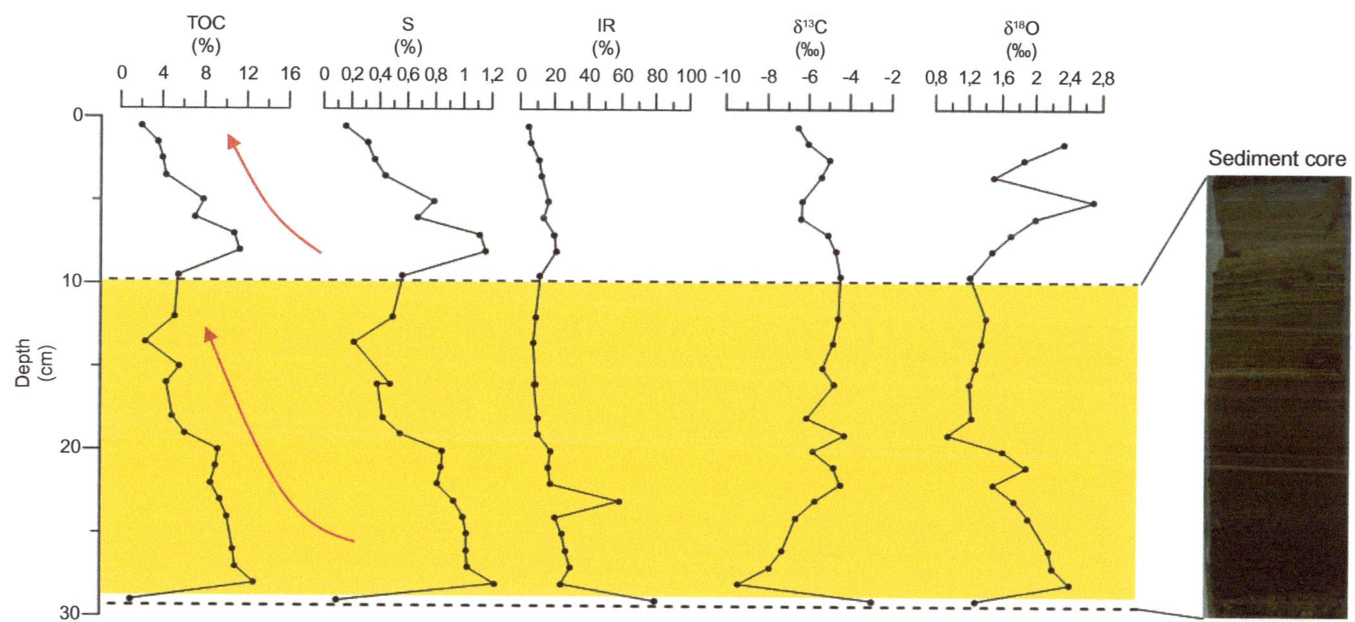

Fig. 12 Total Organic Carbon (TOC), Sulfur Content (S), Insoluble residue (IR) and inorganic Carbon and Oxygen stable isotopes for sediments from Lagoa Vermelha (RJ, Brazil). Interpretation of geochemical an isotopic data suggests more saline and anoxic condition in the bottom of the studied section (dark sediment layers) with a dilution with coarse terrestrial material toward the top (white sediment layers).

The extreme ^{12}C enrichment associated with higher values of TOC and sulfur (between 25 and 30 cm) in the lagoon have been associated with the development of anaerobic microbial communities of sulfate reducing bacteria (Vasconcelos et al., 1997). Although, as also observed by Vasconcelos et al. (1997) in other section of Lagoa Vermelha, carbon isotopic composition never reached ^{13}C depleted values to evoke only organic derived carbonate ions. In this sense, other carbon sources was possible contributing with the final $\delta^{13}C_{carbonate}$ composition. The hypothesis of the influence of Sulfur reducing bacteria affecting the carbon isotopic composition in Lagoa Vermelha is also supported by the identification of a halotolerant sulfate-reducing bacterium, *Desulfovibrio brasiliensis* (Warthmann et al., 2005)

5. Final considerations

The examples presented in this chapter demonstrate the application bulk-rock stable isotopes analyses in stratigraphic studies in Brazilian sections. The use of bulk-rocks has proved to be efficient and an excellent tool in the study of the extensive recordings of carbonaceous rocks of the Brazilian platform and, although diagenetic effects may be affecting the isotopic signals, in the Brazilian basins, the use of such technique allows to establish excellent interpretations mainly in terms of salinity and temperature variations, as already discussed by Azevedo et al. (1997).

For the isotopic study of organic matter, we emphasize that the integrated study using simpler techniques such as elemental analyses of carbon and sulfur, as well as Rock-Eval, may be interesting in the construction of paleoenvironmental interpretations, such as identifying intervals with better or worse preservation of matter organic or even an evaluation on thermal evolution.

Also, in the study of organic matter, biomarkers can provide a molecular view on their composition and thermal evolution. Further, from the observed variations observed paleoenvironmental changes can be deduced supporting the interpretation of the stable isotopes.

Acknowledgments

The authors are grateful for the support provided by Capes (Coordenação de Aperfeiçoamento de Pessoal de Nível Superior), CNPq (Conselho Nacional de Pesquisa), FINEP (Financiadora de Estudos e Projeto) and DNPM (Departamento Nacional de Produção Mineral) and ANP (Agência Nacional do Petróleo) by the access to samples. This study was financed in part by the Coordenação de Aperfeiçoamento de Pessoal de Nível Superior-Brasil (CAPES)- Finance Code 001.

References

Alferes, C.L.F., Rodrigues, R., Pereira, E., 2011. Geoquímica orgânica aplicada à Formação Irati, na área de São Mateus do Sul (PR), Brasil. Geochim. Bras. 25, 47–54.

Arai, M., 2014. Aptian/Albian (Early Cretaceous) paleogeography of the South Atlantic: a paleontological perspective. Braz. J. Geol. 44, 339–350.

Araújo, L.M., 2001. Análise da expressão estratigráfica dos parâmetros de geoquímica orgânica e inorgânica nas seqüências deposicionais Irati. PhD thesis, vol. 2. Instituto de Geociências, Universidade Federal do Rio Grande do Sul, p. 307.

Azevedo, R.L.M., Rodrigues, R., TAKAKI, T., 1997. Aplicação dos dados isotópicos de carbono e oxigênio, de rocha-total, em estudos paleoambiental. vol. 11. Boletim de Geociências da Petrobras, pp. 19–32.

Bastos, L.P.H., Pereira, E., Cavalcante, D.C., Rodrigues, R., 2014. Estratigrafia química aplicada à Formação Codó nos furos de sondagem UN-24-PI e UN-37-PI (Aptiano/Albiano da Bacia do Parnaíba). vol. 22. Boletim de Geociências da Petrobras, pp. 290–312.

Bergamaschi, S., 1999. Análise estratigráfica do Siluro–Devoniano (formações Furnas e Ponta Grossa) da Sub-bacia de Apucarana, Bacia do Paraná, Brasil. Unpublished PhD. thesis, Universidade de São Paulo. Instituto de Geociências, São Paulo, p. 167.

Boersma, A., Shackleton, N., Hall, M., Given, Q., 1979. Carbon and oxygen isotope records at DSDP site 384 (North Atlantic) and some Paleocene paleotemperatures and carbon isotope variations in the Atlantic Ocean. In: Initial Reports of the Deep Sea Drilling Project. vol. 34. U.S. Government Printing Office, Washington, DC, pp. 695–717.

Brandes, J.A., Devol, A.H., 2002. A global marine-fixed nitrogen isotopic budget: implications for Holocene nitrogen cycling. Global Biogeochem. Cycles 16 (4), 67–71.

Canfield, D.E., Glazer, A.N., Falkowski, P.G., 2010. The evolution and future of Earth's nitrogen cycle. Science 330 (6001), 192–196.

Carvalho, M.A., Bengtson, P., Lana, C.C., 2016. Late Aptian (cretaceous) paleoceanography of the South Atlantic Ocean inferred from dinocyst communities of the Sergipe Basin, Brazil. Paleoceanography 31, 2–26.

Craig, H., 1953. The geochemistry of the stable carbon isotopes. Geochim. Cosmochim. Acta 3 (2–3), 53–92.

Dean, W.E., Arthur, M.A., Claypool, G.E., 1986. Depletion of 13 C in cretaceous marine organic matter: source, diagenetic, or environmental signal? Mar. Geol. 70, 119–157.

Dias, A.E.S., 2004. Estratigrafia química e identificação de intervalos potencialmente geradores no devoniano inferior da Formação Ponta Grossa. 62 f. Monograph. (Faculdade de Geologia)—Universidade do Estado do Rio de Janeiro, p. 61.

Dias, A.E.S., Pereira, E., Rodrigues, R., 2003. Carbon and oxygen isotope data from Cenomanian—Turonian boundary in the Laranjeira area, Sergipe—Alagoas Basin, Brazil. In: Simposio Sul-Americano de Geologia Isotópica—IV SSAGI, 2003, Salvador-vol. I. pp. 337–340. Short Papers.

Emiliani, C., 1955. Pleistocene paleotemperatures. J. Geol. 63, 538–578.

Epstein, S., Buchsbaum, R., Lowenstam, H.A., Urey, H.C., 1953. Revised carbonate water isotopic temperature scale. Geol. Soc. Am. Bull 64, 1315–1326.

Espitalié, J., Laporte, J.L., Madec, M., Marquis, F., Leplat, P., Paulet, J., 1977. Méthcde rapide de caractérization dês roches mères, de leur potenciel pétrolier et de leur degree d'évolution. 32, Revue Institute Français de Pétrole, pp. 23–43.

Estrella, G.O., Gaglianone, P.C., Melo, M.R., 1984. The Espirito Santo Basin-Brazil: source rock characterization and petroleum habitat. In: Petroleum Geochemistry and Basin Evaluation. 35, AAPG Memoir, pp. 253–271.

Faure, K., Cole, D., 1999. Geochemical evidence for lacustrine microbial blooms in the vast Permian Main Karoo, Paraná, Falkland Islands and Huab basins of southwestern Gondwana. Palaeogeogr. Palaeoclimatol. Palaeoecol. 152, 189–213.

Freeman, K.H., Hayes, J.M., 1992. Fractionation of carbon isotopes by phytoplankton and estimates of ancient CO2 levels. Global Biogeochem. Cycles 6 (2), 185–198.

Gingerich, P.D., 2003. Mammalian responses to climate change at the Paleocene-Eocene boundary: Polecat Bench record in the northern Bighorn Basin, Wyoming. Geol. Soc. Am. Spec. Pap. 369, 463–478.

Grice, K., Schouten, S., Nissenbaum, A., Charrach, J., Damsté, J.S.S., 1998. Isotopically heavy carbon in the C21 to C25 regular isoprenoids in halite-rich deposits from the Sdom formation, Dead Sea Basin, Israel. Org. Geochem. 28 (6), 349–359.

Gruber, N., Galloway, J.N., 2008. An earth-system perspective of the global nitrogen cycle. Nature 451 (7176), 293.

Hayes, J.M., 1993. Factors controlling 13C contents of sedimentary organic compounds: principles and evidence. Mar. Geol. 113, 111–125.

Hsü, K.J., McKenzie, J.A., 1985. A "Strangelove" ocean in the earliest tertiary. In: Sundquist, E.T., Broecker, W.S. (Eds.), The Carbon Cycle and Atmospheric CO2: Natural Variations Archean to Present. In: vol. 32. American Geophysical Monographs, Washington, DC, pp. 487–492.

Jahren, A.H., Arens, N.C., Sarmiento, G., Guerrero, J., Amundson, R., 2001. Terrestrial record of methane hydrate dissociation in the early cretaceous. Geology 29, 159–162.

Jenkyns, H.C., Gale, A.S., Corfield, R.M., 1994. Carbon-and oxygen-isotope stratigraphy of the English Chalk and Italian Scaglia and its palaeoclimatic significance. Geol. Mag. 131 (1), 1–34.

Junium, C.K., Arthur, M.A., 2007. Nitrogen cycling during the Cretaceous, Cenomanian-Turonian oceanic anoxic event II. Geochem. Geophys. Geosyst. 8 (3).

Kennett, J.P., Stott, L.D., 1990. Proteeus and proto-oceanus: ancestral Paleogene oceans as revealed from Antarctic stable isotopic results; ODP Leg 113. In: Proceedings of the Ocean Drilling, Scientific Results. vol. 113, pp. 865–879.

Kuypers, M.M., van Breugel, Y., Schouten, S., Erba, E., Sinninghe Damsté, J.S., 2004. N2-fixing cyanobacteria supplied nutrient N for Cretaceous oceanic anoxic events. Geology 32, 853–856.

Laut, L., Martins, M.V.A., Frontalini, F., Ballalai, J.M., Belart, P., Habib, R., Fontana, L.F., Clemente, I.M.M.M., Lorini, M.L., Filho, J.G.M., Laut, V.M., Figueiredo, M.S.L., 2017. Assessment of the trophic state of a hypersaline-carbonatic environment: Vermelha Lagoon (Brazil). PLoS One 12 (9), e0184819.

Li, L., Lollar, B.S., Li, H., Wortmann, U.G., Lacrampe-Couloume, G., 2012. Ammonium stability and nitrogen isotope fractionations for NH4+-NH3(aq)-NH3(gas)systems at 20-70°C and pH of 2-13: applications to habitability and nitrogen cycling in low-temperature hydrothermal systems. Geochim. Cosmochim. Acta 84, 280–296.

Livsey, C.M., Babila, T.L., Robinson, M.M., Bralower, T.J., 2019. The planktonic foraminiferal response to the Paleocene-Eocene thermal maximum on the Atlantic coastal plain. Mar. Micropaleontol. 146, 39–50.

Méhay, S., Keller, C.E., Bernasconi, S.M., Weissert, H., Erba, E., Bottini, C., Hochuli, P.A., 2009. A volcanic CO2 pulse triggered the Cretaceous Oceanic Anoxic Event 1a and a biocalcification crisis. Geology 37 (9), 819–822.

Mello, M.R., Telnaes, N., Gaglianone, P.C., Chicarelli, M.I., Brassel, S., Maxwell, J.R., 1988. Organic geochemical characterization of depositional palaeoenvironments of sources rocks and oils in Brazilian marginal basins. Org. Geochem. 13, 31–45.

Merico, A., Tyrrell, T., Wilson, P.A., 2008. Eocene/Oligocene Ocean de-acidification linked to Antarctic glaciation by sea-level fall. Nature 452 (7190), 979.

Meyers, P.A., 1997. Organic geochemical proxies of paleoceanographic, paleolimnologic, and paleoclimatic processes. Org. Geochem. 27, 213–250.

Milani, E., Zalán, P., 1999. An outline of the geology and petroleum systems of the Paleozoic interior basins of South America. Episodes 22 (3), 199–205. Journal of International Geoscience.

Miller, K.G., Wright, J.D., Fairbanks, R.G., 1991. Unlocking the ice house: Oligocene-Miocene oxygen isotopes, eustasy, and margin erosion. J. Geophys. Res. Solid Earth 96 (B4), 6829–6848.

O'Leary, M.H., 1988. Carbon isotopes in photosynthesis. Bioscience 38, 328–336.

Pereira, E., Aguiar, A.P.O., Brasil, F.A.F., Blazutti, D., 2007. Estratigrafia de Alta Resolução e o registro icnofaciológico. In: Ismar de Souza Carvalho; Antonio Carlos Siqueira Fernandes, Icnologia. SBG, São Paulo, pp. 32–39.

Pereira, E., Rodrigues, R., Bergamaschi, S., Souza, M.S.P., 2010. Caracterização Químioestratigráfica do Devoniano Inferior da Bacia do Paraná. In: Memorias—Museu e Laboratorio Mineralogico e Geologico, Faculdade de Ciencias. vol. 14. Universidade do Porto, pp. 349–353.

Peters, K.E., Walters, C.C., Moldowan, J.M., 2005. The Biomarker Guide, second ed. Cambridge University Press.

Rau, G.H., Takahashi, T., Des Marais, D.J., Repeta, D.J., Martin, J.H., 1992. The relationship between d13C of organic matter and [CO2 (aq)] in ocean surface water: data from a JGOFS site in the northeast Atlantic Ocean and a model. Geochim. Cosmochim. Acta 56, 1413–1419.

Rodrigues, R., 2005. Chemostratigraphy. In: Koutsoukos, E.A.M. (Ed.), Applied Stratigraphy. Topics in Geobiology. In: vol. 23. Springer, New York, pp. 165–178.

Rodrigues, R., Oliveira, L.C.V., Carvalho, M.D., Kousmann, R.O., 1998. Anomalous carbon and oxygen signature in upper Maastrichtian Calcilutite, Campos Basin, Brazil. In: 5th International Conference on Gas in Marine Sediments, 1998, Bologna, 1998, pp. 146–147.

Rodrigues, R., Pereira, E., Bergamaschi, S., Alferes, C.L.F., 2010. Carbon isotope stratigraphy of organic matter: Irati formation, lower Permian of Paraná Basin. In: Proceedings of VII SSAGI–South American Symposium on Isotope Geology, pp. 522–525.

Rodrigues, R., 1995. A geoquímica Orgânica na Bacia do Parnaíba. Unpublished PhD thesis, Universidade Federal do Rio Grande do Sul, Porto Alegre.

Salamy, K.A., Zachos, J.C., Cruz, S., 1999. Latest Eocene-Early Oligocene climate change and Southern Ocean fertility: inferences from sediment accumulation and stable isotope data. Palaeogeogr. Palaeoclimatol. Palaeoecol. 145, 61–77.

Santos, R.V., Souza, P.A., de Alvarenga, C.J.S., Dantas, E.L., Pimentel, M.M., de Oliveira, C.G., de Araújo, L.M., 2006. Shrimp U-Pb zircon dating and palynology of bentonitic layers from the Permian Irati Formation, Paraná Basin, Brazil. Gondwana Res. 9, 456–463.

Schidlowski, M., Gorzawski, H., Dor, I., 1992. Experimental hypersaline ponds and model environments for stromatolite formation. 2. Isotopic biogeochemistry. In: Schidlowski, M., Golubic, S., Kimberley, M.M., McKirdy, D.M., Trudinger, P.A. (Eds.), Early Organic Evolution; Implications for Mineral and Energy Resources. Springer, Heidelberg, pp. 494–508.

Schmitz, B., Keller, G., Stenvall, O., 1992. Stable isotope and foraminiferal changes across the cretaceous–tertiary boundary at Stevns Klint, Denmark: arguments for long term oceanic instability before and after bolide-impact event. Palaeogeogr. Palaeoclimatol. Palaeoecol. 96, 233–260.

Scholle, P.A., Arthur, M.A., 1980. Carbon isotope fluctuations in Cretaceous pelagic limestones: potential stratigraphic and petroleum exploration tool. AAPG Bull. 64 (1), 67–87.

Schulte, P., et al., 2010. The Chicxulub asteroid impact and mass extinction at the cretaceous–Paleogene boundary. Science 327, 1214–1218.
Shackleton, N., 1967. Oxygen isotope analyses and Pleistocene temperatures re-assessed. Nature 215 (5096), 15.
Shackleton, N. J., Kennet, J. P., 1975. Paleotemperature history of the Cenozoic and the initiation of Antarctic glaciation: oxygen and carbon isotope analysis in DSDP 277, 279 and 281. Initial Reports of the Deep Sea Drilling Project 29, 743–755.
Sinninghe Damste, J.S., Kenig, F., Koopmans, M.P., Koster, J., Schouten, S., Hayes, J.M., de Leeuw, J.W., 1995. Evidence for gammacerane as an indicator of water column stratification. Geochim. Cosmochim. Acta 59, 1895–1900.
Stiller, M., Rounick, J.S., Shasha, S., 1985. Extreme carbon-isotope enrichments in evaporating brines. Nature 316 (6027), 434.
Svensen, H., Planke, S., Malthe-sørenssen, A., 2004. Release of methane from a volcanic basin as a mechanism for initial Eocene global warming. Nature 429, 3–6.
Takaki, T., Rodrigues, R., 1984. Isótopos estáveis do carbono e oxigênio dos calcários como indicadores paleoambientais—Bacias de Campos, Santos e Espírito Santo. In: Anais do XXXIII Congresso Brasileiro de Geologia, Rio de Janeiro, pp. 4750–4762.
Takaki, T., Rodrigues, R., 1993. Isotopic evidence of the Cenomanian–Turonian anoxic event in the Sergipe-Alagoas Basin, Brazil. In: Third Latin American Congress on Organic Geochemistry, extended abstract, pp. 141–143.
Thomas, E., 2007. Cenozoic mass extinctions in the deep sea: what perturbs the largest habitat on earth? In: Monechi, S., Coccioni, R., Rampino, M.R. (Eds.), Large Ecosystem Perturbations: Causes and Consequences. In: vol. 424. The Geological Society of America Special Paper, pp. 1–24.
Tyson, R.V., 2012. Sedimentary Organic Matter: Organic Facies and Palynofacies. Springer Science & Business Media.
Vasconcelos, C., McKenzie, J.A., 1997. Microbial mediation of modern dolomite precipitation and diagenesis under anoxic conditions (Lagoa Vermelha, Rio de Janeiro, Brazil). J. Sediment. Res. 67 (3), 378–390.
Villa, G., Fioroni, C., Persico, D., Roberts, A.P., Florindo, F., 2014. Middle Eocene to Late Oligocene Antarctic glaciation/deglaciation and Southern Ocean productivity. Paleoceanogr. Paleoclimatol. 29 (3), 223–237.
Wang, R.L., 1998. Acyclic isoprenoids—molecular indicators of bacterial activity in contemporary and ancient Chinese saline/hypersaline environments. Hydrobiologia 381, 59–76.
Warthmann, R., Vasconcelos, C., Sass, H., McKenzie, J.A., 2005. Desulfovibrio brasiliensis sp. nov., a moderate halophilic sulfate-reducing bacterium from Lagoa Vermelha (Brazil) mediating dolomite formation. Extremophiles 9 (3), 255–261.
Weissert, H., Joachimski, M., Sarnthein, M., 2008. Chemostratigraphy. Newsl. Stratigr. 42 (3), 145–179.

Further reading

Heimdal, T.H., Callegaro, S., Svensen, H.H., Jones, M.T., Pereira, E., Planke, S., 2019. Evidence for magma—evaporite interactions during the emplacement of the Central Atlantic Magmatic Province (CAMP) in Brazil. Earth Planet. Sci. Lett. 506, 476–492.

CHAPTER FIVE

Stable-isotope chemostratigraphy of Urgonian-type platform carbonates: Time to be cautious?

Anthony J.-B. Tendil[a,*], Camille Frau[b], Philippe Léonide[a], François Fournier[a], Jean R. Borgomano[a], Cyprien Lanteaume[a], Jean-Pierre Masse[a], Jean-Paul Rolando[c]

[a]UM 34 CEREGE CNRS (UMR 7630)—IRD (UMR 161), Aix-Marseille Université (Centre Saint-Charles), Marseille cedex 03, France
[b]Groupement d'Intérêt Paléontologique, Science et Exposition, Toulon, France
[c]Total S.A., Centre Scientifique and Technique Jean Féger, Pau, France
*Corresponding author: e-mail address: anthony.tendil@gmail.com

Contents

Stratigraphy & Timescales, Volume 4
ISSN 2468-5178
https://doi.org/10.1016/bs.sats.2019.08.001

Abstract

In recent years, $\delta^{13}C$ and $\delta^{18}O$ chemostratigraphy has become a powerful tool used to solve controversial results using different biological time markers from shallow-marine carbonate environments and establish long-distance correlations across proximal and distal settings. However, its reliability was recently questioned due to the impact of diagenesis on stable isotope records. The ammonite-calibrated, platform-to-basin transect of the late Barremian–early Aptian Urgonian Provence platform is made particularly relevant to challenge chemostratigraphy-based correlation methods in shallow carbonate series at different scales. The origin, preservation and stratigraphic potential of isotopic signatures are herein discussed along 11 shallow-marine and basinal sections. In the proximal settings of the Provence domain, tight carbonates—either resulting from muddy depositional textures or early cementation—likely preserve original marine signatures. Contrasting observations arise from porous carbonates whose stable isotope signal is heavily affected by micrite recrystallization (i.e., microporosity genesis) during a mid-Cretaceous tectonic uplift event. Locally, the C-isotope signal of microporous Urgonian limestones can be controlled by the vertical distribution of porosity. Although basinal series are more prone to preserve primary geochemical signatures, extreme values for carbon and oxygen isotopes are observed in correspondence with fine-grained bioclastic material input originating from the adjacent platform, following the latest Barremian demise of the rudist-dominated platform. This study suggests that stable isotope chemostratigraphy may not constitute a suitable tool for calibrating and correlating Urgonian shallow-marine series.

1. Introduction

The global Carbon (C) cycle can be reconstructed through the geological times using the variations of C-isotope ratios between the main carbon reservoirs, i.e., atmosphere, hydrosphere, lithosphere and biosphere (Falkowski et al., 2000). Time residence of paired organic matter (C_{org}) and carbonate carbon (C_{carb}) is estimated to last 100 ky in the atmosphere–ocean system (Berner, 2006; Kump and Arthur, 1999). This temporal resolution is relevant to track and constrain major environmental perturbations and biological crises at different scales, in deep and shallow-marine carbonates, and continental ones as well.

In the past decades, the worldwide massive deposition of organic-rich shales during the Cretaceous (=Oceanic Anoxic Events—Schlanger and Jenkyns, 1976) gained attraction among the geochemist community since OAEs are associated to major C-isotope excursions (Menegatti et al., 1998). Pioneering studies focused on the $\delta^{13}C$ signal of deep carbonates which are less affected by diagenesis (e.g., Jenkyns et al., 1994; Lini et al., 1992). OAE carbon signatures of deep carbonates expose a stepped negative, then positive, excursion in the isotope profile that might serve as a powerful correlative tool at regional and global scales (e.g., Menegatti et al., 1998; Scholle and Arthur, 1980). Such disturbances in the C-isotope profile can be traced toward slope successions, but there, absolute values may locally differ from the original marine signatures due to their location at the interface of two diagenetical domains (i.e., pelagic vs neritic; Weissert et al., 2008). The use of C-isotope signals has been finally generalized in platform settings (e.g., Amodio and Weissert, 2017; Buonocunto et al., 2002; D'Argenio et al., 2004; Grötsch, 1998; Immenhauser et al., 2003; Jenkyns, 1995; Jenkyns and Wilson, 1999; Parente et al., 2007; Vahrenkamp, 1996, 2010; Wissler et al., 2003; Yamamoto et al., 2013). According to these authors, C-isotope signals show comparable trends between deep and platform settings, notwithstanding that platform carbonates are more subject to complex diagenesis. C-isotope profiles were, therefore, tentatively used to establish and/or refine stratigraphic platform-to-basin correlations.

In recent years, few studies questioned the use of C-isotope chemostratigraphy in platform settings given that shallow-marine carbonates are prone to meteoric diagenesis which alters the C-isotope original signal (Brlek and Glumac, 2014; Fouke et al., 1996; Hamon and Merzeraud, 2007; Joachimski, 1994; Swart, 2015). Metastable mineralogies (i.e., aragonite and high-Mg calcite) and local paleogeographic–paleoclimatic conditions are among the main controlling factors in resetting C-isotope values. Furthermore, past carbonate sedimentary systems are intrinsically characterized by temporal discontinuities which result in the incompleteness of the C-isotope record (Strasser, 2015). Application of C-isotope chemostratigraphy, therefore, requires a fine understanding of platform-to-basin diagenesis to estimate degrees of preservation/alteration of primary isotopic signatures (Godet et al., 2016; Huck et al., 2017; Léonide et al., 2014; Swart and Oehlert, 2018). In line with the previous studies, the present contribution aims to investigate bulk isotopic signatures along a platform-to-basin transect of Barremian–Aptian age (i.e., Urgonian-type Provence platform) and to question the reliability of C-isotope chemostratigraphy in this time interval.

2. Generalities on the Barremian–Aptian transition and chemostratigraphic approach

The Barremian–Aptian period recorded several significant episodes of environmental changes reflected by massive deposition of organic-rich deposits in deep basins and major disturbances in the global C-isotope cycle (Jenkyns, 2010). The most significant episode occurs in the mid-early Aptian and was referred to as the OAE 1a (Arthur et al., 1990). It consists of a distinct negative shift (C3 isotope segment sensu Menegatti et al., 1998; A3–A4 *pro parte* sensu Wissler et al., 2003) followed by a positive excursion (C4–C7 isotope segments sensu Menegatti et al., 1998; Ap6–Ap7 sensu Herrle et al., 2004) and a return to pre-OAE carbon values (C8 or Ap8 isotope segments). The negative excursion is either related to the massive CO_2 degassing and/or methane hydrate destabilization during the build-up of the Ontong Java and Manihiki Plateaus Large Igneous Province (e.g., Hesselbo et al., 2000; Jenkyns and Wilson, 1999; Méhay et al., 2009; Menegatti et al., 1998; Naafs et al., 2016; Renard et al., 2005). Subsequent massive deposition and burial of organic-rich matter deposits would be responsible for the positive carbon excursion (Jenkyns, 1980; Scholle and Arthur, 1980). Signature of the OAE 1a is recorded worldwide in carbonates and in organic carbon of marine (Menegatti et al., 1998) and lacustrine (Zhang et al., 2016) origin.

Severe biotic feedbacks are reported during the Barremian–Aptian period from different depositional settings ranging from shallow environments to the open ocean. In shallow-marine environments, the demise of the Urgonian (rudist-dominated) platform ecosystem along the North and South Tethyan margins was so far considered as a progressive phenomenon starting at the Barremian–Aptian transition and culminating in the mid-early Aptian in concert with the OAE 1a (Föllmi, 2008, 2012). The Tethyan-wide switch from rudistid biota toward various environmental-stress adapted carbonate platform communities (i.e., orbitolinid-dominated deposits; *Lithocodium–Bacinella* microencrusters) is seen as an early harbinger of OAE 1a-related environmental perturbation (Amodio and Weissert, 2017; Bonvallet et al., 2019; Burla et al., 2008, 2009; Di Lucia et al., 2012; Föllmi and Godet, 2013; Graziano, 2013; Graziano and Raspini, 2017; Huck et al., 2010, 2011, 2012, 2013, 2014, 2017; Millán et al., 2009, 2011; Phelps et al., 2015; Pictet et al., 2015; Stein et al., 2012a,b; Vincent et al., 2010; Wissler et al., 2003; Yamamoto et al., 2013). The culmination of the OAE 1a in the mid-early Aptian would be concomitant to a widespread drowning phase in numerous platforms of the Tethys (Huck et al., 2011, 2014 and references therein). The above-mentioned authors assumed that C-isotope signals from Tethyan platforms can been physically correlated with specific chemostratigraphic segments of the OAE 1a-related $\delta^{13}C_{carb}$ curve. As such, high-resolution C-isotope stratigraphy has afforded a parallel framework to the controversial shallow-marine biozonations (i.e., rudists, dasycladales, orbitolinids and echinoids) and strengthened a scenario linking the OAE 1a-related events and platform demise (Huck and Heimhofer, 2015; Huck et al., 2011, 2014).

In southeast France, the final demise of the Urgonian-type rudistid platforms has been re-assigned to the latest Barremian by means of ammonite biostratigraphy (Frau et al., 2017, 2018a; Tendil et al., 2018). The temporal mismatch between the collapse of the Urgonian Provence platform ecosystem and the OAE 1a-related spreading and/or culmination stages, therefore, contrasts with the Huck and collaborators' scenario (op. cit.). This implies rethinking the chronology of events and underlying causes, and questions the reliability of the C-isotope chemostratigraphic approach between platform and basin.

3. Geological settings

During the Barremian–early Aptian period, the hemipelagic Vocontian Basin (Southeast France) was characterized by the deposition of rhythmic limestone-marl alternations (=Barremian–Bedoulian Limestones Fm.) along the northern margin of the Mediterranean Tethys. A gradual restriction of the Vocontian settings occurred through this time interval, in response to the main prograding phases of the surrounding Urgonian carbonate platforms of Provence, Bas-Vivarais and Subalpine margins (Fig. 1A) (Masse, 1993). In the late early Aptian, a regional deepening led to the drowning of the peri-Vocontian Urgonian platforms and the deposition of a marl-dominated regime extending up into the Albian (=Blue Marls Fm.). This formation records major OAEs and secondary related events (Bréhéret, 1997; Dauphin, 2002).

A revised chronostratigraphic framework for the Urgonian peri-Vocontian platforms was detailed by Frau et al. (2016, 2017, 2018a) and Tendil et al. (2018) through the identification of eight key ammonite bioevents. These bioevents provided solid time lines of regional extent and allowed the definition of depositional sequences through the Provence domain against the Standard Mediterranean Ammonite Scale (SMAS sensu

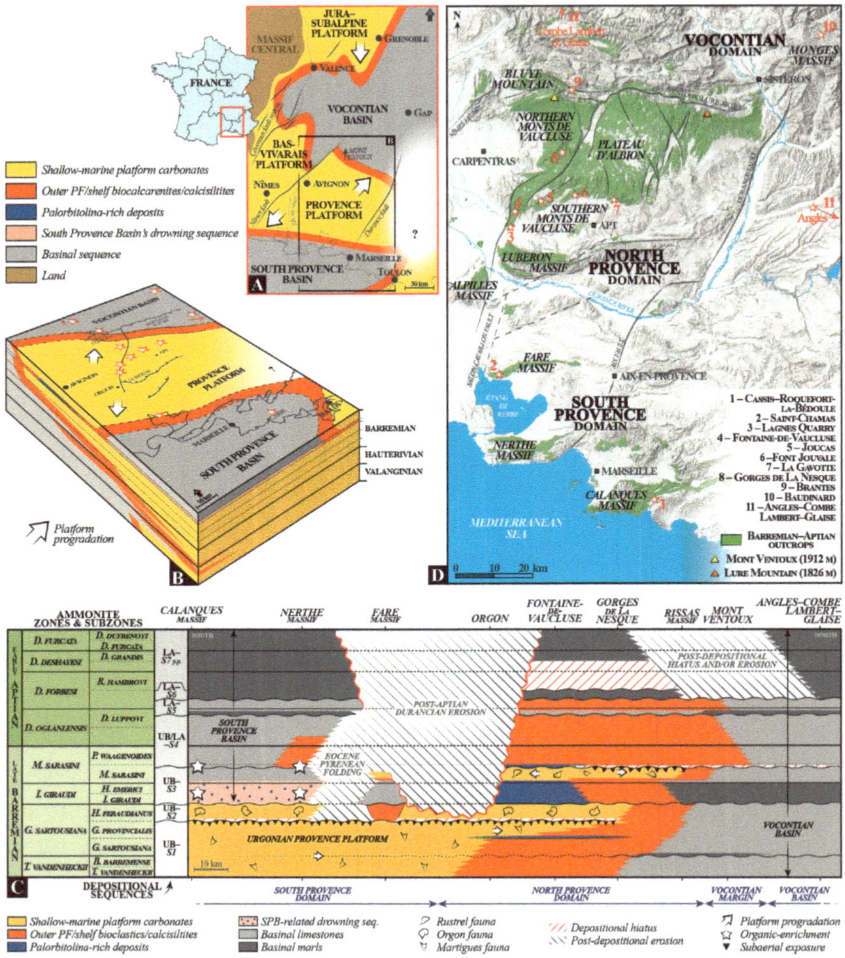

Fig. 1 (A) Latest Barremian paleogeography reconstruction of southeast France showing the joint Urgonian platforms enclosing the Vocontian Basin; (B) Stratigraphic cross section of the Urgonian Provence platform from Valanginian to Barremian times. The topmost paleogeographical map represents the Provence platform during the latest Barremian (lower *M. sarasini* Zone). (C) Lithostratigraphic framework of the upper Barremian–lower Aptian sedimentary succession across the Provence domain. Ammonite age-calibration is after Frau et al. (2018a). *SPB*, South Provence Basin. (D) Simplified geological map displaying the Barremian–Aptian deposits in Provence with locations of the studied sections. *Panel (A): Modified after Ferry, S., Grosheny, D., 2013, Excursion dans le bassin subalpin (20–23 September 2013). Turbidites et brèches carbonatées (Jurassique et Crétacé du bassin subalpin). Excursion du Groupe Français du Crétacé, Série "Excursion", 107.*

Reboulet et al., 2018). Following the installation of a rudistid, *Agriopleura*-bearing accumulation during the (?) early Barremian in southern Provence, the regional evolution of the Provence platform is depicted as follows (Fig. 1B and C):

- UB–*S1/S2* sequences: northward progradation of the rudistid platform (*Agriopleura* and requieniid–monopleurid episodes) toward the northern Provence (Monts de Vaucluse) during the early late Barremian (i.e., ? *T. vandenheckii* and *G. sartousiana* zones interval) in transition with oobioclastic–cherty limestones basinward;
- UB–*S3a* sequence: regional drowning phase during the *I. giraudi* Zone marked by mass occurrence of orbitolinid and echinoid communities in the Monts de Vaucluse and adjacent areas (*Palorbitolina–Heteraster* guide level sensu Masse and Fenerci-Masse, 2011) coeval with the deepening of the intra-shelf South Provencal Basin in the Martigues–Marseille region;
- UB–*S3b* sequence: recovery of a rudistid platform (caprinid-dominated) in northern Provence recording an overall NE–SW bidirectional progradation during the lower *M. sarasini* Zone (i.e., North Provence platform sensu Masse, 1976);
- UB/LA–*S4a* sequence: subsequent demise of caprinid biota which marks the termination of the Urgonian-type rudist regime in the latest Barremian at the *M. sarasini–P. waagenoides* subzonal boundary. At this time, a *Palorbitolina*-rich cover locally developed in the North Provence domain;
- UB/LA–*S4b* and LA–*S5* sequences: deposition of oobioclastic–cherty limestones in the North Provence platform during the latest Barremian to earliest Aptian (upper *M. sarasini* to lower *D. forbesi* zones interval) coeval with calcareous-argillaceous sedimentation in both the South Provence and Vocontian basins;
- LA–*S6/7* sequences: regional deepening expressed by a marly regime through the late early Aptian (upper *D. forbesi* Zone) to the Albian.

It is noteworthy that the reconstruction of the South and North Provence transition is hampered by post-depositional tectonic events (Durancian uplift, Pyrenean and Alpine orogens, Oligocene extension; e.g., Bestani et al., 2016; Masse and Philip, 1976) and related erosion stages which affected Urgonian massifs in the Bouches-du-Rhône department (Fig. 1C).

4. Materials and methods

Recent works by Frau et al. (2017, 2018a) and Tendil et al. (2018) on the Urgonian-type Provence platform provided a platform-to-basin (~180 km), ammonite-based stratigraphic model used in the present work. Direct correlations of Urgonian platform series to Barremian unit-stratotype of Angles (Vocontian Basin) and the early Aptian (former "Bédoulien") unit-stratotype series of Cassis–Roquefort-la-Bédoule (South Provence Basin) strengthen the stratigraphic value of such transect.

A total of 11 stratigraphic sections has been considered for investigating the geochemical signal along the Provence domain (Fig. 1D). Seven stratigraphic sections, including four original localities, have been sampled during the course of this work. These are the Lagnes Quarry core, Joucas, Font Jouvale, Gavotte core, Gorges de La Nesque, Brantes and Baudinard sections. Additional dataset comes from previous published works. This includes, from south to north, the Roquefort-la-Bédoule (Kuhnt et al., 2000), Saint-Chamas (Embry, 2005), Fontaine-de-Vaucluse (Bastide, 2014; Léonide et al., 2014), and the composite Angles–Combe Lambert–Glaise sections (Bodin et al., 2009; Godet, 2006; Westermann et al., 2013). Note that the decimeter-thick Cassis and Martigues sections (Léonide et al., 2014) together with the LSBB (Low-Noise Underground Laboratory) core section (Cochard, 2018) are cited in the discussion part only.

Original oxygen and carbon stable isotopes bulk analyzes have been carried out on 1570 rock samples at the GeoZentrum Nordbayern (Friedrich-Alexander-Universität Erlangen-Nürnberg, Germany). Special attention was given to avoid large sparitic cements filling voids or recrystallized shells. Sampling space was tightened below and above discontinuities when possible. Carbonate powders were reacted with 100% phosphoric acid at 70 °C using a Gasbench II connected to a ThermoFisher Delta V Plus mass spectrometer. All values are reported in per mil relative to V-PDB. Values were normalized with the reference material NBS 19 standards. Values are reported in ‰ PDB (Vienna Pee Dee Belemnite). Reproducibility was checked by replicate analysis of laboratory standards and was $\pm 0.06‰$ (1σ) for oxygen isotopes and $\pm 0.05‰$ (1σ) for carbon isotopes. Values are reported in Supplementary Table S1 in the online version at https://doi.org/10.1016/bs.sats.2019.08.001.

A nonexhaustive compilation of $\delta^{13}C$ and $\delta^{18}O$ published values was made to constrain the global marine carbonate composition of late Barremian–early Aptian interval. Based on this dataset, marine values ranging from $-4.25‰$ to $-1‰$ for the $\delta^{18}O$ and from 1‰ to 3‰ for the $\delta^{13}C$ are used as a referential for the marine calcite of the studied interval (Fig. 2).

Porosity measurements of the Lagnes Quarry and Font Jouvale sections were performed on 1 × 1, 1 × ¾ or 1 × ½ in. plug. Dry bulk density was calculated from the dry mass and measured cylinder volume. Grain densities were

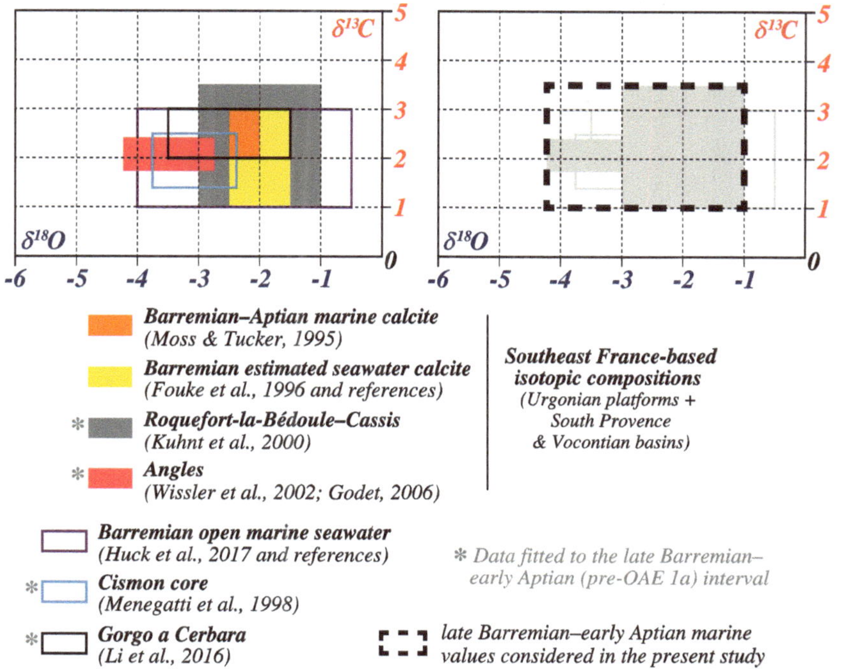

Fig. 2 Compilation of published stable isotopes values used as reference marine water compositions for the late Barremian–early Aptian time interval. The dashed black rectangle indicates the marine values considered in the present work.

measured using a PoroPerm Prod AP-125-004-0 VNCI TECHNOLOGIE. Total porosity was calculated from dry and grain density. Porosity values are compiled in Supplementary Table S2 in the online version at https://doi.org/10.1016/bs.sats.2019.08.001.

5. Geochemical and petrophysical results

5.1 Cassis–Roquefort-la-Bédoule (Fig. 3)

South to Marseille, the so-called Calanques Massif houses the best-developed, Urgonian innermost platform series of southern Provence, cropping out along km-wide cliffs. This Urgonian succession dips east toward the Cassis–Roquefort-la-Bédoule area (~16 km east from Marseille), which corresponds to the depot-center of the South Provence Basin. The latter being known to hosts the early Aptian (former "Bédoulien") unit-stratotype series. The studied part of the section exposes a basinal record of the UB–*S3* to

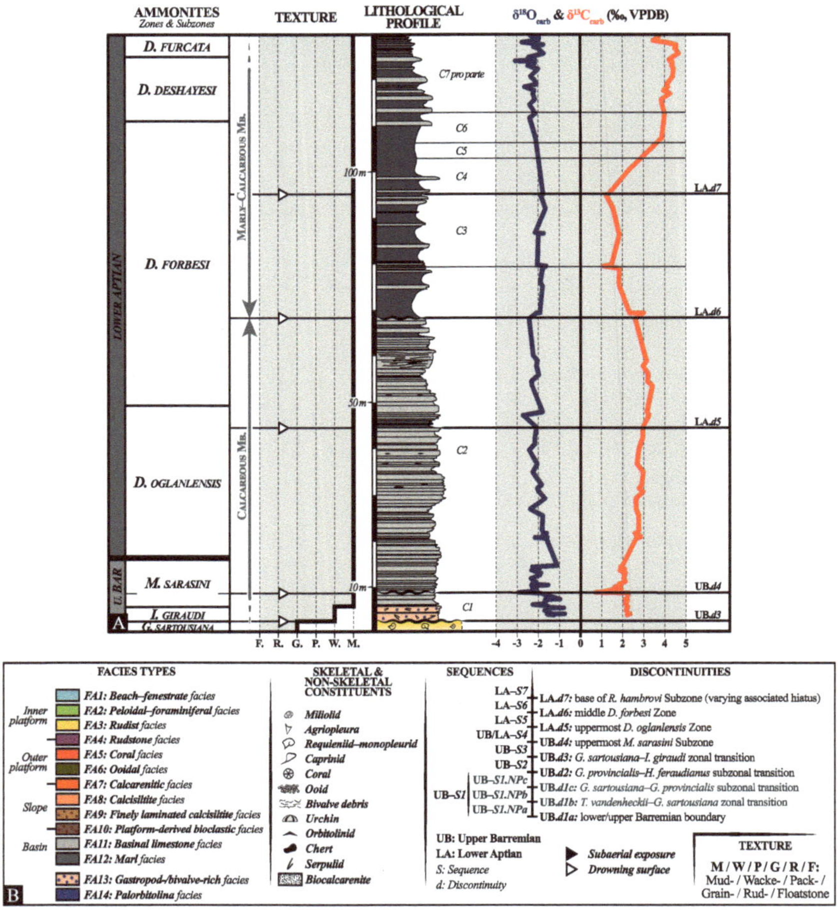

Fig. 3 (A) Litho-, biostratigraphy and bulk-rock stable isotope analyzes of the **Cassis–Roquefort-la-Bédoule** field section. C1–C7 relate to the isotopic segments defined by Menegatti et al. (1998). (B) Facies types, biological–sedimentological symbols and stratigraphic nomenclature used throughout the present contribution. *Panel (A): data from Frau, C., Tendil, A.J.-B., Lanteaume, C., Masse, J.-P., Pictet, A., Bulot, L.G., Luber, T., Redfern, J., Borgomano, J., Léonide, P., Fournier, F., Massonnat, G., 2018a. Late Barremian–early Aptian ammonite bioevents from the Urgonian-type series of Provence (Southeast France): regional stratigraphic correlations and implications for the peri-Vocontian carbonate platforms. Cretaceous Res. 90, 222–253; Frau, C., Bulot, L.G., Delanoy, G., Moreno-Bedmar, J.A., Masse, J.-P., Lanteaume, C., Tendil, A.J.-B.T., 2018b. The candidate Aptian GSSP at Gorgo a Cerbara (Central Italy): an alternative interpretation of the bio-, litho- and chemostratigraphic markers. Newsl. Stratigr. 51, 311–326. Panel (B): data from Tendil, A.J.-B., Frau, C., Léonide, P., Fournier, F., Borgomano, J.R., Lanteaume, C., Masse, J.-P., Massonnat, G., Rolando, J.-P., 2018. Platform-to-basin anatomy of a Barremian–Aptian Tethyan carbonate system: new insights into the regional to global factors controlling the stratigraphic architecture of the Urgonian Provence platform (Southeast France). Cretaceous Res. 91, 382–411.*

LA–S7 sequences. Detailed litho-, biostratigraphy and sedimentology can be found in Tendil et al. (2018, supplementary material, p. 3).

5.1.1 Stable isotope data

Kuhnt et al. (2000) examined the oxygen and carbon isotopic composition (120 bulk samples) of the Cassis–Roquefort-la-Bédoule basinal succession. $\delta^{18}O$ values vary between −2.94‰ and −0.75‰ while $\delta^{13}C$ values range between +0.70‰ and 4.66‰. According to Kuhnt et al. (2000), the entire succession suffered from low diagenetic overprint. Both oxygen and carbon isotope signals match the marine values (Fig. 4A) and could, therefore, be

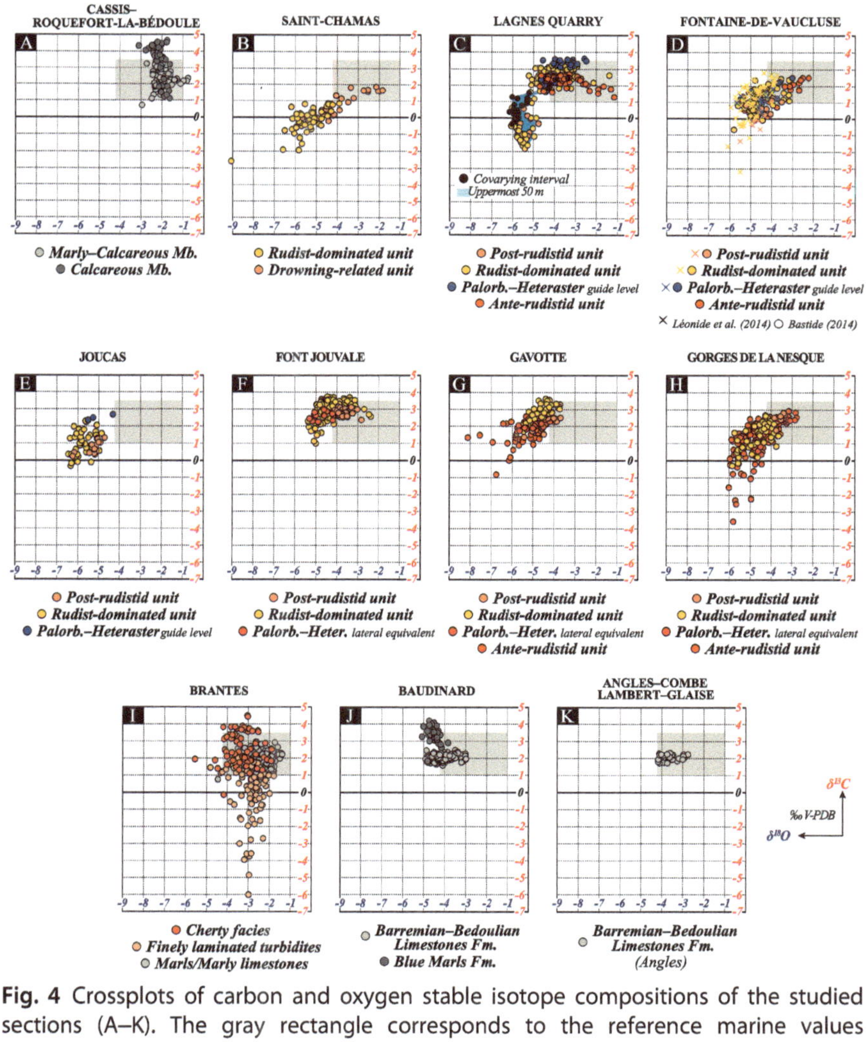

Fig. 4 Crossplots of carbon and oxygen stable isotope compositions of the studied sections (A–K). The gray rectangle corresponds to the reference marine values defined in Fig. 2.

considered as preserved. Frau et al. (2018b) correlated the trend of the Cassis–Roquefort-la-Bédoule carbon isotope curve to the successive segments (i.e., C1–C7) of the OAE 1a-related carbon cycle defined by Menegatti et al. (1998).

The latest Barremian beds are represented by C-isotope values around +2.25‰ while the corresponding oxygen-isotope signal scatters from −3‰ to −0.70‰ (C1 segment *pro parte*). Within this succession, a marked negative excursion of both $\delta^{18}O$ (−2.94‰) and the $\delta^{13}C$ (+0.70‰) signatures is observed in Bed 45. The early Aptian limestone-dominated sequence (Calcareous Mb.) records a slight $\delta^{13}C$ positive trend up to +3.42‰ in Bed 113 (Segment C2), then followed by decreasing values reaching a level of +3‰ in Bed 129. More negative values (around +1‰) are then observed at the base of the overlying marl-dominated sequence (Marly–Calcareous Mb.). This excursion correlates to the C3 segment and corresponds to the OAE 1a-related negative excursion. Above, a prominent $\delta^{13}C$ increase (C4 Segment) marks the onset of a positive trend (up to +4.66‰; C6 and C7 *pro parte* segments), only interrupted by a transient negative excursion (C5 segment). An overall opposite evolution of the $\delta^{18}O$ curve can be identified throughout the rest of the succession, with maximum values (~−1.70‰) reached during the latest early Aptian.

5.2 Saint-Chamas (Fig. 5)

The Saint-Chamas field section is located in the Fare Massif which forms a wide anticline running east to west from Éguilles to Saint-Chamas (Bouches-du-Rhône). We thereafter focus on the UB–*S3* Sequence which therein represents the southward termination of the caprinid-bearing Urgonian platform. Detailed litho-, biostratigraphy and sedimentology can be found in Tendil et al. (2018, supplementary material, p. 10).

5.2.1 Stable isotope data

Embry (2005) provided stable isotopes analyzes from 77 samples of the Saint-Chamas section. The oxygen-isotope signal oscillates between −9.17‰ and −1.86‰ while the carbon-isotope values vary between −3.78‰ and +1.87‰. A striking feature of the Saint-Chamas section relates to the covariant changes in carbon and oxygen isotope composition. Together with the overall offset of $\delta^{18}O$ and $\delta^{13}C$ signatures compared to marine values (Fig. 4B), a strong diagenetic resetting of primary isotopic signatures can be suggested throughout the entire Saint-Chamas succession.

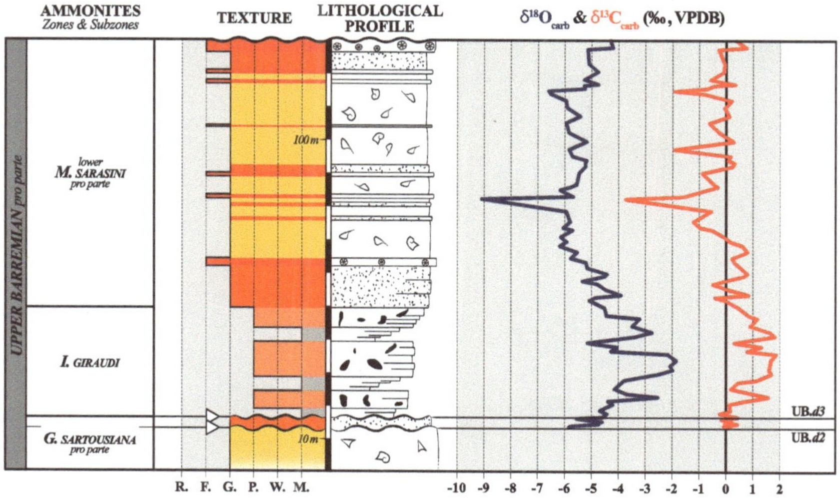

Fig. 5 Litho-, biostratigraphy and bulk-rock stable isotope analyzes of the **Saint-Chamas** field section. See Fig. 3B for explanation of symbols, color legend and stratigraphic nomenclature. *Data from Embry, J.C., 2005. Paléoécologie et architecture stratigraphique en haute résolution des plates-formes carbonatées du Barrémien–Aptien de la Néo-Téthys (Espagne, Suisse, Provence–Vercors). Impact respectif des différents facteurs de contrôle. Doctoral thesis, Institut Français du Pétrole, 298 pp.*

5.3 Lagnes Quarry (Fig. 6)

The Lagnes Quarry borehole section was drilled in the nominative quarry, located ~2 km south to Lagnes (Vaucluse) in the southwesternmost part of the Monts de Vaucluse. The studied part of the section exposes an inner platform record of the UB–*S1* to UB/LA–*S4* sequences *pro parte*. Detailed litho-, biostratigraphy and sedimentology can be found in Tendil et al. (2018, supplementary material, p. 20).

5.3.1 Stable isotope data

272 bulk stable isotopic measurements were performed throughout the 246.50 m-thick core. Isotope values vary between −1.83 and +3.61‰ in $\delta^{13}C$, and −6.16 and −1.14‰ in $\delta^{18}O$, which differ from the original marine signal (Fig. 4C).

Basal calcarenites of the UB–*S1* Sequence are marked by relatively constant $\delta^{13}C$ values (~2.50‰) while their $\delta^{18}O$ signatures oscillate between −4.35 and −3.30‰. The transition to the overlying serpulid-dominated facies is characterized by opposite isotopic trends, i.e., positive shifts for the $\delta^{18}O$ (amplitude of 1.30‰) and negative shifts for the $\delta^{13}C$

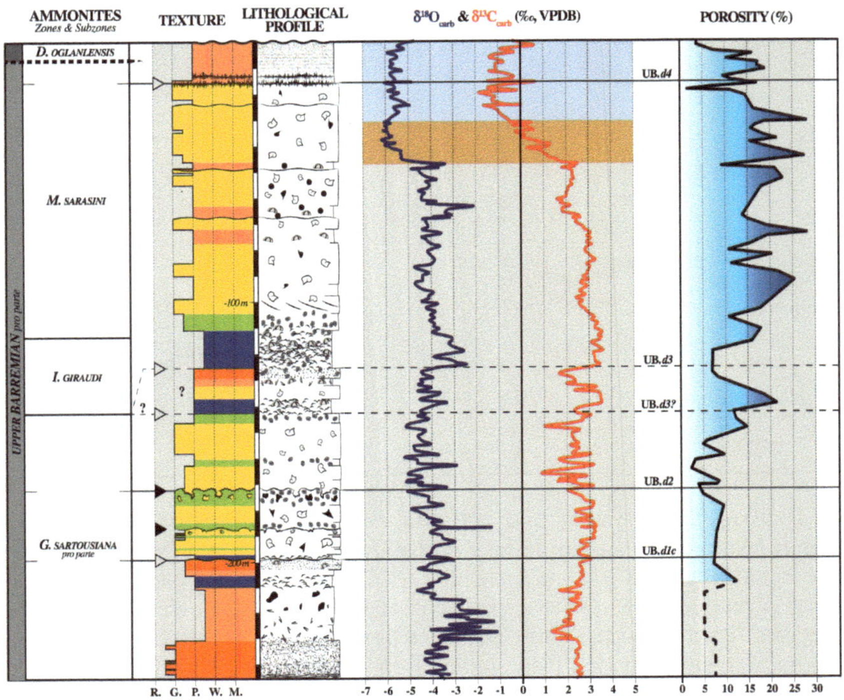

Fig. 6 Litho-, biostratigraphy, bulk-rock stable isotope analyzes and porosity evolution of the **Lagnes Quarry core** section. The highlighted orange area indicates the covarying $\delta^{18}O$ and $\delta^{13}C$ signatures (see discussion Section 6.1). See Fig. 3B for explanation of symbols, color legend and stratigraphic nomenclature.

(amplitude of 0.60‰). An overall positive trend of the $\delta^{13}C$ signal (up to 3.20‰) is then observed up to the *Agriopleura*-bearing unit. In this interval, oxygen-isotope values scatter from −4.70 to −1.15‰. The erosive discontinuity at the base of the second orbitolinid-rich level (−199.20 m;= UB.*d1c*) records positive shifts for both $\delta^{18}O$ (amplitude of ~1‰) and $\delta^{13}C$ (amplitude of ~0.50‰) values. Above a discrete emersive surface, the foraminiferal–peloidal interval located at −187 m is characterized by a low-amplitude (~1‰) negative shift of the C-isotope values while the $\delta^{18}O$ signal exhibits a large positive spike up to −1.37‰. The top of the *Agriopleura*-bearing unit, represented by an erosive and emersive surface (= UB.*d2*; see also Fig. 8D in Tendil et al., 2018), is marked by negative shifts for both $\delta^{18}O$ (amplitude of 1.50‰) and $\delta^{13}C$ (amplitude of 1.10‰). In the overlying requieniid–monopleurid unit (UB–*S2* Sequence), the C-isotope values vary between +1 and +2‰ while their oxygen counterparts fluctuate between −4 and −5‰. Slightly more positive values of the

$\delta^{13}C$ (~3.5‰) and the $\delta^{18}O$ (~−3‰) represent the *Palorbitolina–Heteraster* guide level (−126 to −112 m; UB–*S3a* Sequence). The remainder of the Urgonian succession (UB–*S3b* and UB/LA–*S4 pro parte* sequences) is defined by an overall negative trend of the $\delta^{13}C$ signatures. This is notably expressed in the uppermost 50 m where the values decrease from +2.50 to values near −1.80‰. A comparative trend in also observed in the $\delta^{18}O$ signal, except the presence of distinct positive shifts recorded in fine-grained, coral- and echinoid-rich intervals. Oxygen-depleted values (between −6‰ and −5‰) are found in the topmost part of the succession. These correspond to the most depleted isotopic signatures of the Lagnes Quarry section compared to the original marine values.

5.3.2 Porosity data

Porosity of the Lagnes Quarry borehole section was measured from 67 samples. Porosity values range from 1.62% to 28.09%. The lower part of the succession (up to −150 m) is characterized by values less than 10%. A gradual increase of the porosity is then observed from the topmost part of the requieniid–monopleurid unit (UB–*S2* Sequence). Maximum values are observed in the caprinid-bearing unit (UB–*S3b* Sequence) within which porosity often exceeds 15%. Microporosity is the dominant pore type at Lagnes and can be associated to vuggy/moldic porosity.

5.4 Fontaine-de-Vaucluse (Fig. 7)

The study area is located in the southwestern part of the Monts de Vaucluse along the 7 km-long cliff, extending between Valescure and Lagnes (Vaucluse). The section corresponds to the reference inner platform succession for the northern Urgonian Provence platform and exposes the UB–*S1* to UB/LA–*S4* sequences *pro parte*. Detailed litho-, biostratigraphy and sedimentology can be found in Tendil et al. (2018, supplementary material, p. 17).

5.4.1 Stable isotope data

Oxygen and carbon stable isotopes were initially measured by Léonide et al. (2014) throughout the Fontaine-de-Vaucluse succession from 79 whole rock samples. Isotopic compositions vary between −6.09‰ and −3.73‰ for the $\delta^{18}O$ and between −3.18‰ and +2.78‰ for the $\delta^{13}C$. A parallel work by Bastide (2014) later provided 112 stable isotopes measurements. The corresponding oxygen- and carbon-isotope values scatter, respectively, from −5.8‰ to −2.22‰ and from −0.67‰ to +2.52‰. Léonide et al. (2014) and Bastide (2014) $\delta^{18}O$ and $\delta^{13}C$ curves show comparable values

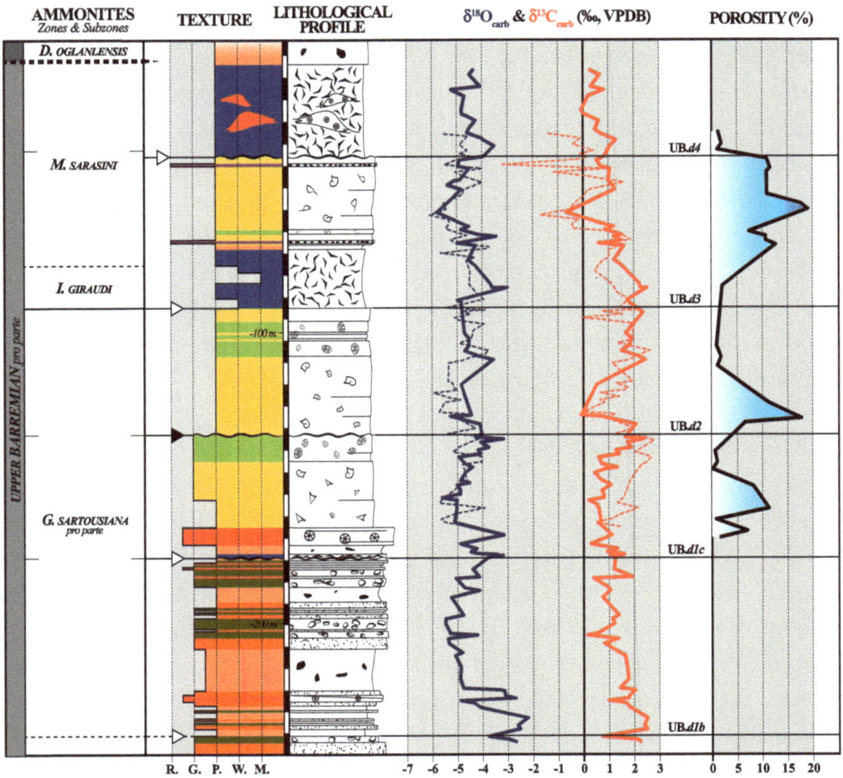

Fig. 7 Litho-, biostratigraphy, bulk-rock stable isotope analyzes and porosity evolution of the **Fontaine-de-Vaucluse** field section. Full lines correspond to the isotopic data from Bastide (2014) while dashed lines correspond to those from Léonide et al. (2014). See Fig. 3B for explanation of symbols, color legend and stratigraphic nomenclature. *Data from Bastide, F., 2014. Synthèse de l'évolution de la plateforme Urgonienne (Barrémien tardif à aptien précoce) du Sud-Est de la France: Faciès, micropaléontologie, géochimie, géométries, paléotectonique et géomodélisation. Doctoral thesis, Université de Grenoble, 300 pp and Léonide, P., Fournier, F., Reijmer, J.J.G., Vonhof, H.B., Borgomano, J., Dijk, J., Rosenthal, M., van Goethem, M., Cochard, J., Meulenaars, K., 2014. Diagenetic patterns and pore space distribution along a platform to outer-shelf transect (Urgonian limestone, Barremian–Aptian, SE France). Sediment. Geol. 306, 1–23.*

and isotopic evolution despite some local variations as discussed below. Only sporadic few oxygen and carbon isotopic compositions (i.e., from certain oobioclastic and rudist beds) conform to the reference marine values (Fig. 4D).

In the upper part of the oobioclastic-dominated unit (UB–*S1* Sequence), an overall decreasing trend of both $\delta^{18}O$ (from $\sim -2.5‰$ to $\sim -5‰$) and

$\delta^{13}C$ (from ~+2.5‰ to ~+1‰) is identified. Marked negative and positive spikes are also recognized in this interval. Above the UB.*d1c* drowning discontinuity, slightly more positive oxygen- and carbon-isotope values are observed (amplitude of ~0.5‰). The overlying *Agriopleura*-bearing unit (topmost part of the UB–*S1* Sequence) is represented by a gradual positive trend of the stable isotopes compositions, respectively from ~−5‰ to ~−3‰ for the $\delta^{18}O$ and from ~+0.5‰ to ~+2.5‰ for the $\delta^{13}C$. Negative shifts of the $\delta^{18}O$ (amplitude of ~1‰) and $\delta^{13}C$ (amplitude of ~1‰) signatures are detected below the UB.*d2* emersive discontinuity. The following requieniid–monopleurid-dominated unit (UB–*S2* Sequence) is characterized at its base by a decreasing trend of the stable isotopes as seen in the $\delta^{13}C$ whose compositions reach ~0‰. Increasing carbon-isotope values (up to ~+2.5‰) are then observed up to the middle part of the *Palorbitolina–Heteraster* guide level (UB–*S3a* Sequence *pro parte*). From the middle part of the *Palorbitolina–Heteraster* guide level up to the last *Palorbitolina*-rich event (UB/LA–*S4a* Sequence), the oxygen- and carbon-isotope profiles show a decreasing trend, respectively from −3‰ to −4.5‰ and from ~+2.5‰ to ~0.5‰. In the caprinid-bearing unit (UB–*S3b* Sequence), depleted carbon (around −1‰) and oxygen (~−6‰) values are identified within a 20m-thick interval. The strongly depleted carbon values (up to −3‰) observed in the topmost part of the caprinid-bearing unit by Léonide et al. (2014) fall in an interval rich in biogenic components (i.e., coral–rudist rudstone facies). Reexamination of the samples from this level indicates that diagenetically altered debris have been analyzed and could explain the negative $\delta^{13}C$ excursions. This is not the case in the $\delta^{13}C$ profile of Bastide (2014) in which carbon-isotope signatures exhibit positive values.

5.4.2 Porosity data

The vertical distribution of porosity at Fontaine-de-Vaucluse was published by Léonide et al. (2014). The dataset consists of 27 measurements which display porosity values ranging from 0.12% to 19.14%. Porous intervals (porosity values >10%) are only documented in the rudist-dominated units (UB–*S1 pro parte*, UB–*S2* and UB–*S3b* sequences). Cemented carbonates (porosity values <5%) can either be found in rudistid (i.e., below UB.*d2* and UB.*d3* discontinuities) or *Palorbitolina*-rich units (i.e., UB–*S3* and UB/LA–*S4* sequences). Microporosity appears as the dominant pore type at Fontaine-de-Vaucluse.

5.5 Joucas (Fig. 8)

The Joucas field section is located in the Véroncle Combe, ~2.7 km west-northwest to Joucas (Vaucluse), in the south-central part of the Monts de Vaucluse. This inner platform section spans the uppermost UB–*S2* to UB/LA–*S4* sequences. Detailed litho-, biostratigraphy and sedimentology can be found in Tendil et al. (2018, supplementary material, p. 23).

5.5.1 Stable isotope data

A carbon and oxygen-isotope analysis of 66 outcrop samples was performed at Joucas section. The $\delta^{18}O$ ranges between −6.45‰ and −4.32‰ while the $\delta^{13}C$ varies between −0.33‰ and +2.67‰. None of these isotopic compositions plot within the data field of Barremian–Aptian marine values (Fig. 4E).

The topmost part of the requieniid–monopleurid unit (UB–*S2* Sequence) exhibits a negative trend of both $\delta^{18}O$ (minimum value −6.06‰) and $\delta^{13}C$ (minimum value of +1.37‰) signatures. A marked positive shift of the stable isotopes is then identified in the overlying, meter-thick *Palorbitolina–Heteraster* guide level (UB–*S3a* Sequence). There, oxygen values reach up −4.32‰ and carbon values are around +2.5‰. Above, the caprinid-bearing unit (UB–*S3b* Sequence) is characterized by $\delta^{18}O$ signatures scattering from −4.8‰ to

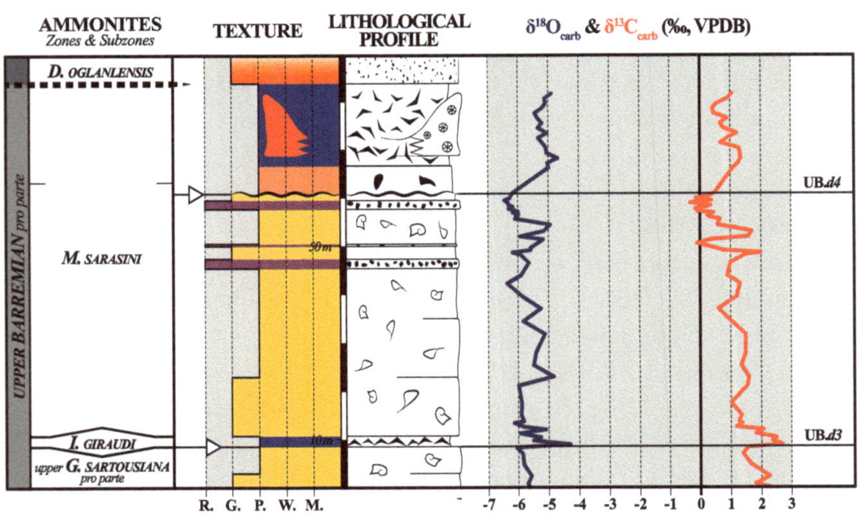

Fig. 8 Litho-, biostratigraphy and bulk-rock stable isotope analyzes of the **Joucas** field section (lithostratigraphic data from Léonide et al., 2012). See Fig. 3B for explanation of symbols, color legend and stratigraphic nomenclature.

$-6.5‰$ and an overall negative trend of the $\delta^{13}C$. In this interval, negative carbon values (up to $-0.33‰$) correlate to coral–rudist rudstone beds, as observed below the UB.*d4* drowning discontinuity. The transition to *Palorbitolina*-rich and bioclastic limestones (UB/LA–*S4* Sequence) is represented by more positive signatures with $\delta^{18}O$ values near $-5‰$ and $\delta^{13}C$ values around $+1‰$.

5.6 Font Jouvale (Fig. 9)

The Font Jouvale field succession is located southeast to the hamlet of Font Jouvale (Vaucluse) in south-central part of the Monts de Vaucluse. It includes the uppermost UB–*S1* to UB/LA–*S4* sequences of inner platform settings. Detailed litho-, biostratigraphy and sedimentology can be found in Tendil et al. (2018, supplementary material, p. 32).

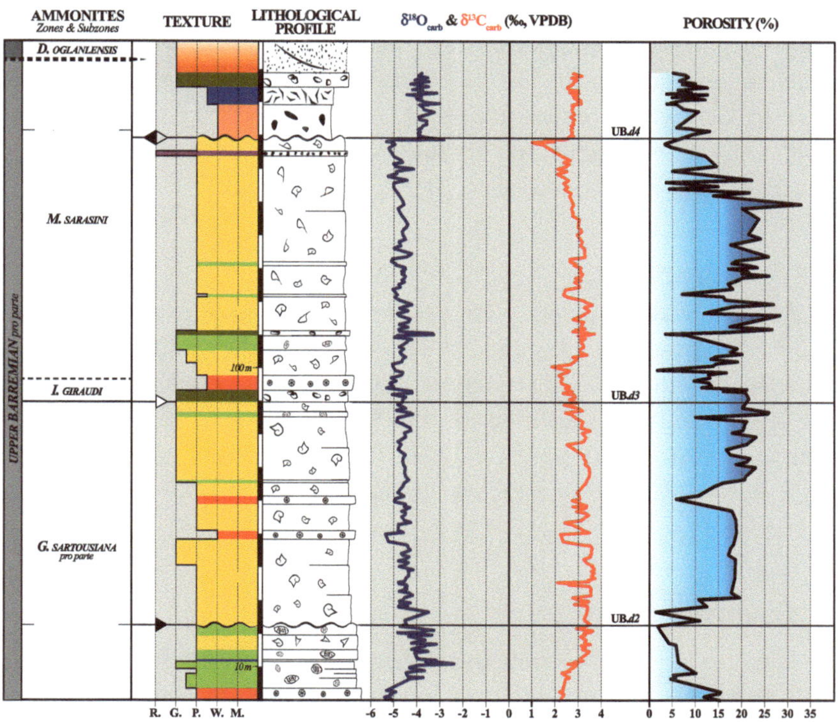

Fig. 9 Litho-, biostratigraphy, bulk-rock stable isotope analyzes and porosity evolution of the **Font Jouvale** field section (lithostratigraphic data from Léonide et al., 2012). See Fig. 3B for explanation of symbols, color legend and stratigraphic nomenclature.

5.6.1 Stable isotope data

298 samples collected along the Font Jouvale outcrop were analyzed for their $\delta^{18}O$ and $\delta^{13}C$ values. Oxygen-isotope values fluctuate between −5.38‰ and −2.39‰ while carbon-isotope values vary between +0.99‰ and +3.74‰. Most of these signatures differ from the marine record (Fig. 4F), except those located below the UB.*d2* emersive discontinuity and above the UB.*d4* composite surface.

Increasing values of both $\delta^{18}O$ and $\delta^{13}C$, respectively from −5.3‰ to ~−3.8‰ and from +2.3‰ to +3.5‰, are recorded in the peloidal and *Agriopleura*-bearing limestones (UB–*S1* Sequence *pro parte*) at the base of the Font Jouvale section. Above, oxygen-isotope values display a relative constant pattern around −4.5‰ through the entire rudistid succession. The corresponding carbon-isotope signatures oscillate between +2‰ and +3.5‰. In this interval, two distinct $\delta^{13}C$ negative trends can be identified. The first represents a ~30 m-thick interval beginning in the upper part of the requieniid–monopleurid unit (UB–*S2* Sequence) and terminates slightly above the oolitic–coral beds (UB–*S3a* Sequence). The second negative excursion starts in the upper 50 m of the caprinid-bearing unit (UB–*S3b* Sequence) and culminates just below the UB.*d4* composite discontinuity where $\delta^{13}C$ reaches compositions of +1‰. More positive $\delta^{18}O$ (~−4‰) and $\delta^{13}C$ (~+2.8‰) signatures are then recognized in the overlying cherty and *Palorbitolina*-rich limestones (UB/LA–*S4* Sequence).

5.6.2 Porosity data

146 porosity measurements were performed at Font Jouvale. Values are comprised between 1.38% and 28.91%. Most of the rudistid succession (i.e., UB–*S2* and UB–*S3a* sequences) displays porosity values higher than 10%. Minimum values are tied to muddy, peloidal-rich facies at the base of the section (topmost part of the UB–*S1* Sequence) or more distal facies (e.g., coral, ooidal, cherty and *Palorbitolina* facies). Microporosity predominates at Font Jouvale.

5.7 Gavotte (Fig. 10)

The Gavotte borehole section was drilled near the Emergency Exit of the LSBB, in the vicinity of the Rustrel village (Vaucluse). It represents an eastward equivalent of the Font Jouvale succession, but includes a thicker UB–*S1* Sequence. Detailed litho-, biostratigraphy and sedimentology can be found in Tendil et al. (2018, supplementary material, p. 37).

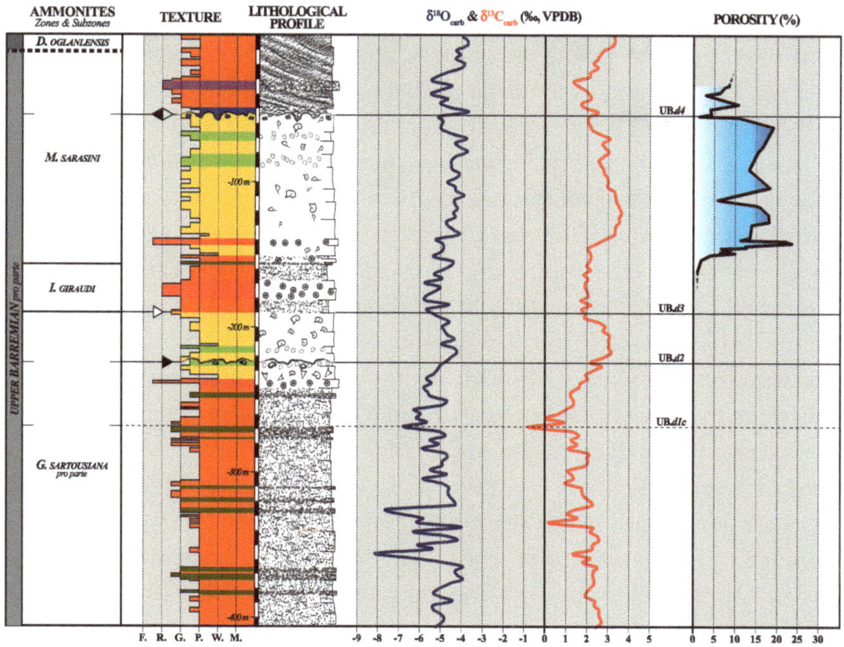

Fig. 10 Litho-, biostratigraphy, bulk-rock stable isotope analyzes and porosity evolution of the **Gavotte borehole** section. Porosity data derive from the near Les Antennes outcropping section (Léonide et al., 2014). See Fig. 3B for explanation of symbols, color legend and stratigraphic nomenclature.

5.7.1 Stable isotope data

134 bulk-isotopic measurements were made from cuttings samples collected from the Gavotte borehole. The $\delta^{13}C$ of the Gavotte sections scatters from −0.81 to +3.63‰ while the $\delta^{18}O$ scatters from 8.12 to −3.69‰. Most of these values differ from the marine values (Fig. 4G).

From the base of the section to −270 m, the oobioclastic unit exhibits an overall negative trend of the C-isotope values (from +2.70‰ to ~+1‰). Two marked negative shifts (amplitude of ~2‰) are observed at ~−333 and −270 m. The oxygen-isotope values display a comparable evolution, except between −360 and −320 m where the $\delta^{18}O$ signal fluctuates between −8.12 and −4‰. Then, isotopic signatures gradually increase until the bottom part of the requieniid–monopleurid unit at −220 m. A slight negative spike of the $\delta^{13}C$ (amplitude of 0.5‰) can nevertheless be detected below the exposure surface (UB.*d2*) which caps the *Agriopleura*-bearing unit (UB–*S1* Sequence). Above this surface, $\delta^{13}C$ stabilizes around +3‰ but then decreases to +2‰ toward the top of the requieniid–monopleurid unit

(UB–*S2* Sequence). A slight negative trend (amplitude of ~1‰) of the $\delta^{18}O$ signal is recognized in this interval. The apparition of coral–oobioclastic facies (UB.*d3*)—lateral variations of the *Palorbitolina–Heteraster* guide level (UB–*S3a* Sequence) according to Tendil et al. (2018)—coincides with the onset of a 50 m-thick interval where $\delta^{13}C$ displays values around +2‰. Increasing C-isotope values (from 2‰ to 3.60‰) are observed following the recovery of the Urgonian rudistid regime (UB–*S3b* Sequence). However, the remaining 70 m of this caprinid-bearing sequence are characterized with decreasing values to +2‰. A gradual positive evolution of the $\delta^{18}O$ curve (from −5.60‰ to −4‰) is noticed from the UB.*d3* discontinuity. The topmost part of the UB–*S3b* Sequence displays a 15 m-thick interval where both $\delta^{18}O$ and $\delta^{13}C$ signatures shift to more negative values, respectively −5‰ and +2‰, below the emersive UB.*d4* discontinuity. Slightly more positive $\delta^{18}O$ (−3.7‰) and $\delta^{13}C$ (+2.50‰) values are identified in the overlying *Palorbitolina*-rich unit (UB/LA–*S4b* Sequence). These signatures then decrease and reach compositions of −5.5‰ for the $\delta^{18}O$ and +1.2‰ for the $\delta^{13}C$ in the basal part of the oobioclastic unit before increasing again in the remaining of the succession.

Masse et al. (1999) produced a composite $\delta^{13}C$ curve of the Rustrel (Les Antennes)–Lagarde d'Apt–Murs region while Léonide et al. (2014) studied the $\delta^{18}O$ and $\delta^{13}C$ responses of the Rustrel (Les Antennes) section. Comparable isotopic trends can be described between those localities and the Gavotte one.

5.7.2 Porosity data

Previous work by Léonide et al. (2014) provided a porosity log of the outcropping Rustrel section located ~3 km east to the Gavotte borehole. Based on 33 measurements, porosity values fluctuate between 0.30% and 23.15%. The caprinid-bearing unit (UB–*S3b* Sequence) displays porosity values exceeding 10% while outer platform and *Palorbitolina*-rich limestones (i.e., UB–*S3b* and UB/LA–*S4b* sequences *pro parte*) appear more cemented (porosity values <10%). As for the Lagnes Quarry, Fontaine-de-Vaucluse and Font Jouvale localities, the Rustrel section is dominated by microporosity.

5.8 Gorges de La Nesque (Fig. 11)

The sampling was done through the UB–*S1* to lowermost UB/LA–*S4* sequences along the deep river Gorges de La Nesque (Rocher du Cire cliff and Lauze locality), situated ~7 km east to Villes-sur-Auzon (Vaucluse, SE

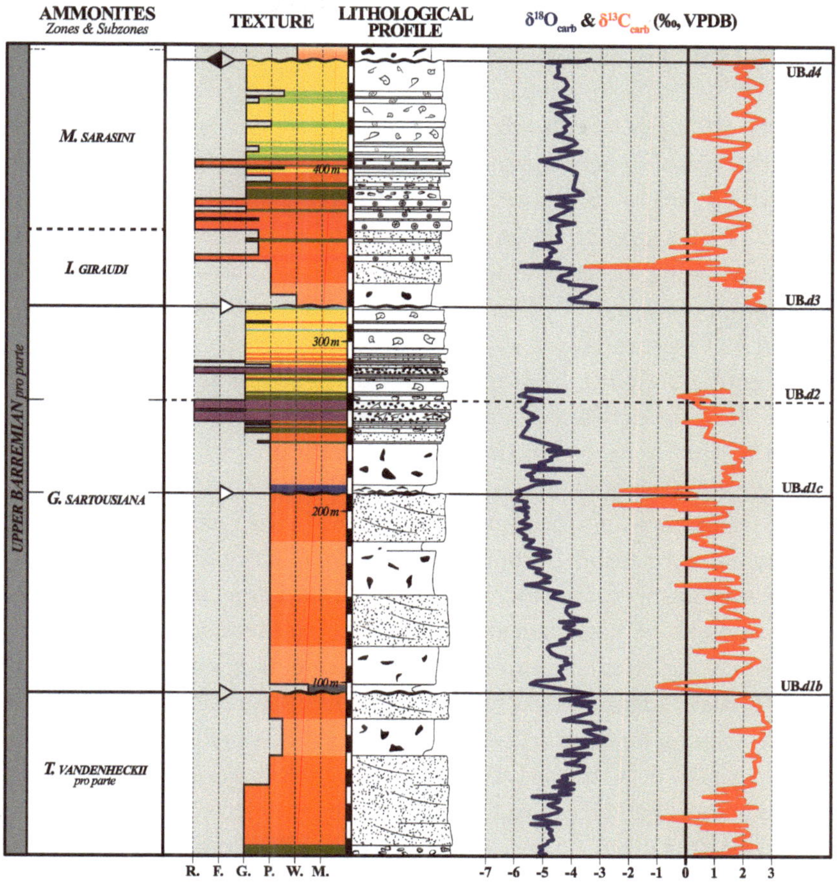

Fig. 11 Litho-, biostratigraphy and bulk-rock stable isotope analyzes of the **Gorges de La Nesque** field section (lithostratigraphic data from Léonide et al., 2012). See Fig. 3B for explanation of symbols, color legend and stratigraphic nomenclature.

Monts de Vaucluse). This succession is located at the transition between the inner and outer platform settings. Detailed litho-, biostratigraphy and sedimentology can be found in Tendil et al. (2018, supplementary material, p. 45).

5.8.1 Stable isotope data

267 stable isotopes measurements were performed. Oxygen-isotope values fluctuate between −6.01‰ and −2.77‰ while carbon-isotope values are comprised between −3.56‰ and +2.93‰. Only few $\delta^{18}O$ and $\delta^{13}C$

signatures, mostly from oobioclastic–cherty limestones, match those of the marine record of the Barremian–Aptian interval (Fig. 4H).

The $\delta^{18}O$ and $\delta^{13}C$ profiles derived from the Rocher du Cire cliff section (UB–*S1* Sequence) show comparable trends. This is such a case at the base of the section where a gradual increase in isotopic compositions can be observed, i.e., from −5‰ to −3.5‰ for the $\delta^{18}O$ and from 0.5‰ to +2‰ for the $\delta^{13}C$. Decreasing values are then identified between the UB.*d1b* and UB.*d1c* drowning discontinuities, with oxygen and carbon values reaching respectively −6‰ and ∼−2‰. It is worth noting the presence of strongly depleted $\delta^{13}C$ signatures (up to −2‰) in the marly and the orbitolinid-rich beds directly overlying the latter discontinuities. The topmost part of the Rocher du Cire cliff section is first represented by more positive $\delta^{18}O$ (∼−4.5‰) and $\delta^{13}C$ (∼+1.5‰) in cherty limestones before being more negative, respectively ∼−5.5‰ and ∼+0.8‰, in oobioclastic facies.

The isotopic compositions at the base of the Lauze section present a stepwise decrease of carbon-isotope values observed throughout the UB–*S3a* Sequence, from ∼+2.5‰ to values below 0‰. Large amplitude negative excursions (up to −3.5‰) are identified within this sequence in oobioclastic limestones. Care must be taken when considering these shifts as a major fault system is locally identified. As such, isotopic resetting (e.g., fluid circulation processes) cannot be excluded. A similar pattern, albeit of lesser amplitude, is observed for the $\delta^{18}O$ in the UB–*S3a* Sequence. An overall positive trend of the carbon-isotope values (up to +2.5‰) can be detected in the following caprinid-bearing unit (UB–*S3b* Sequence) while oxygen-isotope values scatter from −3.8‰ to −5.1‰. Below the UB.*d4* composite discontinuity, a meter-thick interval records a negative excursion of both the oxygen- (amplitude of 0.5‰) and carbon-isotope (amplitude of 1‰) signals. Above the surface, more positive $\delta^{18}O$ (∼−3.5‰) and $\delta^{13}C$ (∼+2.6‰) values are identified in the cherty limestones of the UB/LA–*S4b* Sequence.

5.9 **Brantes** (Fig. 12)

The Brantes field section is located on the eastern part of the Bluye Mountain (Vaucluse), which forms a kilometric-wide syncline in front of the north facing side of the Mont Ventoux. This succession corresponds to a slope settings and includes the UB–*S1* to LA–*S5* sequences. Detailed litho-, biostratigraphy and sedimentology can be found in Tendil et al. (2018, supplementary material, p. 67).

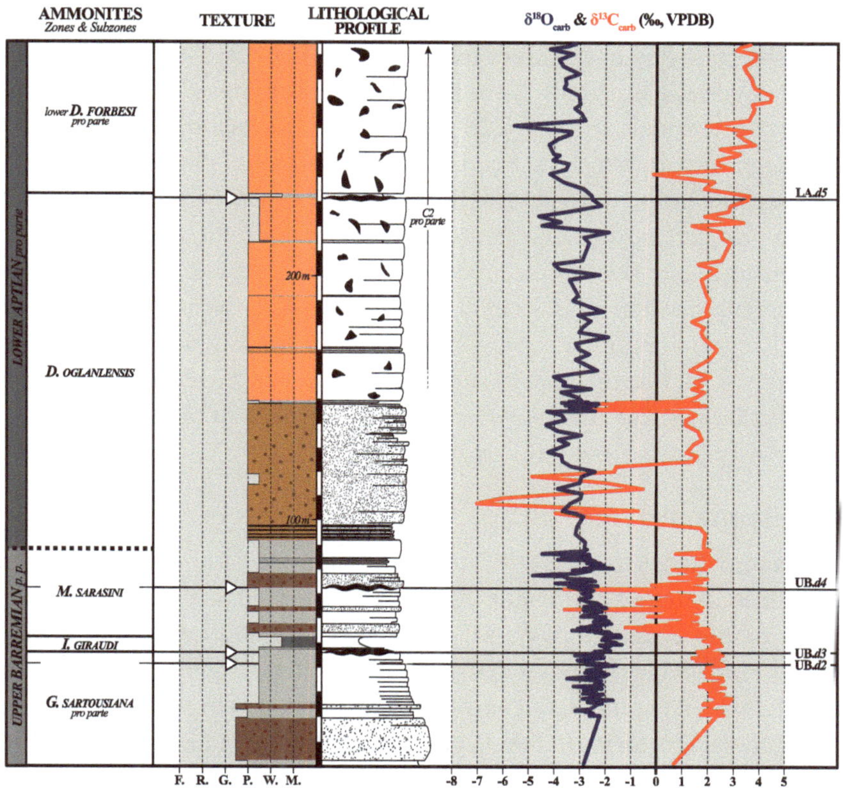

Fig. 12 Litho-, biostratigraphy and bulk-rock stable isotope analyzes of the **Brantes** field section. See Fig. 3B for explanation of symbols, color legend and stratigraphic nomenclature.

5.9.1 Stable isotope data

333 whole-rock measurements of the stable isotope signatures were performed. $\delta^{18}O$ values vary between −5.55‰ and −1.38 while $\delta^{13}C$ values scatter from −7.02‰ to +4.50‰. Most of those isotopic signatures conform well to the marine values of the Barremian/Aptian boundary interval (Fig. 4I).

Marl-limestones alternations at the base of the section (UB–*S1* and UB–*S2* sequences) along with the *Heteroceras* Marls (UB–*S3a* Sequence) are characterized by oscillating $\delta^{13}C$ values between +1.50‰ and +3.00‰. Extreme negative $\delta^{13}C$ spikes (up to −3.60‰) are identified in the overlying platform-derived biocalcarenites (UB–*S3b* Sequence). Even more C-isotope depleted values (up to −7‰) are observed in the laminated calcisiltites of the UB/LA–*S4* Sequence *pro parte* (100–125 m). These negative signatures are encompassed by $\delta^{13}C$ values varying between +1 and +2‰ both in the marl-limestones alternations above the UB.*d4* discontinuity and in the upper part of the turbiditic interval (125–145 m). The corresponding $\delta^{18}O$ curve exhibits an overall decreasing trend (from ∼−2‰ to ∼−4‰) since the base of the section. Strongly negative $\delta^{13}C$ values (up to −3‰) are also recorded in the topmost part of the laminated calcisiltites unit and the overlying meter-thick, marly unit. By contrast, more positive $\delta^{18}O$ values (up to −2.5‰) are recognized within this interval. Above, a gradual positive excursion of the $\delta^{13}C$ (from +1.50‰ to +∼4‰) occurs in the bedded, chert-bearing calcilutites (UB/LA–*S4 pro parte* and LA–*S5* sequences) while the $\delta^{18}O$ fluctuates between −2‰ and −4.50‰. It merely reflects part of the long-lasting, pre-OAE 1a positive trend (C2 segment *pro parte*).

5.10 Baudinard (Fig. 13)

The Baudinard field section is located in the Monges massif which represents the highest point of the Digne Prealps (Alpes-de-Haute-Provence). It is linked to the Esparron NW–SE syncline whose southwest termination (i.e., La Pinée Ridge) exposes a complete Vocontian-type basinal succession of Barremian–Aptian age, unconformably overlain by the Red Molasse Fm. of Oligocene age. Detailed litho-, biostratigraphy and sedimentology can be found in Tendil et al. (2018, supplementary material, p. 73).

5.10.1 Stable isotope data

200 whole-rock stable isotopic measurements were performed. The Baudinard section displays values varying between −5.08‰ and −2.98‰ for the $\delta^{18}O$ and +1.57‰ and +4.19‰ for the $\delta^{13}C$. These values match well those of the marine values, even though some $\delta^{18}O$ values are slightly depleted (Fig. 4J).

The 64 m-thick Barremian–Bedoulian Limestones Fm. (= UB–*S1* to LA–*S5* sequences equivalents) is defined by $\delta^{13}C$ values close to +2‰ while the oxygen-isotope profile varies between −5‰ and −3‰. Above, the Blue Marls Fm. (= LA–*S7* Sequence equivalent) exhibits more positive carbon-isotope values (between +3‰ and +4‰) and a relatively stable pattern of the $\delta^{18}O$ (around −4.5‰). None geochemical feature is recognized at the UB. *d4* and LA.*d6–7* discontinuities, although hiatuses of variable duration are identified by means of ammonites (Frau et al., 2018a,b). As such, only pre- (C2 segment *pro parte*) and post-OAE 1a (C7 segment *pro parte*) isotopic segments are recognized.

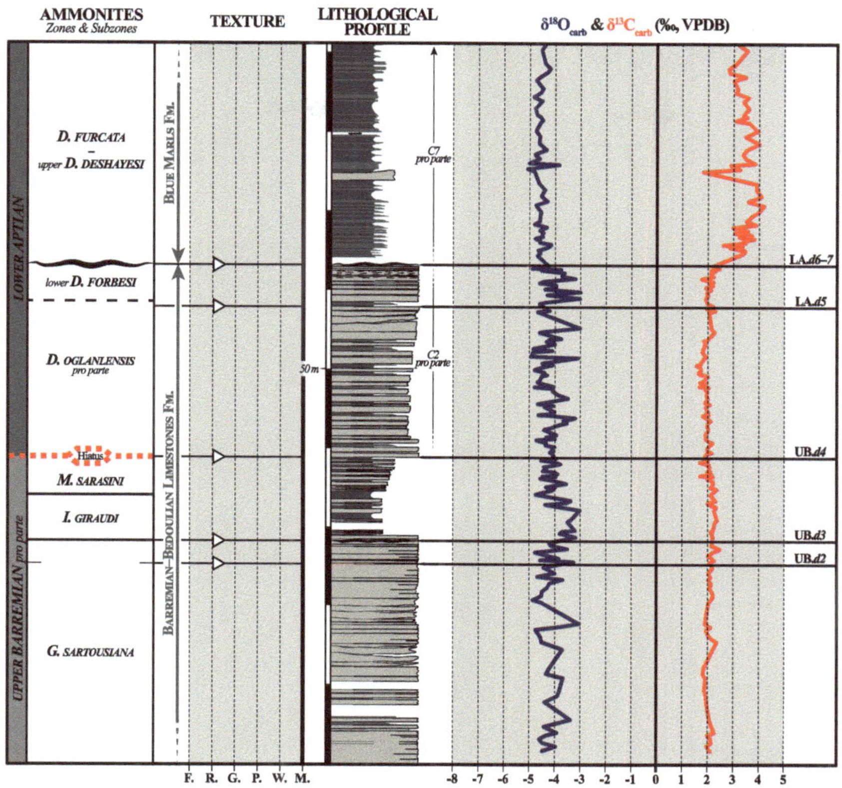

Fig. 13 Litho-, biostratigraphy and bulk-rock stable isotope analyzes of the **Baudinard** field section. See Fig. 3B for explanation of color legend and stratigraphic nomenclature.

5.11 Angles–Combe Lambert–Glaise (Fig. 14)

The composite Angles–Combe Lambert–Glaise field succession is based on the study of three outcrops located in the Vocontian Basin, as described below:

- The Angles section is situated ~10 km north of Castellane, along the departmental road D33. The Angles succession provides the most complete Barremian–Aptian succession of the Vocontian Basin and has been chosen as the stratotype for the Barremian (Busnardo, 1965). The late Barremian to early Aptian series (*T. vandenheckii* to *D. forbesi pro parte* zones) consists of ammonite-bearing, limestone-dominated beds and marly interbeds (i.e., the Barremian–Bedoulian Limestones Fm). A distinct marly interval, tied to the *Heteroceras* Marls sensu Paquier (1900), characterizes the *I. giraudi* Zone of late Barremian age. At its base,

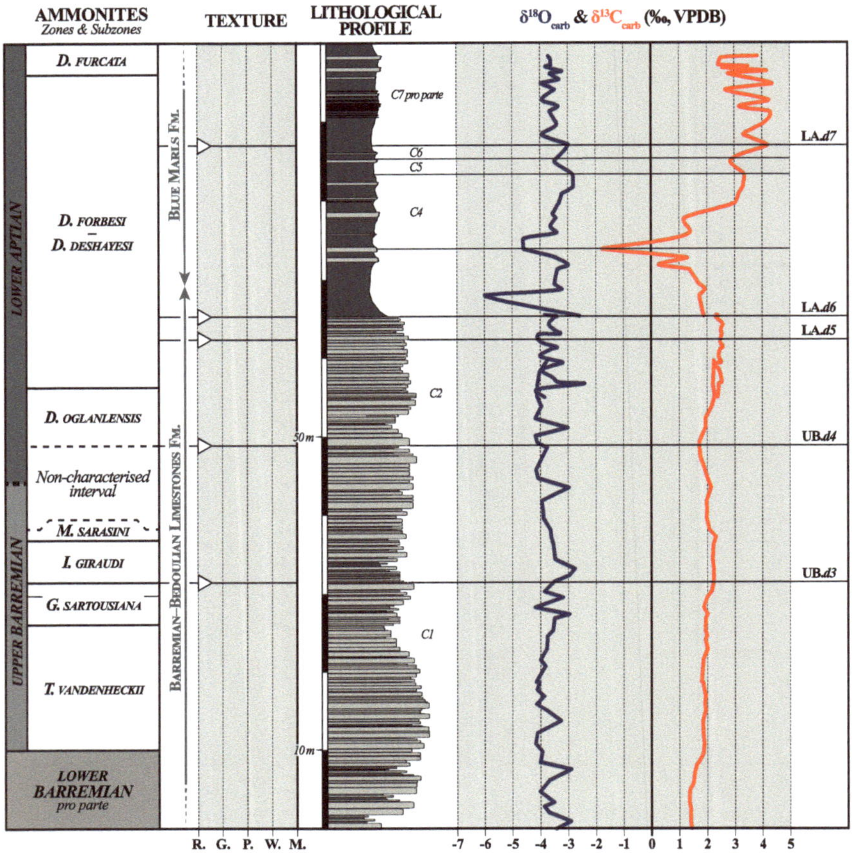

Fig. 14 Litho-, biostratigraphy and bulk-rock stable isotope analyzes of the **Angles–Combe Lambert–Glaise composite** field section. C1–C7 relate to the isotopic segments defined by Menegatti et al. (1998). See Fig. 3B for explanation of color legend and stratigraphic nomenclature. *Data from Godet, A., 2006. The evolution of the Urgonian platform in the western Swiss Jura realm and its interactions with palaeoclimatic and palaeoceanographic change along the Northern Tethyan Margin (Hauterivian-earliest Aptian), Doctoral thesis, Université de Neuchâtel, 405 pp.; Bodin, S., Fiet, N., Godet, A., Matera, V., Westermann, S., Clément, A., Janssen, N.M.M., Stille, P., Föllmi, K.B., 2009. Early Cretaceous (late Berriasian to early Aptian) palaeoceanographic change along the northwestern Tethyan margin (Vocontian Trough, southeastern France): $\delta^{13}C$, $\delta^{18}O$ and Sr-isotope belemnite and whole-rock records. Cretaceous Res. 30, 1247–1262 and Westermann, S., Stein, M., Matera, V., Fiet, N., Fleitmann, D., Adatte, T., Föllmi, K.B., 2013. Rapid changes in the redox conditions of the western Tethys Ocean during the early Aptian oceanic anoxic event. Geochim. Cosmochim. Ac. 121, 467–486.*

a ferruginous firmground (UB.*d3*), comparable to those of Brantes and Baudinard sections, can be identified.

- ~10 km south of Die, the 12 m-thick Combe Lambert section correlates to the topmost part of the Barremian–Bedoulian Limestones Fm. and exhibits decimeter-thick hemipelagic beds interbedded with centimeter-thick marly intervals (Delanoy, 1997).
- The Glaise section is located ~4 km north of Veynes. The 35 m-thick marly succession corresponds to the base of the Aptian–Albian Blue Marls Formation in which centimetric turbiditic layers are recognized (Bréhérét, 1997). A decametric interval records the deposition of discrete, OAE 1a-related black shales.

5.11.1 Stable isotope data

A reference isotopic curve from the composite Angles–Combe Lambert–Glaise section was reconstructed by Huck et al. (2011) based on the works of Godet (2006), Bodin et al. (2009) and Westermann (2010). Oxygen-isotope values range between −5.98‰ and −2.40‰ while carbon-isotope signatures vary between −1.76‰ and 4.29‰. The entire set of stable isotope measurements from the limestone-dominated sequence conforms to the reference marine values as illustrated in Fig. 4K.

A relatively stable pattern of the carbon-isotope profile (~+1.8‰) is observed at the base of the late Barremian (UB–*S1* Sequence equivalent). A slight increase of $\delta^{13}C$ values (up to ~+2.2‰) is then detected during the *H. feraudianus* Subzone (UB–*S2* Sequence equivalent). These signatures remain near ~+2.2‰ during the *I. giraudi* Zone (UB–*S3a* Sequence equivalent) before decreasing to values around +1.7‰ at the Barremian–Aptian boundary discontinuity (UB.*d4*). Above, a positive trend of the $\delta^{13}C$ is identified throughout the limestone-dominated sequence (UB/LA–*S4 pro parte* and LA–*S5* sequences equivalent) and reaches values of ~+2.5‰. A prominent negative shift (minimal value of −1.76‰) marks the transition with the marl-dominated regime and is considered as the expression of the OAE 1a-related negative excursion (LA–*S6*/*S7* sequences *pro parte* equivalent). The deposition of the overlying black shales deposits (i.e., the Goguel level) coincides with the onset of a large amplitude positive excursion where carbon values exceed +4‰. According to Frau et al. (2018b), the pattern of the $\delta^{13}C$ curve matches with the C1–C7 segments defined by Menegatti et al. (1998), albeit a significant condensation of the C3 segment.

The oxygen-isotope curve displays a positive trend up to the *I. giraudi* Zone where the $\delta^{18}O$ reaches the value of −2.75‰. An overall decreasing excursion (up to −4.18‰) characterizes the upper part of the limestone-dominated sequence. As for the $\delta^{13}C$ profile, a negative shift (up to −4.6‰) then followed by a positive trend (amplitude of ~1.8‰) is identified during the spreading and culmination of the OAE 1a.

6. Regional features of the stable isotope signatures

The Urgonian Provence platform chronostratigraphic model used in the present work is based on the recognition of (i) shallow-marine and deep biological markers (rudists, dasycladales, echinoids and ammonites) and (ii) remarkable stratigraphic surfaces of regional significance as detailed in Frau et al. (2017, 2018a) and Tendil et al. (2018). This framework (Fig. 1C) provides stratigraphic constraints for detailed investigation of the spatial and temporal variations of the $\delta^{13}C$–$\delta^{18}O$ isotopic signatures as represented in Fig. 15. Key observations are detailed as follow:

6.1 Variability in $\delta^{13}C$–$\delta^{18}O$ absolute values

The drowning of the requieniid–monopleurid limestones during the *I. giraudi* Zone (UB–*S3a* sequence) corresponds to a conspicuous change in carbonate production through the Provence domain. Mass occurrence of *Palorbitolina lenticularis* and *Heteraster oblongus* communities are represented by $\delta^{13}C$ values >3‰ at **Lagnes Quarry core** section (Fig. 6). At **Fontaine-de-Vaucluse** section (Fig. 7), such facies is associated to a $\delta^{13}C$ which scatters from +1.2‰ to +2.5‰ (data from Bastide, 2014) and +0.4‰ to +1.8‰ (data from Léonide et al., 2014) while at **Joucas** section (Fig. 8), C-isotope values are characterized by values near +2.5‰. The *Palorbitolina–Heteraster* guide level grades laterally to coral–oobioclastic–cherty limestones. Corresponding $\delta^{13}C$ signal fluctuates between +2.2‰ and +3‰ at **Font Jouvale** section (Fig. 9) and between −3.5‰ and +2.7‰ at **Gorges de La Nesque** section (Fig. 11), or is marked by relative constant values around +2‰ at **Gavotte borehole** section (Fig. 10). Similar observations can be made for the $\delta^{18}O$ signal whose values oscillate around −5‰ at Gavotte, between −4‰ and −3‰ at Lagnes Quarry and **Baudinard** (Fig. 13), and −2‰ at **Brantes** (Fig. 12) during the UB–*S3a* sequence.

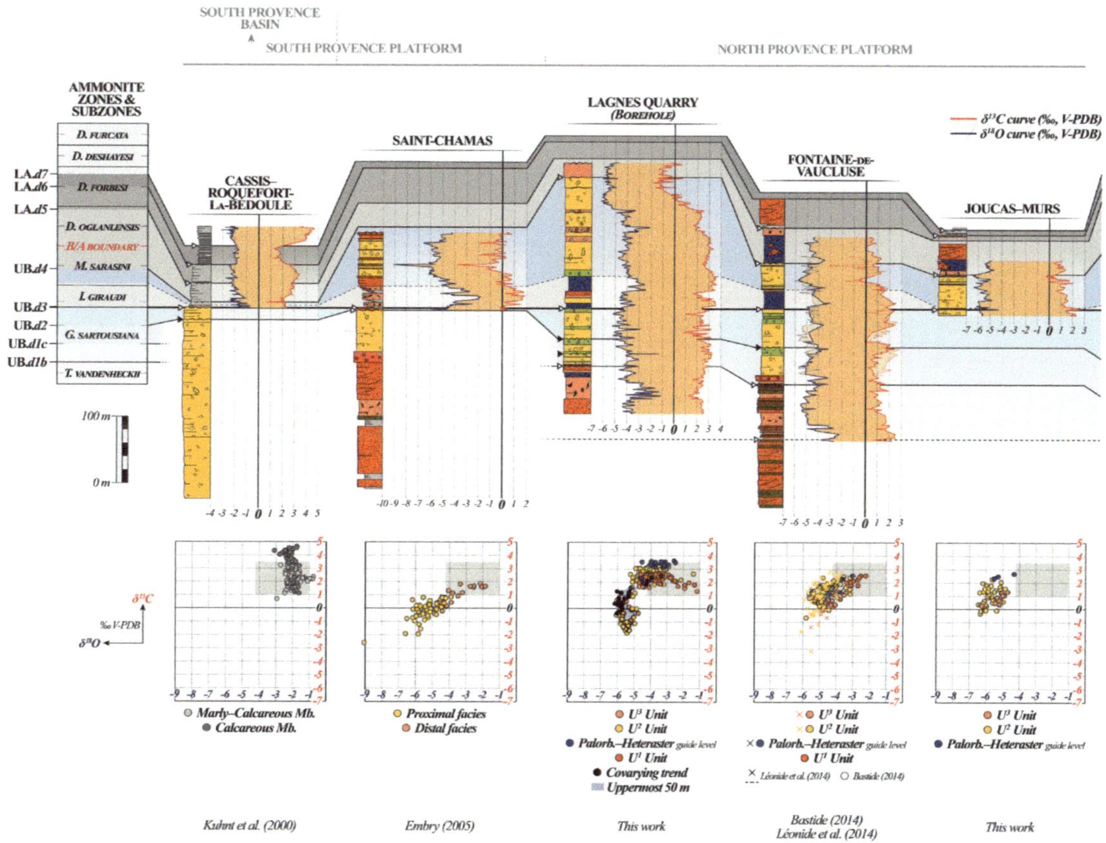

Fig. 15 Regional correlations of the δ18O and δ13C signatures following the platform-to-basin chronostratigraphic model of Tendil et al. (2018) and associated crossplots. See Fig. 3B for explanation of symbols, color legend and stratigraphic nomenclature.

(Continued)

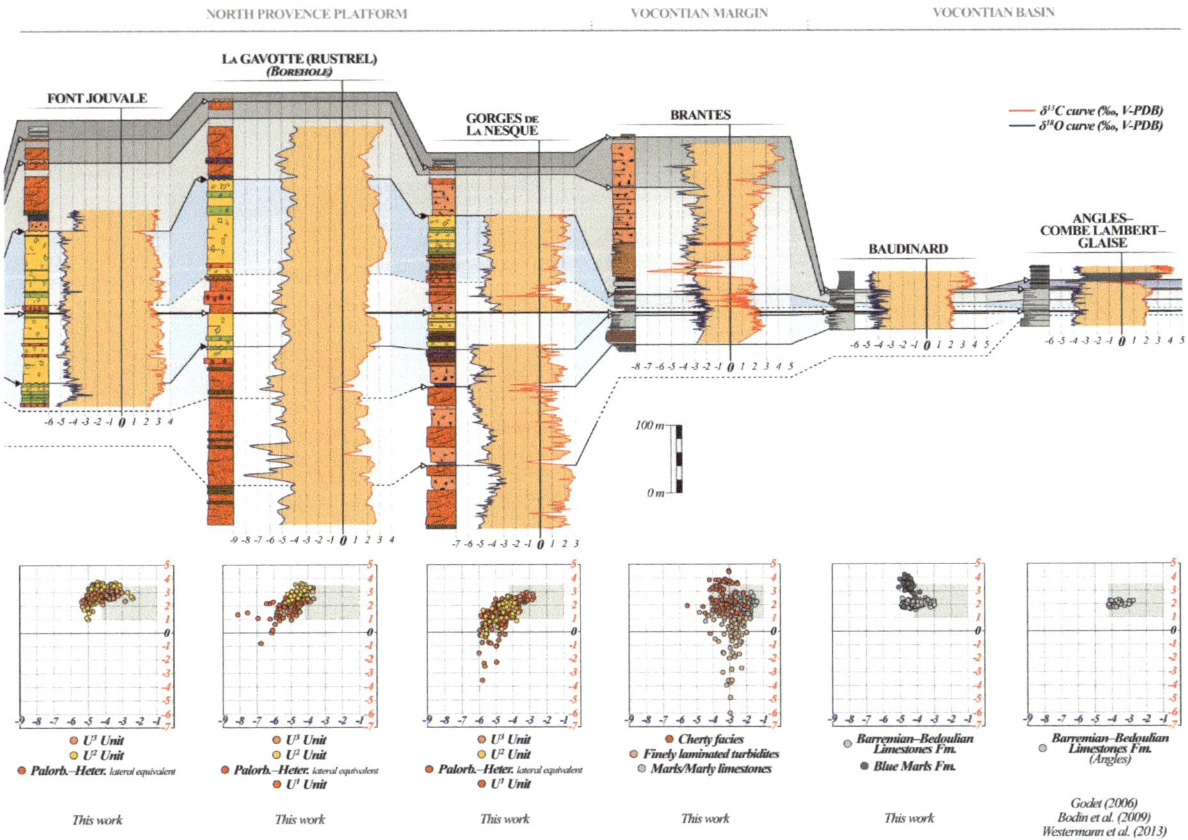

NORTH PROVENCE PLATFORM
VOCONTIAN MARGIN
VOCONTIAN BASIN
FONT JOUVALE
LA GAVOTTE (RUSTREL)
(BOREHOLE)
GORGES DE LA NESQUE
BRANTES
BAUDINARD
ANGLES–COMBE LAMBERT–GLAISE
δ13C curve (‰, V-PDB)
δ18O curve (‰, V-PDB)
100 m
0 m
U3 Unit
U2 Unit
Palorb.–Heter. lateral equivalent
U1 Unit
Cherty facies
Finely laminated turbidites
Marls/Marly limestones
Barremian–Bedoulian Limestones Fm.
Blue Marls Fm.
Barremian–Bedoulian Limestones Fm. (Angles)
This work
Godet (2006)
Bodin et al. (2009)
Westermann et al. (2013)

Fig. 15—Cont'd

6.2 Variability in $\delta^{13}C$–$\delta^{18}O$ trends

During the UB–*S3a* sequence, $\delta^{13}C$ values around +2.3‰ are recognized in the marly interval of the Vocontian margin (**Brantes** section; Fig. 12) and Basin (**Baudinard** and **Angles** sections; Figs. 13 and 14). Comparable signatures are observed in the drowning sequence of the South Provence Basin (**Cassis–Roquefort-la-Bédoule** section; Fig. 3). A similar stable pattern also occurs at the **Gavotte borehole** section (Fig. 10) as previously noted. Elsewhere, the $\delta^{13}C$ profiles in contemporaneous intervals are either marked by a decreasing trend (from +3‰ to +2.5‰ at **Font Jouvale** section; Fig. 9) or by a wide scattering of values (e.g., from −3.5‰ to +2.7‰ at **Gorges de La Nesque** section; Fig. 11).

The following rudist-rich, UB–*S3b* sequence exhibits fluctuating $\delta^{13}C$ evolution trends. In the **Lagnes Quarry core** section (Fig. 6), a stepwise decreasing trend is reported throughout the sequence while at **Gorges de La Nesque** section an overall positive trend can be detected. Others examples include the **Font Jouvale** and **Gavotte** sections where the C-isotope trends are first positive and then become negative. In distal settings, the **Brantes** section shows erratic carbon isotopic excursions (between −3.6‰ and +2‰) while a relative constant pattern is identified in both the **South Provence** and **Vocontian** basinal sections at that time (Figs. 3, 13, and 14).

6.3 Local perturbations of the $\delta^{13}C$ signal

In the North Provence platform settings, majority of C-isotope profiles are represented by positive values although negative $\delta^{13}C$ values sporadically occur. This is such a case for the **Lagnes Quarry core** (Fig. 6), **Fontaine-de-Vaucluse** (Fig. 7) and **Gorges de La Nesque** (Fig. 11) sections where negative $\delta^{13}C$ values (up to −3.6‰) are documented within decameter-thick intervals. In Vocontian slope settings, the finely laminated turbidites outcropping at **Brantes** (Fig. 12) are marked by extremely negative carbon values (up to −7‰). Based on the platform-to-basin stratigraphic model herein used, no contemporaneous negative derives can be identified elsewhere in the Provence domain at that time.

In the **Lagnes Quarry core** section, a positive covariance ($R^2 \sim 0.72$) between $\delta^{18}O$ and $\delta^{13}C$ values is documented within a 17 m-thick interval at the top of the UB–*S3b* Sequence. A comparable covariant trend ($R^2 \sim 0.72$) is seen at the **Saint-Chamas** section (Fig. 5) across the whole succession. It is worth mentioning the presence of major, post-Urgonian epikarstic surfaces—respectively of mid-Cretaceous (Saint-Chamas section; Masse, 1976) and Burdigalian (Lagnes Quarry section) ages—atop those sections.

6.4 Expansion/condensation of the stable isotope signal

In all basinal sections studied in this work (**Cassis–Roquefort-la-Bédoule, Angles–Combe Lambert–Glaise, Baudinard**), the isotopic trends of the whole OAE 1a-related signal are recognizable (C1–C8) and show comparable absolute values from contemporaneous basins elsewhere (Frau et al., 2018a). Two main perturbation events occur and show contrasting expressions:

(1) The transition between the *M. sarasini* and *P. waagenoides* subzones at **Cassis–Roquefort-la-Bédoule** is marked by a strong negative oxygen and carbon isotope shift (~1.5‰), seen in paired organic matter and carbonate carbon (Fig. 3). This excursion, here referred to as the Intra-Sarasini Negative Excursion (ISNE), falls at a serpulid-rich, wavy discontinuity which gives evidence of a short-termed, *syn*-depositional hiatus (Frau et al., 2018a). As noticed by the latter authors, the ISNE may result from seawater enrichment by isotopically-light, dissolved inorganic carbon derived from intense weathering of the hinterland during the emersion of the Urgonian Provence platform. According to Frau et al. (2018a), the ISNE is also recorded in the Vocontian Basin (**Angles** section; Fig. 14) but its amplitude is lowered (~1.5‰). By contrast, the hiatus associated to the corresponding discontinuity is of greater duration (including part of the Barremian–Aptian boundary interval—see discussion in Frau et al., 2018b, p. 320).

(2) The OAE 1a-related negative excursion (C3 segment) is distinctly recorded at **Cassis–Roquefort-la-Bédoule** along a 3 m-thick marly succession. Absolute values decrease by, at least, ~2‰ compared to pre-OAE levels. By contrast, the record of the OAE 1a-related negative excursion from Vocontian basinal record is lacking (**Baudinard**; Fig. 13) or complete (**Glaise**; Fig. 14). At **Baudinard**, a hiatus including the C2–C7 segments *pro parte* is documented. This omission surface is widespread through the Vocontian Basin and marked by a *Rhizocorallium* hard ground on top of the Barremian–Bedoulian Limestone Fm. (Cotillon, 2000). Overlying levels of the Blue Marls Fm. are locally slumped during the deposition of the C7 segment *pro parte*. At **Glaise**, the OAE 1a-related signal is complete but markedly condensed

Urgonian Provence platform which developed before the Pyreneo-Provencal tectonic inversion (Lamarche et al., 2012). Stable isotope analyzes performed on these sparite cements indicate massively depleted carbon values between −1‰ and −7‰ (Cochard, 2018), comparable to those recorded in the Brantes turbidites. In the case these calcite sparites reflect a single cementation episode across the northern Provence platform, this would indicate a major contribution of the latter calcite in the $\delta^{13}C$ profile of the Brantes section as well as its telogenetic origin. An alternative explanation consists in an early calcitization of the sponge spicules in a deep marine environment undersaturated with respect to silice and where bacterial activity would be intense, as documented in late Cretaceous spicule-rich turbidites in the Polish Outer Carpathians (Bąk et al., 2015). As a result, chemostratigraphy applies in slope settings but local sedimentary and/or diagenetical disturbances can mask the original marine carbon isotope signal (Fig. 19E and F).

7.3 Geochemical record from basin settings

The Barremian–Aptian isotopic trends record two major perturbations of the global carbon cycle (i.e., latest Barremian ISNE and mid-early Aptian OAE 1a), both characterized by a short-lived, prominent negative excursion then followed by a long-lasting, positive trend in the isotopic composition of paired organic matter and carbonate carbon. These isotopic disturbances are, in both cases, associated to the deposition of organic-rich black shales (Kuhnt et al., 2000; Stein et al., 2012a; Westermann et al., 2013).

The ISNE has been either tied to a hydrate methane destabilization event (Renard et al., 2005) or a change toward a warm and humid climate leading to the mobilization of terrestrial organic matter (Stein et al., 2012a). In any case, these changes were related to global warming conditions due to high rate of volcanically-induced pCO_2 at the time of the first main development phase of the Pacific Large Igneous Provinces (i.e., Ontong Java–Manihiki–Nova Canton trough system). As further evidence, the Pacific LIPs likely start during the latest Barremian following the temporal recalibration of the CM0 magnetozone (Frau et al., 2018b). Besides, the high-amplitude shift of the ISNE in the South Provence Basin (Fig. 3) may result from seawater enrichment by isotopically-light, dissolved inorganic carbon derived from intense weathering of the hinterland during the final emersion of the North Provence rudistid platform (Frau et al., 2018a). The ISNE seems to be associated to a minor sedimentary hiatus compared to **Cassis–Roquefort-la-Bédoule** section. Maximum of condensation occurs during the deposition of the C3 segment which exhibits a high amplitude shift (~5‰) associated to nannoconid-free organic-rich shales.

Post-Urgonian basinal marls of the North Provence platform (= LA–*S6/S7* sequences) correspond to a diachrone formation of mid-early Aptian to Albian age (Frau et al., 2017). According to these authors, the record of the OAE 1a C-isotope signal is flawed in this region and partially obliterated by a *syn*-sedimentary discontinuity including the OAE 1a culmination.

7. Lessons learned from the chemostratigraphy of the Urgonian Provence domain

7.1 Geochemical record from platform settings

Subaerial exposure of carbonate platform environments promote interactions with ^{18}O and ^{13}C-depleted meteoric fluids, which are ultimately reflected as negative excursions of stable isotopes in the sedimentary record. Such depleted carbon values are considered useful for identifying sequence boundaries and, therefore, for correlating sedimentary successions. Nevertheless, the resulting expression of ^{13}C depends on many other factors such as the extension and the duration of the exposure event along with the erosion intensity during the subsequent transgression (e.g., Joachimski, 1994; Railsback et al., 2003). In the Urgonian Provence platform, the most prominent exposure events are observed atop the *Agriopleura*-rich (UB–*S1* Sequence) and caprinid-bearing (UB–*S3b* Sequence) subtidal units (Frau et al., 2017, 2018a; Tendil et al., 2018). Subtle to conspicuous negative shifts of the $\delta^{13}C$ values occur at these levels (see Fontaine-de-Vaucluse, Font Jouvale, Gavotte and Gorges de La Nesque sections; Figs. 7, 9–11). In the innnermost settings of the Urgonian Provence platform, numerous subaerial exposure events terminate the peritidal cycles made either of rudist grainstones or miliolid-, peloid-, oncoid-, gastropod-rich wackestones (Fenerci-Masse et al., 2005; Léonide et al., 2014; Masse et al., 2001, 2003). Two different types of peritidal cycles are thereafter identified:

(1) The meter-thick peritidal cycle (e.g., Cassis type section; Fig. 16A).

This cycle shows very low porosity values (<5%) either due to the micritic depositional texture (muddy facies; Fig. 17A) or the early marine cementation (grainy facies). A relatively constant pattern of stable isotopes is identified in those carbonates (Figs. 16A and 19A). The top of such peritidal cycle is characterized by intertidal to supratidal

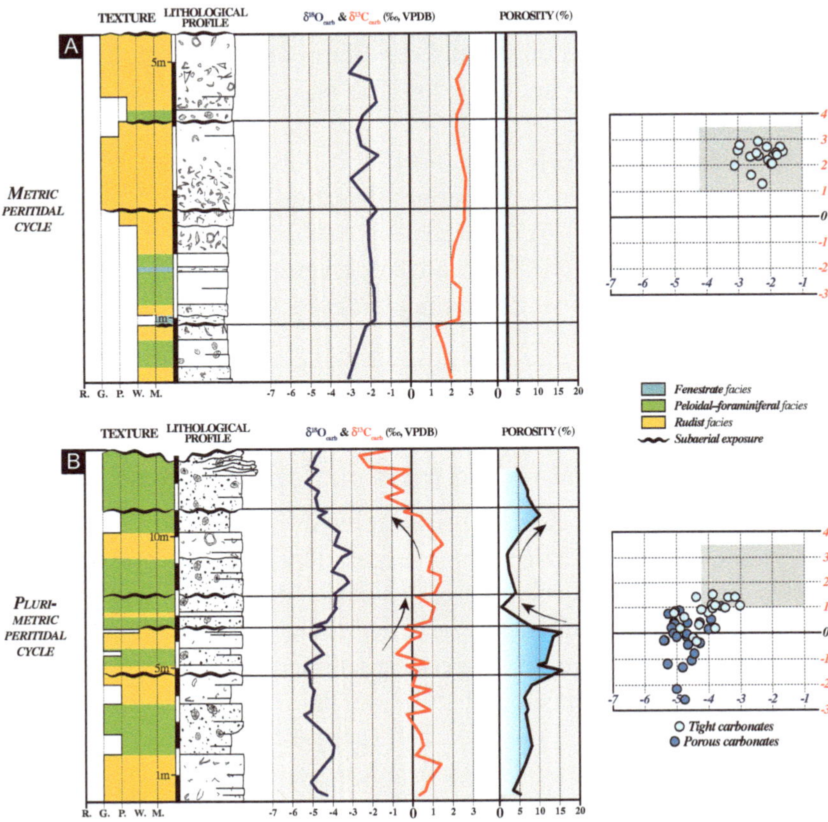

Fig. 16 Lithostratigraphy, bulk-rock stable isotope analyzes and porosity evolution of innermost platform carbonate series. (A) Metric peritidal cycle (Cassis). Complete early marine and/or early phreatic meteoric cementation promotes the preservation of an apparent primary isotopic signal. (B) Pluri-metric peritidal cycle (Martigues). Note that stable isotope values are more depleted in microporous limestones than in cemented ones (i.e., below cycle tops), as indicated by the black arrows. The gray rectangle corresponds to the reference marine values defined in Fig. 2. See Fig. 3B for explanation of symbols, color legend and stratigraphic nomenclature. *Pabel (A): data from Léonide, P., Fournier, F., Reijmer, J.J.G., Vonhof, H.B., Borgomano, J., Dijk, J., Rosenthal, M., van Goethem, M., Cochard, J., Meulenaars, K., 2014. Diagenetic patterns and pore space distribution along a platform to outer-shelf transect (Urgonian limestone, Barremian–Aptian, SE France). Sediment. Geol. 306, 1–23; Panel (B): data from Léonide, P., Fournier, F., Reijmer, J.J.G., Vonhof, H.B., Borgomano, J., Dijk, J., Rosenthal, M., van Goethem, M., Cochard, J., Meulenaars, K., 2014. Diagenetic patterns and pore space distribution along a platform to outer-shelf transect (Urgonian limestone, Barremian–Aptian, SE France). Sediment. Geol. 306, 1–23.*

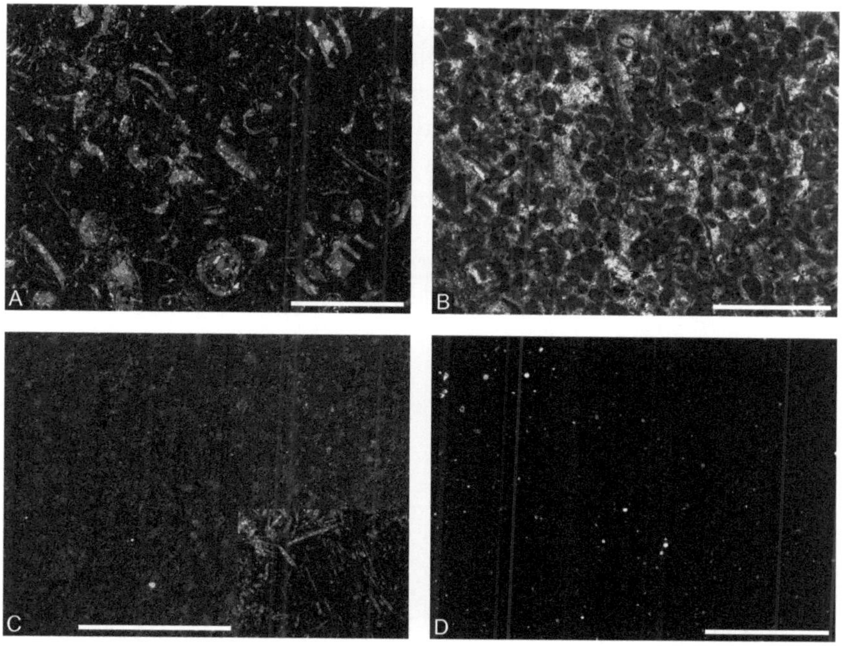

Fig. 17 (A) Gastropod-rich wackestone, with porosity less than 5% (Cassis section). (B) Microporous peloidal grainstone, with porosity greater than 10% (Martigues section). (C) Sponge spicule-rich packstone (Brantes section). Note the contrast between the orange to red luminescent micritic matrix and the non-luminescent sparite crystals. (D) Typical basinal mudstone (Baudinard section). Scale bar is 500 μm.

facies in which dissolution features occur (Léonide et al., 2014). The absence of negative shifts in relation with these subaerial evidences can be explained by the short-termed duration of exposure events which prevents the effect of meteoric waters and the development of mature paleosoils. Similar interpretation is made by Joachimski (1994) in Purbeckian limestones from the Swiss–French Jura domain. In meter-thick peritidal cycle, the $\delta^{18}O$ and $\delta^{13}C$ signatures seemingly conform to those of the marine reference values and could be of interest for chemostratigraphic correlations (Fig. 16A).

(2) The decameter-thick peritidal cycle (e.g., Martigues type section; Fig. 16B).

The presence of both cemented and porous intervals in this type of peritidal cycle suggest a more complex diagenetical evolution. This differentiation likely illustrates the infiltration depth of meteoric fluids and the ensuing early cementation of cycle tops, below subaerial facies (Léonide et al., 2014). At Martigues, the isotopic response reflects the diagenetic pattern as evidenced by the slightly more negative carbon-isotope values in tight units (Figs. 16B and 19A).

In more distal part of the inner platform, a comparable relationship between the diagenetical evolution (i.e., resulting porosity) and the expression of the stable isotopes is reported in both decameter-thick (Rustrel section; Fig. 18A) (Cochard, 2018) and pluri-decameter-thick

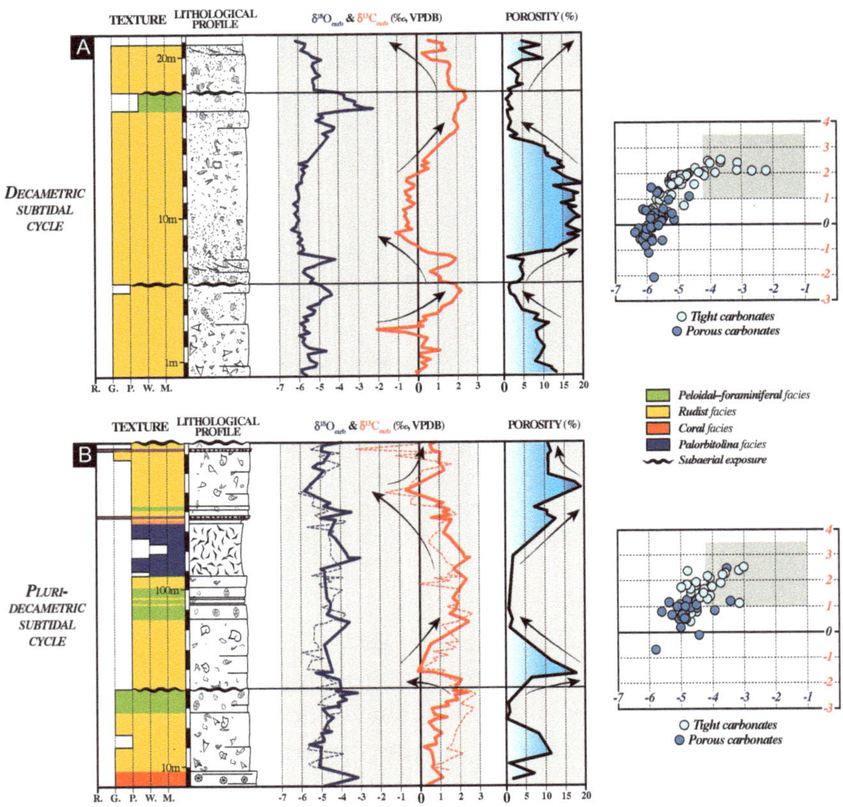

Fig. 18 Lithostratigraphy, bulk-rock stable isotope analyzes and porosity evolution of inner platform carbonate series. (A) Decametric subtidal cycle (Rustrel, LSBB). (B) Pluri-decametric subtidal cycle (Fontaine-de-Vaucluse). As for the Martigues section (Fig. 16B), microporous limestones are frequently represented by more negative stable isotope values than in cemented ones. The gray rectangle corresponds to the reference marine values defined in Fig. 2. See Fig. 3B for explanation of symbols, color legend and stratigraphic nomenclature. *Panle (A): data from Cochard, J., 2018. Analyses des propriétés réservoirs d'une série carbonatée microporeuse fracturée. Doctoral thesis, Aix-Marseille Université, 198 pp; Panel (B): data from Léonide, P., Fournier, F., Reijmer, J.J.G., Vonhof, H.B., Borgomano, J., Dijk, J., Rosenthal, M., van Goethem, M., Cochard, J., Meulenaars, K., 2014. Diagenetic patterns and pore space distribution along a platform to outer-shelf transect (Urgonian limestone, Barremian–Aptian, SE France). Sediment. Geol. 306, 1–23.*

subtidal cycles (Fontaine-de-Vaucluse section; Fig. 18B) (Léonide et al., 2014). Such diagenetical control prevents, therefore, any use of isotopic curves for correlating sections in inner platform settings.

In the Martigues, Rustrel and Fontaine-de-Vaucluse examples, porous intervals are dominated by microporosity (Fig. 17B). Depleted oxygen and carbon isotope values of microporous micrites support an important meteoric influence on the isotopic record (Léonide et al., 2014). In the study area, the development of microporosity mainly occurred during the mid-Cretaceous Durancian tectonic uplift and is thought to have accentuated the porosity development (Godeau et al., 2018; Léonide et al., 2014). The stable isotope signatures of the Martigues, Rustrel and Fontaine-de-Vaucluse sections mostly differ from the marine values, except in some tight units where the $\delta^{18}O$ and $\delta^{13}C$ values are heavier (Figs. 16B and 18A and B). The depletion of $\delta^{18}O$ and $\delta^{13}C$ values at these three localities may result from the prevalent subaerial conditions occurring during the Durancian event (Fig. 19B).
However, the latter diagenetical–geochemical relationship can be locally inverted, as seen at Font Jouvale (Fig. 9) and Gavotte (Fig. 10) sections. At Font Jouvale, the highly microporous rudist-dominated units are characterized by the most positive C-isotope values (between +3 and +4‰) of the whole Provence platform domain. Comparable observations are made in the microporous caprinid-bearing unit (UB–*S3b* Sequence) of the Gavotte borehole section which is represented by more positive $\delta^{13}C$ signatures than the underlying cemented units (*Palorbitolina–Heteraster* guide level lateral equivalent *pro parte*). Based on point counting method, Tendil (2018, Fig. 32) suggested that the carbon-isotope signatures within the caprinid-bearing unit of Gavotte roughly mimic the vertical evolution of sparite proportions. Indeed, the vertical increase of sparite proportions (from ~5 to ~15%) between −140 m and −90 m is reflected by a negative trend in the $\delta^{13}C$ values (from ~+3.5 to ~+2.5‰). Above, slightly more positive carbon-isotope values (~+3‰) are documented within a 20 m-thick interval (from −90 to −70 m) which records a decrease in sparite proportions (from ~15 to ~3%). The topmost part of the caprinid-bearing unit is represented by the most depleted carbon values (~+2‰) of the whole UB–*S3b* Sequence which may result from both the input of meteoric fluids below the UB.*d4* subaerial discontinuity and the increase of sparite proportions. Such blocky sparite is dated to the Turonian Stage (Godeau et al., 2018) and may imply the non-preservation of the primary isotopic signal in the case of the Gavotte borehole section (Fig. 19C).

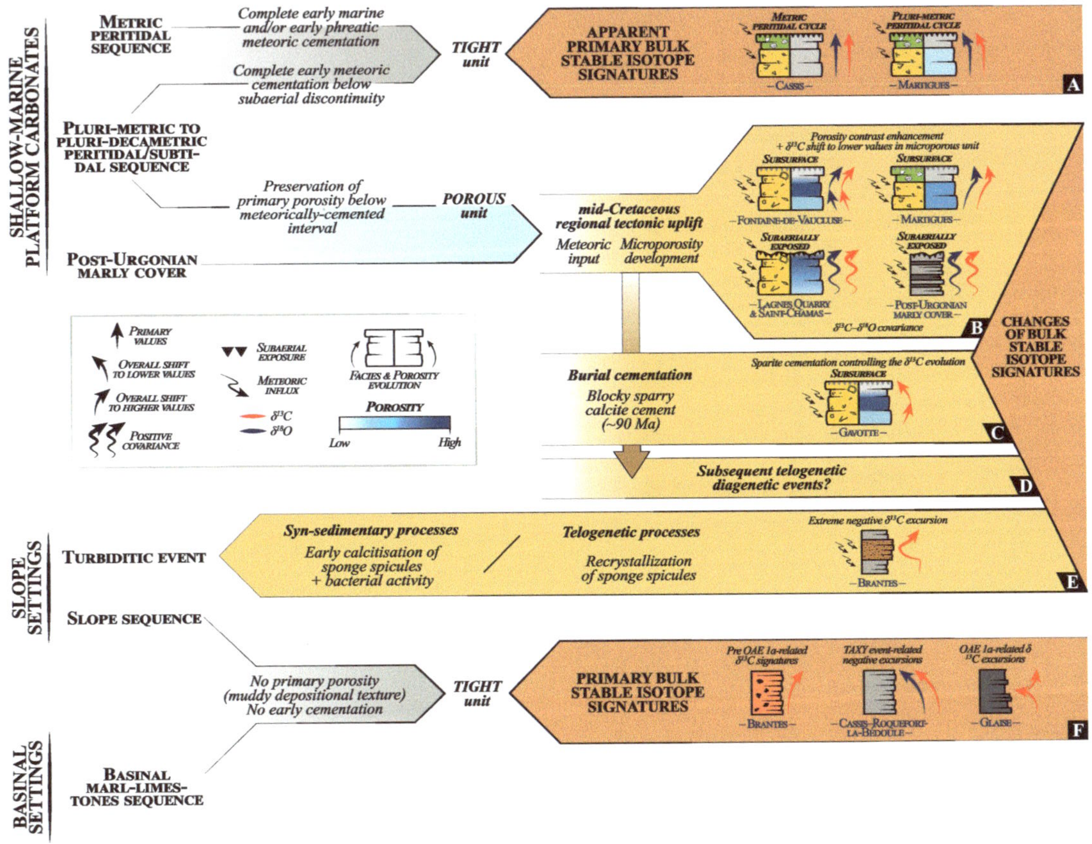

Fig. 19 Conceptual evolution model representing the carbon- and oxygen-isotope responses of carbonates according to the main diagenetical events from the Provence domain. See discussion in the text for details and Fig. 3B for explanation of symbols and color legend.

At the Lagnes Quarry, the caprinid-bearing unit (UB–*S3b* Sequence) exhibits porosity values up to 28% (Fig. 6). The corresponding $\delta^{13}C$ signatures scatter from −2‰ to +3.5‰. A prominent shift toward more negative values is observed in the topmost 50 m of the section. The base of this interval is represented by a positive covariance ($R^2 \sim 0.75$) between the $\delta^{18}O$ and $\delta^{13}C$. According to Allan and Matthews (1982), such covariance would reflect a diagenetical alteration in the mixing zone. However, as discussed by Swart and Oehlert (2018), carbonate dissolution processes predominate in the mixing zone and hampers the development of a characteristic geochemical imprint. Variable intensities of diagenetical processes (i.e., recrystallization, neomorphism and cementation) within the phreatic meteoric zone would rather result in the covariance between the $\delta^{18}O$ and $\delta^{13}C$ signatures according to Swart and Oehlert (2018). As no major *syn*-sedimentary exposure of the caprinid-bearing unit is reported in this part of the platform (Léonide et al., 2012; Masse and Fenerci-Masse, 2011, 2013; Tendil et al., 2018), the regional Durancian uplift of mid-Cretaceous age (Masse and Philip, 1976) constitutes the main hypothesis to explain the development of a large extent meteoric domain. The observation of more positive carbon-isotope values toward the base of the caprinid-bearing unit combined with an overall, slight decreasing of the microporosity likely indicates a vertical decreasing of the meteoric alteration. The Durancian erosion of the sedimentary cover has probably strengthened the contribution of the meteoric-related processes in the Lagnes Quarry area where porosity values are among the highest in the Provence domain (Tendil, 2018).

The truncation of the Lagnes Quarry Urgonian series by an unconformity of Burdigalian age testifies of a polyphased erosion of the post-caprinid unit. The regional uplift of the Provence domain following the tectonic Pyreneo-Provencal inversion during the Eocene may have also played a role in the sedimentary truncation and in the reactivation of meteoric fluids circulations (Fig. 19D). Nevertheless, the amalgam of several subaerial exposure events does not necessarily imply a modification of the isotopic signal due to the mineralogical stabilization which occurred during the first emersive period (Swart and Oehlert, 2018). In the Lagnes Quarry core section, the topmost part of the carbon-isotope profile may represent the Durancian-related geochemical responses of the Urgonian carbonates. To some extent, this could also be the case for other Provence localities affected by the Durancian event.

In the southern part of the Urgonian Provence platform, both the Martigues (Fig. 16B) and Saint-Chamas (Fig. 5) sections are also truncated by an erosion surface originating from the Durancian uplift. At Martigues section, only a negative shift of the $\delta^{13}C$ is identified below the surface, while at Saint-Chamas, the $\delta^{18}O$ and $\delta^{13}C$ expose a covariant trend ($R^2 \sim 0.71$) throughout the topmost 100 m (at least) of the Urgonian succession (Figs. 5 and 19B). The vertical organization of the depositional sequences, i.e., pluri-metric peritidal cycle at Martigues and decametric subtidal cycle at Saint-Chamas, may have played a key role in the distribution of meteoric lenses which ultimately led to isotopic perturbations. A covariant trend is not exclusive to Urgonian carbonates as it is also reported in the post-Urgonian marly cover of the northern part of the Provence platform ($R^2 \sim 0.62$–0.80 in LA–*S6*/S7 sequences; Fig. 19B) (Frau et al., 2017). These marls were probably close to the alteration front of the Durancian exposure as Albian deposits did not exceed a few tens of meters in the close vicinity (Joseph et al., 1987, 1989), hence their isotopic resetting. The Durancian tectonic uplift appears, therefore, as a major event in the diagenetical history of the Barremian–Aptian deposits of the Provence domain and led to important perturbations of the stable isotopes record, preventing the recognition of regional isotopic patterns.

7.2 Geochemical record from slope settings

The southern margin of the Vocontian Basin records the sedimentary influence of the near Urgonian Provence platform through the Barremian/Aptian boundary interval (Tendil et al., 2018). This is such a case in the Brantes section where massive or finely-laminated turbidites are interbedded between basinal marly limestones and cherty calcisiltites which record a primary, pre-OAE 1a $\delta^{13}C$ signal (Fig. 12). At Brantes, the earliest Aptian turbidites exhibit strongly depleted $\delta^{13}C$ values (up to $-7‰$; Fig. 12). Such negative excursions are not observed elsewhere in the deep Vocontian and shallow-marine Provence settings (Fig. 15). Higher proportions of syntaxial and calcite sparite cements are identified from point counting performed on selected thin sections (Tendil, 2018, fig. 33). Most of the calcite derives from the recrystallization of sponge spicules whose accumulation can be very important in platform-derived deposits. Cathodoluminescence analysis shows a luminescent micrite (Fig. 17C) which suggests a strong diagenetical alteration and contrasts with the non-luminescent sparite. This sparite shares common luminescence characteristics with sparite cements observed at Rustrel (*Agriopleura*-rich unit; Cochard, 2018). Of interest is that the Rustrel sparite cements occlude the most recent set of fractures reported in the at the transition between the *M. sarasini* and *P. waagenoides* subzones. Coeval hiatus, but of greater duration occurs in the Vocontian Basin at that time (Frau et al., 2018b) which also exhibits a negative carbon shift at a sedimentary break spanning part of the Barremian/Aptian boundary interval (Fig. 14). Increasing *syn*-sedimentary tectonics (i.e., onset of the Iberia rotation) together with relative sea level drop of high amplitude during the latest Barremian likely contributed to the demise of the Urgonian rudistid platform, although genetic relationships need further refinement (Tendil et al., 2018).

The OAE 1a has a good record in both the South Provence and Vocontian basins but with contrasting expressions (see Section 6; Figs. 3 and 14). As largely discussed in the literature (Jenkyns, 2010 and references therein), enhanced primary productivity and/or oxygen depletion in oceans leading to a massive burial of both marine and terrestrial organic matter during the OAE 1a would reflect the paroxysm of Pacific LIPs activity. The worldwide recognition of these characteristic C-isotope signatures during the mid-early Aptian prove the preservation of an original signal in muddy basinal facies (Fig. 17D). Nevertheless, the OAE 1a record can be lacking either due to local hiatus development (Frau et al., 2017) or regional resedimentation event (Cotillon, 2000).

In summary, major C-isotope disturbance events typically reflect global changes in the atmosphere–ocean system and prove useful for chemostratigraphic purposes (Fig. 19F). However, specific segments of the carbon-isotope profile could show contrasting expressions depending on local to regional tectonosedimentary context (Fig. 19B–E).

8. Conclusion

The geochemical investigation undertaken in this work highlights the wide scattering of both $\delta^{18}O$ (from −9.08‰ to −0.75‰) and $\delta^{13}C$ (from −7.02‰ to +4.66‰) values along the platform-to-basin transect of the Provence domain. Most of the stable isotopes recorded in Urgonian platform carbonates differ from the Barremian–Aptian original marine signal and their absolute values and overall isotopic trends cannot be regionally recognized.

Our analysis of the Urgonian platform series shows that the O- and C-isotope signatures are closely related to the vertical stacking pattern of depositional sequences. In the innermost platform settings, the $\delta^{18}O$ and $\delta^{13}C$ of early cemented, meter-thick peritidal cycles apparently conform to the marine values despite successive, short-termed subaerial exposures.

In thicker peritidal to subtidal cycles, the stable isotopes record commonly mimics the diagenetic imprint of platform carbonates (occurrence of microporous and tight units). Microporous limestones are frequently represented by more negative values than in cemented platform carbonates, thus reflecting the massive input of meteoric waters during the mid-Cretaceous Durancian tectonic uplift of the Provence region which led to the microporosity enhancement and preservation. Such geochemical response is not unique as varying cement proportions and diagenetic processes directly below Durancian-truncated surfaces locally add complexity to the isotopic signal interpretation. There, microporous units are either characterized by more positive $\delta^{13}C$ values or even covariant $\delta^{18}O$–$\delta^{13}C$ profiles.

In slope settings, the marly limestone successions are interbedded with thick turbiditic, spicule-rich intervals. These spicules were heavily affected by telogenetic meteoric fluids which induced the diagenetic resetting of the turbidites-related C-isotope signatures. The original perturbations of the global carbon cycle during the Barremian–Aptian transition are distally preserved in the South Provence and Vocontian basinal sections.

Building from these results, the interpretation of the stable isotopes record remains a challenging issue and great caution is necessary in applying chemostratigraphy in carbonate environments. The high-resolution temporal calibration of the Urgonian Provence platform series reveals that the C-isotope stratigraphy tool cannot be applied for refining the regional chronostratigraphic framework. At best, the interpretation of isotopic compositions of Urgonian carbonates must be considered as a hypothesis-generating approach when defining local to global scales stratigraphic correlations and/or reconciling established biostratigraphic schemes. Through this contribution, the authors aim to encourage stimulating discussions toward a further understanding of the processes controlling the geochemical signatures of carbonates.

Acknowledgments

This work is a contribution to the Ph.D thesis of the first author (A.J.-B.T.). It is part of the ALBION R&D Project initiated by TOTAL S.A. whose financial support is gratefully acknowledged. We are grateful to the editor Michael Montenari for having invited us to contribute to Elsevier's 'Stratigraphy & Timescales' volume. Project managers Joanna Collett and Denny Mansingh are also thanked for their kind help. We warmly thank Dominique Sandri-Mouze (LHOIST Group) and the Laboratoire Souterrain à Bas Bruit (LSBB) staff for granting us access to the Lagnes Quarry and LSBB facilities, respectively. Alix Loumbangoy assisted the first author in sampling the Lagnes Quarry core. Last but not least, the first two authors wish to express their warmest thanks to their families for their help and continuous support during the preparation of this work.

References

Allan, J.R., Matthews, R.K., 1982. Isotope signatures associated with early meteoric diagenesis. Sedimentology 29, 797–817.

Amodio, S., Weissert, H., 2017. Palaeoenvironment and palaeoecology before and at the onset of Oceanic Anoxic Event (OAE) 1a: reconstructions from Central Tethyan archives. Palaeogeogr. Palaeoclimatol. Palaeoecol. 479, 71–89.

Arthur, M.A., Brumsack, H.-J., Jenkyns, H.C., Schlanger, S.O., 1990. Stratigraphy, geochemistry, and paleoceanography of organic carbon-rich Cretaceous sequences. In: Ginsburg, R., Beaudoin, B. (Eds.), Cretaceous Resources, Events, and Rhythms. In: NATO ASI Series C, 304, Kluwer Academic, pp. 75–119.

Bąk, M., Górny, Z., Bąk, K., Wolska, A., Stożek, B., 2015. Successive stages of calcitization and silicification of Cenomanian spicule-bearing turbidites based on microfacies analysis, Polish Outer Carpathians. Ann. Soc. Geol. Pol. 85, 187–203.

Bastide, F., 2014. Synthèse de l'évolution de la plateforme Urgonienne (Barrémien tardif à aptien précoce) du Sud-Est de la France: Faciès, micropaléontologie, géochimie, géométries, paléotectonique et géomodélisation. Doctoral thesis, Université de Grenoble, p. 300.

Berner, R.A., 2006. GEOCARBSULF: a combined model for Phanerozoic atmospheric O_2 and CO_2. Geochim. Cosmochim. Acta 70, 5653–5664. A Special Issue Dedicated to Robert A. Berner.

Bestani, L., Espurt, N., Lamarche, J., Bellier, O., Hollender, F., 2016. Reconstruction of the Provence Chain evolution, southeastern France. Tectonics 35, 1506–1525.

Bodin, S., Fiet, N., Godet, A., Matera, V., Westermann, S., Clément, A., Janssen, N.M.M., Stille, P., Föllmi, K.B., 2009. Early Cretaceous (late Berriasian to early Aptian) palaeoceanographic change along the northwestern Tethyan margin (Vocontian Trough, southeastern France): $\delta^{13}C$, $\delta^{18}O$ and Sr-isotope belemnite and whole-rock records. Cretac. Res. 30, 1247–1262.

Bonvallet, L., Arnaud-Vanneau, A., Arnaud, H., Adatte, T., Spangenberg, J.E., Stein, M., Godet, A., Föllmi, K.B., 2019. Evolution of the Urgonian shallow-water carbonate platform on the Helvetic shelf during the late Early Cretaceous. Sediment. Geol. 387, 18–56.

Bréhérét, J.-G., 1997. L'Aptien et l'Albien de la Fosse vocontienne (des bordures au bassin). Evolution de la sédimentation et enseignements sur les évènements anoxiques. Publ. Soc. Géol. Nord 25, 614.

Brlek, M., Glumac, B., 2014. Stable isotopic ($\delta^{13}C$ and $\delta^{18}O$) signatures of biogenic calcretes marking discontinuity surfaces: a case study from Upper Cretaceous carbonates of central Dalmatia and eastern Istria, Croatia. Facies 60, 773–788.

Buonocunto, F.P., Sprovieri, M., Bellanca, A., D'Argenio, B., Ferreri, V., Neri, R., Ferruzza, G., 2002. Cyclostratigraphy and high-frequency carbon isotope fluctuations in Upper Cretaceous shallow-water carbonates, southern Italy. Sedimentology 49, 1321–1337.

Burla, S., Heimhofer, U., Hochuli, P.A., Weissert, H., Skelton, P., 2008. Changes in sedimentary patterns of coastal and deep-sea successions from the North Atlantic (Portugal) linked to Early Cretaceous environmental change. Palaeogeogr. Palaeoclimatol. Palaeoecol. 257, 38–57.

Burla, S., Oberli, F., Heimhofer, U., Wiechert, U., Weissert, H., 2009. Improved time control on Cretaceous coastal deposits: new results from Sr isotope measurements using laser ablation. Terra Nova 21, 401–409.

Busnardo, R., 1965. Rapport sur l'étage Barrémien. In: Colloque sur le Crétacé inférieur (Lyon, Septembre 1963). 34, BRGM, pp. 161–169. Mém.

Cochard, J., 2018. Analyses des propriétés réservoirs d'une série carbonatée microporeuse fracturée. Doctoral thesis, Aix-Marseille Université, p. 198.

Cotillon, P., 2000. Sea bottom current activity recorded on the southern margin of the Vocontian basin (southeast France) during the Lower Aptian. Evidence for a climatic signal. Bull. Soc. Geol. Fr. 181, 3–18.

D'Argenio, B., Ferreri, V., Weissert, H., Amodio, S., Buonocunto, F.P., Wissler, L., 2004. A multidisciplinary approach to global correlation and geochronology. The Cretaceous shallow-water carbonates of southern Apennines, Italy. In: D'Argenio, B., Fischer, A.G., Premoli Silva, I., Weissert, H., Ferreri, V. (Eds.), Cyclostratigraphy: Approaches and Case Histories. In: vol. 81. SEPM Special Publication, pp. 103–122.

Dauphin, L., 2002. Litho-, bio-, et chronostratigraphie comparées dans le Bassin Vocontien à l'Aptien. Doctoral thesis, Université de Lille, p. 451.

Delanoy, G., 1997. Biostratigraphie des faunes d'Ammonites à la limite Barrémien–Aptien dans la région d'Angles–Barrême–Castellane. Etude particulière de la famille des Heteroceratina Spath, 1922 (Ancyloceratina, Ammonoidea). Ann. Mus. Hist. Nat. Nice 12, 1–270.

Di Lucia, M., Trecalli, A., Mutti, M., Parente, M., 2012. Biochemostratigraphy of the Barremian-Aptian shallow-water carbonates of the southern Apennines (Italy): pinpointing the OAE1a in a Tethyan carbonate platform. Solid Earth 3, 1–28.

Embry, J.C., 2005. Paléoécologie et architecture stratigraphique en haute résolution des plates-formes carbonatées du Barrémien–Aptien de la Néo-Téthys (Espagne, Suisse, Provence–Vercors). Impact respectif des différents facteurs de contrôle. Doctoral thesis, Institut Français du Pétrole, p. 298.

Falkowski, P., Scholes, R.J., Boyle, E., Canadell, J., Canfield, D., Elser, J., Gruber, N., Hibbard, K., Högberg, P., Linder, S., Mackenzie, F.T., Iii, B.M., Pedersen, T., Rosenthal, Y., Seitzinger, S., Smetacek, V., Steffen, W., 2000. The global carbon cycle: a test of our knowledge of earth as a system. Science 290, 291–296.

Fenerci-Masse, M., Masse, J.-P., Pernarcic, E., 2005. Quantitative stratigraphy of rudist limestones and its bearing on spatial organisation of rudist communities: the Late Barremian, Urgonian, sequences of Provence (S.E. France). Palaeogeogr. Palaeoclimatol. Palaeoecol. 215, 265–284.

Föllmi, K.B., 2008. A synchronous, middle Early Aptian age for the demise of the Helvetic Urgonian platform related to the unfolding oceanic anoxic event 1a (Selli event). Rev. Paléobiol. 27, 461–468.

Föllmi, K.B., 2012. Early Cretaceous life, climate and anoxia. Cretac. Res. 35, 230–257.

Föllmi, K.B., Godet, A., 2013. Palaeoceanography of Lower Cretaceous Alpine platform carbonates. Sedimentology 60, 131–151.

Fouke, B.W., Everts, A.-J.W., Zwart, E.W., Schlager, W., Smalley, P.C., Weissert, H., 1996. Subaerial exposure unconformities on the Vercors carbonate platform (SE France) and their sequence stratigraphic significance. J. Geol. Soc. Lond. Spec. Publ. 104, 295–319.

Frau, C., Delanoy, G., Masse, J.-P., Lanteaume, C., Tendil, A.J.-B., 2016. New Heteroceratidae (Ammonoidea) from the late Barremian deepening succession of Marseille (Bouches-du-Rhône, France). Acta Geol. Pol. 66, 205–225.

Frau, C., Pictet, A., Spangenberg, J., Masse, J.-P., Tendil, A.J.-B.T., Lanteaume, C., 2017. New insights on the age of the post-Urgonian marly cover of the Apt region (Vaucluse, SE France) and its implications on the demise of the North Provence carbonate platform. Sediment. Geol. 359, 44–61.

Frau, C., Tendil, A.J.-B., Lanteaume, C., Masse, J.-P., Pictet, A., Bulot, L.G., Luber, T., Redfern, J., Borgomano, J., Léonide, P., Fournier, F., Massonnat, G., 2018a. Late Barremian–Early Aptian ammonite bioevents from the Urgonian-type series of Provence (Southeast France): regional stratigraphic correlations and implications for the peri-Vocontian carbonate platforms. Cretac. Res. 90, 222–253.

Frau, C., Bulot, L.G., Delanoy, G., Moreno-Bedmar, J.A., Masse, J.-P., Lanteaume, C., Tendil, A.J.-B.T., 2018b. The candidate Aptian GSSP at Gorgo a Cerbara (Central Italy): an alternative interpretation of the bio-, litho- and chemostratigraphic markers. Newsl. Stratigr. 51, 311–326.

Godeau, N., Deschamps, P., Guilhou, A., Léonide, P., Tendil, A., Gerdes, A., Hamelin, B., Girard, J.-P., 2018. U-Pb dating of calcite cement and diagenetic history in microporous carbonate reservoirs: case of the Urgonian Limestone, France. Geology 46 (3), 247–250.

Godet, A., 2006. The Evolution of the Urgonian Platform in the Western Swiss Jura Realm and Its Interactions With Palaeoclimatic and Palaeoceanographic Change Along the Northern Tethyan Margin (Hauterivian-Earliest Aptian). Doctoral thesis, Université de Neuchâtel, p. 405.

Godet, A., Durlet, C., Spangenberg, J.E., Föllmi, K.B., 2016. Estimating the impact of early diagenesis on isotope records in shallow-marine carbonates: a case study from the Urgonian Platform in western Swiss Jura. Palaeogeogr. Palaeoclimatol. Palaeoecol. 454, 125–138.

Graziano, R., 2013. Sedimentology, biostratigraphy and event stratigraphy of the Early Aptian oceanic anoxic event (OAE1A) in the Apulia Carbonate Platform margin–Ionian basin system (Gargano promontory, southern Italy). Cretac. Res. 39, 78–111.

Graziano, R., Raspini, A., 2017. High-resolution chronostratigraphy of palaeoecologic and isotopic changes in shallow-marine carbonates: deciphering the completeness of the Aptian record in the Apennine carbonate platform (southern Italy). Cretac. Res. 86, 97–128.

Grötsch, J., 1998. Carbon-isotope stratigraphy in shallow-water carbonates: implications for Cretaceous black-shale deposition. Sedimentology 45, 623–634.

Hamon, Y., Merzeraud, G., 2007. C and O isotope stratigraphy in shallow-marine carbonate: a tool for sequence stratigraphy (example from the Lodève region, peritethian domain). Swiss J. Geosci. 100, 71–84.

Herrle, J.O., Kössler, P., Friedrich, O., Erlenkeuser, H., Hemleben, C., 2004. High-resolution carbon isotope records of the Aptian to Lower Albian from SE France and the Mazagan Plateau (DSDP Site 545): a stratigraphic tool for paleoceanographic and paleobiologic reconstruction. Earth Planet. Sci. Lett. 218, 149–161.

Hesselbo, S.P., Gröcke, D.R., Jenkyns, H.C., Bjerrum, C.J., Farrimond, P., Morgans Bell, H.S., Green, O.R., 2000. Massive dissociation of gas hydrate during a Jurassic oceanic anoxic event. Nature 406, 392–395.

Huck, S., Heimhofer, U., 2015. Improving shallow water carbonate chemostratigraphy by means of rudist bivalve sclerochemistry. Geochem. Geophys. Geosyst. 16, 18.

Huck, S., Rameil, N., Korbar, T., Heimhofer, U., Wieczorek, T.D., Immenhauser, A., 2010. Latitudinally different responses of Tethyan shoal-water carbonate systems to the Early Aptian oceanic anoxic event (OAE 1a). Sedimentology 57, 1585–1614.

Huck, S., Heimhofer, U., Rameil, N., Bodin, S., Immenhauser, A., 2011. Strontium and carbon-isotope chronostratigraphy of Barremian–Aptian shoal-water carbonates: Northern Tethyan platform drowning predates OAE 1a. Earth Planet. Sci. Lett. 304, 547–558.

Huck, S., Heimhofer, U., Immenhauser, A., 2012. Early Aptian algal bloom in a neritic proto-North Atlantic setting: harbinger of global change related to OAE 1a? Geol. Soc. Am. Bull. 124, 1810–1825.

Huck, S., Heimhofer, U., Immenhauser, A., Weissert, H., 2013. Carbon-isotope stratigraphy of Early Cretaceous (Urgonian) shoal-water deposits: diachronous changes in carbonate-platform production in the north-western Tethys. Sediment. Geol. 290, 157–174.

Huck, S., Stein, M., Immenhauser, A., Skelton, P.W., Christ, N., Föllmi, K.B., Heimhofer, U., 2014. Response of proto-North Atlantic carbonate platform ecosystems to OAE1a-related stressors. Sediment. Geol. 313, 15–31.

Huck, S., Wohlwend, S., Coimbra, R., Christ, N., Weissert, H., 2017. Disentangling shallow-water bulk carbonate carbon isotope archives with evidence for multi-stage diagenesis: an in-depth component-specific petrographic and geochemical study from Oman (mid-Cretaceous). Depos. Rec. 3 (2), 233–257.

Immenhauser, A., Della Porta, G., Kenter, J.A.M., Bahamonde, J.R., 2003. An alternative model for positive shifts in shallow-marine carbonate $\delta^{13}C$ and $\delta^{18}O$: an alternative model for carbon and oxygen isotope shifts. Sedimentology 50, 953–959.

Jenkyns, H.C., 1980. Cretaceous anoxic events: from continents to oceans. J. Geol. Soc. Lond. 137, 171–188.

Jenkyns, H.C., 1995. Carbon-isotope stratigraphy and paleoceanographic significance of the Lower Cretaceous shallow-water carbonates of Resolution Guyot, Mid-Pacific Mountains. Proc. Ocean Drill. Program Sci. Results 143, 99–104. Ocean Drilling Program College Station.

Jenkyns, H.C., 2010. Geochemistry of oceanic anoxic events: review. Geochem. Geophys. Geosyst. 11 (3), 1–30.

Jenkyns, H.C., Wilson, P.A., 1999. Stratigraphy, paleoceanography, and evolution of Cretaceous Pacific guyots: relics from a greenhouse Earth. Am. J. Sci. 299 (5), 341–392.

Jenkyns, H.C., Gale, A.S., Corfield, R.M., 1994. Carbon- and oxygen-isotope stratigraphy of the English Chalk and Italian Scaglia and its palaeoclimatic significance. Geol. Mag. 131 (1), 1–34.

Joachimski, M.M., 1994. Subaerial exposure and deposition of shallowing upward sequences: evidence from stable isotopes of Purbeckian peritidal carbonates (basal Cretaceous), Swiss and French Jura Mountains. Sedimentology 41 (4), 805–824.

Joseph, P., Cabrol, C., Friès, G., 1987. Blocs basculés et passes sous-marines dans le champ de Banon (France, S.E.) à l'Apto-Albien: une paléotopographie directement contrôlée par la tectonique synsédimentaire décrochante. C.R. Acad. Sci. Ser. IIa Sci. Terre Planets 304 (9), 447–452.

Joseph, P., Beaudoin, B., Friès, G., Parize, O., 1989. Les vallées sous-marines enregistrent au Crétacé inférieur le fonctionnement en blocs basculés du domaine vocontien. C.R. Acad. Sci. Ser. IIa Sci. Terre Planets 309 (10), 1031–1038.

Kuhnt, W., Moullade, M., Masse, J.-P., Erlenkeuser, H., 2000. Carbon isotope stratigraphy of the lower Aptian historical stratotype at Cassis–La Bédoule. Geol. Mediter. 25 (1998), 63–79.

Kump, L.R., Arthur, M.A., 1999. Interpreting carbon-isotope excursions: carbonates and organic matter. Chem. Geol. 161 (1–3), 181–198.

Lamarche, J., Lavenu, A.P.C., Gauthier, B.D.M., Guglielmi, Y., Jayet, O., 2012. Relationships between fracture patterns, geodynamics and mechanical stratigraphy in Carbonates (South-East Basin, France). Tectonophysics 581, 231–245.

Léonide, P., Borgomano, J., Masse, J.-P., Doublet, S., 2012. Relation between stratigraphic architecture and multi-scale heterogeneities in carbonate platforms: the Barremian–lower Aptian of the Monts de Vaucluse, SE France. Sediment. Geol. 265, 87–109.

Léonide, P., Fournier, F., Reijmer, J.J.G., Vonhof, H.B., Borgomano, J., Dijk, J., Rosenthal, M., van Goethem, M., Cochard, J., Meulenaars, K., 2014. Diagenetic patterns and pore space distribution along a platform to outer-shelf transect (Urgonian limestone, Barremian–Aptian, SE France). Sediment. Geol. 306, 1–23.

Lini, A., Weissert, H., Erba, E., 1992. The Valanginian carbon isotope event: a first episode of greenhouse climate conditions during the Cretaceous. Terra Nova 4, 374–384.

Masse, J.-P., 1976. Les calcaires urgoniens de Provence (Valanginien–Aptien inférieur). Stratigraphie, paléontologie, les paléoenvironnements et leur évolution. Doctoral thesis, Université d'Aix–Marseille II, p. 445.

Masse, J.-P., 1993. Valanginian to early Aptian carbonate platforms from Provence (SE France). In: Simo, A., Scott, R.W., Masse, J.-P. (Eds.), Cretaceous Carbonate Platforms. In: Am. Assoc. Pet. Geol. Bull. Mem., vol. 5. , pp. 363–374.

Masse, J.-P., Fenerci-Masse, M., 2011. Drowning discontinuities and stratigraphic correlations in platform carbonates. The late Barremian–Early Aptian record of South-East France. Cretac. Res. 32, 659–684.
Masse, J.-P., Fenerci-Masse, M., 2013. Drowning events, development and demise of carbonate platforms and controlling factors: the Late Barremian–Early Aptian record of Southeast France. Sediment. Geol. 298, 28–52.
Masse, J.P., Philip, J., 1976. Paléogéographie et tectonique du Crétacé moyen en Provence: révision du concept d'isthme durancien. Rev. Géogr. Phys. Géol. Dyn. 18, 49–66.
Masse, J.-P., El Albani, A., Erlenkeuser, H., 1999. Stratigraphie isotopique ($\delta^{13}C$) de l'Aptien inférieur de Provence (SE France): Application aux corrélations plate-forme/bassin. Eclogae Geol. Helv. 92, 259–263.
Masse, J.-P., Fenerci, M., Borgomano, J., 2001. Levelling pattern in peritidal carbonates, Late Barremian from Cassis (Marseille region, SE France). Anatomic implications. Geol. Mediter. 28, 117–120.
Masse, J.-P., Fenerci, M., Pernarcic, E., 2003. Palaeobathymetric reconstruction of peritidal carbonates: Late Barremian, Urgonian, sequences of Provence (SE France). Palaeogeogr. Palaeoclimatol. Palaeoecol. 200, 65–81.
Méhay, S., Keller, C.E., Bernasconi, S.M., Weissert, H., Erba, E., Bottini, C., Hochuli, P.A., 2009. A volcanic CO_2 pulse triggered the Cretaceous Oceanic Anoxic Event 1a and a biocalcification crisis. Geology 37, 819–822.
Menegatti, A.P., Weissert, H., Brown, R.S., Tyson, R.V., Farrimond, P., Strasser, A., Caron, M., 1998. High-resolution $\delta^{13}C$ stratigraphy through the Early Aptian "Livello selli" of the Alpine Tethys. Paleoceanography 13, 530–545.
Millán, M.I., Weissert, H.J., Fernández-Mendiola, P.A., García-Mondéjar, J., 2009. Impact of Early Aptian carbon cycle perturbations on evolution of a marine shelf system in the Basque-Cantabrian Basin (Aralar, N Spain). Earth Planet. Sci. Lett. 287, 392–401.
Millán, M.I., Weissert, H.J., Owen, H., Fernández-Mendiola, P.A., García-Mondéjar, J., 2011. The Madotz Urgonian platform (Aralar, northern Spain): paleoecological changes in response to Early Aptian global environmental events. Palaeogeogr. Palaeoclimatol. Palaeoecol. 312, 167–180.
Naafs, B.D.A., Castro, J.M., Gea, G.A.D., Quijano, M.L., Schmidt, D.N., Pancost, R.D., 2016. Gradual and sustained carbon dioxide release during Aptian Oceanic Anoxic Event 1a. Nat. Geosci. 9, 135.
Parente, M., Frijia, G., Lucia, M.D., 2007. Carbon-isotope stratigraphy of Cenomanian–Turonian platform carbonates from the southern Apennines (Italy): a chemostratigraphic approach to the problem of correlation between shallow-water and deep-water successions. J. Geol. Soc. Lond. 164, 609–620.
Phelps, R.M., Kerans, C., Da-Gama, R.O., Jeremiah, J., Hull, D., Loucks, R.G., 2015. Response and recovery of the Comanche carbonate platform surrounding multiple Cretaceous oceanic anoxic events, northern Gulf of Mexico. Cretac. Res. 54, 117–144.
Pictet, A., Delanoy, G., Adatte, T., Spangenberg, J.E., Baudouin, C., Boselli, P., Boselli, M., Kindler, P., Föllmi, K.B., 2015. Three successive phases of platform demise during the early Aptian and their association with the oceanic anoxic Selli episode (Ardèche, France). Palaeogeogr. Palaeoclimatol. Palaeoecol. 418, 101–125.
Railsback, L.B., Holland, S.M., Hunter, D.M., Jordan, E.M., Diaz, J.R., Crowe, D.E., 2003. Controls on geochemical expression of subaerial exposure in Ordovician limestones from the Nashville Dome, Tennessee, U.S.A. J. Sediment. Res. 73, 790–805.
Reboulet, S., Szives, O., Aguirre-Urreta, B., Barragan, R., Company, M., Frau, C., Kakabadze, M.V., Klein, J., Moreno-Bedmar, J.A., Lukeneder, A., Pictet, A., Ploch, I., Raisossadat, S.N., Vašíček, Z., Baraboshkin, E.J., Mitta, V.V., 2018. Report on the 6th international meeting of the IUGS Lower Cretaceous ammonite working group, the Kilian Group (Vienna, Austria, 20th August 2017). Cretac. Res. 91, 100–110.

Renard, M., de Rafélis, M., Emmanuel, L., Moullade, M., Masse, J.-P., Kuhnt, W., Bergen, J.A., Tronchetti, G., 2005. Early Aptian $\delta^{13}C$ and manganese anomalies from the historical Cassis-La Bédoule stratotype sections (S.E. France): relationship with a methane hydrate dissociation event and stratigraphic implications. Carnets Geol. 24, 18. Article 2005/04 (CG2005_A04).

Schlanger, S.O., Jenkyns, H.C., 1976. Cretaceous oceanic anoxic events: causes and consequences. Neth. J. Geosci. 55 (3–4), 179–184.

Scholle, P.A., Arthur, M.A., 1980. Carbon isotope fluctuations in Cretaceous pelagic limestones: potential stratigraphic and petroleum exploration tool. Am. Assoc. Pet. Geol. Bull. 64, 67–87.

Stein, M., Arnaud-Vanneau, A., Adatte, T., Fleitmann, D., Spangenberg, J.E., Föllmi, K.B., 2012a. Palaeoenvironmental and palaeoecological change on the northern Tethyan carbonate platform during the Late Barremian to earliest Aptian: sedimentary pattern of the Rawil Member. Sedimentology 59, 939–963.

Stein, M., Westermann, S., Adatte, T., Matera, V., Fleitmann, D., Spangenberg, J.E., Föllmi, K.B., 2012b. Late Barremian–Early Aptian palaeoenvironmental change: the Cassis-La Bédoule section, southeast France. Cretac. Res. 37, 209–222.

Strasser, A., 2015. Hiatuses and condensation: an estimation of time lost on a shallow carbonate platform. Depos. Rec. 1, 91–117.

Swart, P.K., 2015. The geochemistry of carbonate diagenesis: the past, present and future. Sedimentology 62 (5), 1233–1304.

Swart, P.K., Oehlert, A.M., 2018. Revised interpretations of stable C and O patterns in carbonate rocks resulting from meteoric diagenesis. Sediment. Geol. 364, 14–23.

Tendil, A.J.-B., 2018. Tectonic, Climatic and Paleoceanographic Controls on the Stratigraphic Architecture of the Urgonian Provence Carbonate Platform (France): Iintegrated Sedimentological, Seochemical and Numerical Approaches. Doctoral thesis, Aix–Marseille University, p. 436.

Tendil, A.J.-B., Frau, C., Léonide, P., Fournier, F., Borgomano, J.R., Lanteaume, C., Masse, J.-P., Massonnat, G., Rolando, J.-P., 2018. Platform-to-basin anatomy of a Barremian–Aptian Tethyan carbonate system: new insights into the regional to global factors controlling the stratigraphic architecture of the Urgonian Provence platform (Southeast France). Cretac. Res. 91, 382–411.

Vahrenkamp, V.C., 1996. Carbon isotope stratigraphy of the Upper Kharaib and Shuaiba formations: implications for the Early Cretaceous evolution of the Arabian Gulf Region. Am. Assoc. Pet. Geol. Bull. 80, 647–661.

Vahrenkamp, V.C., 2010. Chemostratigraphy of the Lower Cretaceous Shu'aiba formation: a δ13C reference profile for the Aptian Stage from the southern Neo-Tethys Ocean. In: van Buchem, F.S.P., Al-Husseini, M.I., Maurer, F., Droste, H.J. (Eds.), Barremian-Aptian Stratigraphy and Petroleum Habitat of the Eastern Arabian Plate. In: 4, GeoArabia Special Publication, pp. 107–137. 2.

Vincent, B., van Buchem, F.S., Bulot, L.G., Immenhauser, A., Caron, M., Baghbani, D., Huc, A.Y., 2010. Carbon-isotope stratigraphy, biostratigraphy and organic matter distribution in the Aptian–Lower Albian successions of southwest Iran (Dariyan and Kazhdumi formations). Geoarabia-Manama 4, 139–197. Special Publication.

Weissert, H., Joachimski, M., Sarnthein, M., 2008. Chemostratigraphy. Newsl. Stratigr. 42, 145–179.

Westermann, S., 2010. Trace-Element and Phosphorus Contents in Sediments Associated With Cretaceous Oceanic Anoxic Events. Doctoral thesis, Université de Lausanne, p. 223.

Westermann, S., Stein, M., Matera, V., Fiet, N., Fleitmann, D., Adatte, T., Föllmi, K.B., 2013. Rapid changes in the redox conditions of the western Tethys Ocean during the early Aptian oceanic anoxic event. Geochim. Cosmochim. Acta 121, 467–486.

Wissler, L., Funk, H., Weissert, H., 2003. Response of Early Cretaceous carbonate platforms to changes in atmospheric carbon dioxide levels. Palaeogeogr. Palaeoclimatol. Palaeoecol. 200, 187–205.
Yamamoto, K., Ishibashi, M., Takayanagi, H., Asahara, Y., Sato, T., Nishi, H., Iryu, Y., 2013. Early Aptian paleoenvironmental evolution of the Bab Basin at the southern Neo-Tethys margin: response to global carbon cycle perturbations across Ocean Anoxic Event 1a. Geochem. Geophys. Geosyst. 14, 1104–1130.
Zhang, X., Zhang, G., Sha, J., 2016. Lacustrine sedimentary record of early Aptian carbon cycle perturbation in western Liaoning, China. Cretac. Res. 62, 122–129.

Further reading

Li, J., Hu, X., Zhao, K., Cai, Y., Sun, T., 2016. Paleoceanographic evolution and chronostratigraphy of the Aptian Oceanic Anoxic Event 1a (OAE1a) to oceanic red bed 1 (ORB1) in the Gorgo a Cerbara section (central Italy). Cretac. Res. 66, 115–128.
Moss, S., Tucker, M.E., 1995. Diagenesis of Barremian-Aptian platform carbonates (the Urgonian Limestone Formation of SE France): near-surface and shallow-burial diagenesis. Sedimentology 42, 853–874.
Wissler, L., Weissert, H., Masse, J.-P., Bulot, L., 2002. Chemostratigraphic correlation of Barremian and lower Aptian ammonite zones and magnetic reversals. Int. J. Earth Sci. 91, 272–279.

CHAPTER SIX

Computer modeling of Zircon ($ZrSiO_4$)—Coffinite ($USiO_4$) solid solutions and lead incorporation: Geological implications

Robert A. Jackson[a,*], Michael Montenari[b]
[a]School of Chemical and Physical Sciences, Keele University, Keele, United Kingdom
[b]School of Geography, Geology and the Environment, Keele University, Keele, United Kingdom
*Corresponding author: e-mail address: r.a.jackson@keele.ac.uk

Contents

Abstract

The structure of zircon, $ZrSiO_4$ is modeled using interatomic potentials. The uranium end-member, coffinite ($USiO_4$) and intermediate solid solutions of zircon and coffinite ($U_xZr_{1-x}SiO_4$) are then modeled, allowing the prediction of lattice parameters as a function of uranium concentration. Finally, possible structures resulting from the radioactive decay of uranium to lead in coffinite are considered.

Stratigraphy & Timescales, Volume 4
ISSN 2468-5178
https://doi.org/10.1016/bs.sats.2019.08.005

1. Introduction

1.1 Zircon

Zircon and coffinite are common minerals of the tetragonal silicate zircon-group, where the position of the tetravalent cation within the crystal lattice is either occupied by the chemical element zirconium (as in zircon, $ZrSiO_4$) or uranium (as in coffinite, $USiO_4$).

Zircon occurs as an ubiquitous accessory mineral in igneous and metamorphic rocks with the capability to incorporate significant amounts of uranium, thorium, hafnium, and a wide range of the rare earth elements (Bea, 1996; Cherniak et al., 1997; Grimes et al., 2007; Heaman et al., 1990; Hoskin and Ireland, 2000; Nardi et al., 2013; O'Hara et al., 2001; Sawka, 1988). The concentrations of the elements U-Pb and U-Th-Pb preserved within zircon crystals play an important geochronological role in the age determination of various magmatic and metamorphic processes and hence contribute significantly toward the elucidation of global processes ranging from the formation of Earth's early continental crust to large-scale orogenic processes (Amelin et al., 1999; Harrison et al., 2005; Wilde et al., 2001). Furthermore, concentrations of minor and trace elements preserved within zircon crystals are being used to effectively discriminate between various magmatic settings. Zircon crystals isolated from ultramafic materials such as kimberlites (with zircon crystals potentially derived from Earth's mantle) have, for example, the tendency to be typically very low in the actinide elements thorium ($Th < 10$ ppm) and uranium ($U < 30$ ppm) and in the REEs (Heaman et al., 1990). Zircon crystals from mafic materials have been demonstrated to be characterized by elevated thorium to uranium ratios ($Th/U > 1$), fractionated REE patterns and unusual high enrichments in scandium (with Sc concentrations varying between 86 and 230 ppm). Zircon crystals from magmatic suites associated with strongly extensional tectonic settings, such as continental rifts, differ fundamentally from the above described in having high ratios of the elements zirconium to hafnium ($Zr/Hf > 60$) and correspondingly low concentrations in the element hafnium ($Hf \leq 6000$–8000 ppm). These geochemical signatures preserved within zircon crystals have proven to be of utmost importance in the tracing and identification of source areas (Heaman et al., 1990).

Aside from their original magmatic and metamorphic country rocks zircon crystals are also commonly found as part of the detrital component in many siliciclastic sediments, were they are being used in provenance studies. Authigenic zircon crystals associated with hydrothermal activity have been described in the literature, but are considered to be very rare (Baruah et al., 1995; Hower et al., 1999; Saxena, 1966).

1.2 Coffinite

First described by Stieff et al. (1956) from the sedimentary deposits of the Colorado Plateaus province, vein deposits in the Copper King mine, Larimer County (Colorado) and asphaltic pellets preserved within sediments in Texas and Oklahoma, the UO^{4+} mineral coffinite was considered to be not much more than an oddity among the plethora of uranium-bearing minerals. It's natural occurrence seemed to be limited to rather insignificant amounts preserved within uranium sandstone deposits (Mesbah et al., 2015), in marked contrast to uraninite (or "pitchblende," UO_2) which was considered to be the main carrier of industrial relevant quantities of uranium.

The drastically growing demand for uranium that was triggered by its use as a fuel for nuclear power generation but also to manufacture nuclear weapons during the 1960s and 70s, sparked an increased interest in uranium exploration and exploitation. The resulting detailed research into uranium ores revealed, that coffinite was a far more widespread uranium mineral than previously assumed. It emerged, that coffinite forms large quantities of uranium-bearing phases, previously considered to be impure uraninite. Furthermore, it has meanwhile been described on a micro- to nanoscale from different geomaterials ranging from organic matter to apatite (Deditius et al., 2008, 2010, 2012) and from various geochemical settings such as the natural nuclear reactor of Oklo (Gabon) and hydrothermal deposits of, e.g., Cigar Lake, Canada (Fayek et al., 1997; Finch and Murakami, 1999; Janeczek, 1999; Janeczek and Ewing, 1992).

The tetragonal mineral coffinite is isostructural with the minerals zircon ($ZrSiO_4$, Pointer et al., 1988) hafnon ($HfSiO_4$, Salt and Hornung, 1967) thorite ($ThSiO_4$, Fuchs and Gebert, 1958) and, according to Mesbah et al. (2015), can form in the zircon structure type ($ASiO_4$) the end-member compositions of the transuranium elements, where A = Zr, Hf, Th, Pa, U, Np, Pu, and Am (Fuchs and Gebert, 1958; Keller, 1963; Williford et al., 2000). The A-site cations are surrounded by eight oxygen atoms, and the AO8 polyhedra form edge-sharing chains parallel to the a-axis. The SiO_4

monomers join the AO8 polyhedra in an alternating, edge-sharing arrangement parallel to the c-axis (Mesbah et al., 2015). The structure has a large inter-polyhedra space parallel to the c-axis that can host molecular water (Taylor and Ewing, 1978). Hydroxyl may also substitute for oxygen, such that the formula is $U(SiO_4)^{1-x}(OH)_{4x}$ (Stieff et al., 1956). Coffinite may incorporate substantial amounts of rare earth elements (REEs) and phosphorus by coupled substitution mechanisms (Finch and Murakami, 1999; Mesbah et al., 2015).

This contribution explores the possibility of in-situ formation of coffinite inclusions within zircon crystals from a theoretical perspective. A potential naturally occurring geochemical environment for such a process might be a hydrothermal setting.

2. Modeling background

Computer modeling has been used as a tool in mineralogical studies for more than 30 years. Examples include the modeling of the properties and behavior of mantle minerals by Parker and Price (1989), the modeling of carbonate minerals by Fisler et al. (2000), and modeling of water interactions with low-coordinated forsterite surfaces by King et al. (2010).

There have been a number of atomistic modeling studies of zircon published in recent years, including Meis and Gale (1998), who modeled uranium and plutonium diffusion in zircon, Akhtar and Waseem (2001), who modeled the zircon structure, intrinsic defects and doping by a wide range of ions, from monovalent to pentavalent, and Chaplot et al. (2002), who calculated the lattice dynamics of the material as a function of temperature. In addition, there have been a number of studies of zircon using quantum mechanical approaches, or a combination of these with atomistic modeling, including papers by Saadoune and de Leeuw (2009) and Saadoune et al. (2009), which focussed on helium diffusion in zircon.

The aim of this chapter is to study zircon and the structures which result from incorporation of uranium into its structure, with a view to predicting the structural changes which occur as a result. In addition, the question of what structures are formed as the uranium in coffinite decays radioactively (eventually to lead) is considered. The results are interpreted from the viewpoint of the considerable geological interest in the material.

3. Computational methodology

The computational approach used in this chapter involves describing the interactions between ions in the crystal structure by the use of interatomic potentials, and then using lattice energy minimisation to predict the structures. The incorporation of dopant ions in the structures is carried out using the mean field method. Each of these approaches will be discussed in detail below.

3.1 Interatomic potentials

Interatomic potentials are simple mathematical functions which describe the interaction between a pair of atoms or ions. In the case of ions, the most commonly used function is the Buckingham potential, supplemented by an electrostatic term, which has the form:

$$V(r) = A\exp(-r/\rho) - Cr^{-6} + \frac{q_1 q_2}{r}$$

In this equation, V(r) is the interaction energy of the two ions, A, ρ and C are potential parameters which have to be calculated and specified, and q_1 and q_2 are the charges on the ions which are separated by a distance r. Ions can be treated as rigid, or ionic polarisability can be taken into account by the shell model, which divides the ion charge between a core and shell connected by a spring (Dick and Overhauser, 1958). Charges can assume formal values (Zr^{4+}, Si^{4+}, O^{2-}) or be assigned partial values; formal charges are used in the calculations presented here.

In the case of the silicon ion in zircon, this ion is in a tetrahedral environment, and to take account of covalent interactions, a three-body term is added:

$$V_{3-\text{body}} = \tfrac{1}{2}k(\theta - \theta_0)^2$$

In this expression, θ_0 is the equilibrium bond angle (109.74° for silicon in a tetrahedral environment), θ is the actual bond angle, and k is a spring constant. The derivation of the parameters used in this work is described by Akhtar and Waseem (2001).

Interatomic potentials can be obtained in two ways—by empirical fitting to structures and crystal properties, and by calculating potential surfaces using quantum mechanical methods. In this paper the potentials employed were obtained by empirical fitting, in which the potential parameters are varied until the minimum energy structure corresponds to the experimental structure.

Table 1 Potential parameters used in this study.

Interaction	A (eV)	ρ (Å)	C (eV Å^6)
$Zr^{4+} - O^{2-}$	3079.84	0.3166	10.50
$Si^{4+} - O^{2-}$	1219.49	0.3199	11.85
$O^{2-} - O^{2-}$	22,764.0	0.149	43.0
$U^{4+} - O^{2-}$	1518.92	0.3821	65.41

(i) Short range potentials (cut-off = 12 Å). (ii) Shell model and three-body terms.
O^{2-}: $Y = -2.39\,e$, $k = 18.41\,eV\,Å^{-2}$.
O-Si-O: $k_b = 9.50\,eV\,rad^{-2}$, $\theta_0 = 109.47°$.

Table 2 Comparison of experimental and calculated lattice parameters for zircon.

Parameter	Exp. (Å)	Calc. (Å)	% diff.
a = b	6.607	6.601	−0.1
c	5.982	6.071	1.5

To model zircon, the interatomic potentials fitted by Akhtar and Waseem (2001) have been adapted, with the minor modification that highly charged cations (i.e., zirconium and silicon) are treated as rigid ions, with only the anions (oxygen in this case) being represented using the shell model. To model uranium incorporation, potentials have been taken from previous studies of uranium dioxide (Jackson et al., 1986). The interatomic potentials used are given in Tables 1, and 2 shows the agreement obtained in reproduction of the experimentally determined zircon lattice parameters (taken from the structure determination by Robinson et al., 1971). Agreement of better than 2% is obtained, giving confidence in the use of this potential in the calculations that follow.

3.2 Lattice energy minimisation

The lattice energy (LE) of a crystal is defined here as the sum of the interactions between its constituent ions. Hence the following expression applies:

$$LE = \Sigma\,(\text{Buckingham potentials} + \text{three} - \text{body terms}) = \Sigma\,(V(r))$$

The crystal structure (lattice parameters and atomic coordinates) is then varied until a minimum value of the lattice energy is obtained, which is assumed to correspond to the experimental structure. This procedure has been widely used to model mineral structures, including a previous paper by one of the authors on topaz (Jackson and Valerio, 2004).

3.3 Modeling dopant ions in the structure: The mean field approach

To model the solid solutions, use has been made of the mean field approximation. When the zircon site is substituted by a certain percentage of uranium, the interaction between that site and the surrounding lattice is obtained by calculating a weighted average of the zircon and uranium contributions to the overall interaction. The same approach can be used when the coffinite structure is considered, and the uranium is progressively replaced by lead.

The coffinite structure has been modeled in two ways: (i) by taking structures from the literature as a starting point (Table 3) and (ii) by starting with the zircon structure and progressively doping the zirconium site with uranium ions (Table 4). The results of both approaches are discussed in the next section.

Table 3 Comparison of experimental and calculated lattice parameters for coffinite.

	Coffinite		
Parameter	**Exp. (Å)**	**Calc (Å)**	**% diff.**
a=b	6.995	6.874	−1.8
c	6.262	6.371	−1.7

Table 4 Zircon lattice parameters as a function of % uranium at the Zr site.

% uranium	a=b (Å)	c (Å)
0 ($ZrSiO_4$)	6.607	5.982
10	6.624	6.099
20	6.647	6.129
30	6.672	6.159
40	6.698	6.190
50	6.724	6.220
60	6.752	6.250
70	6.781	6.281
80	6.811	6.311
90	6.842	6.341
100 ($USiO_4$)	6.874	6.371

4. Results and discussion

4.1 Agreement with experimentally determined structures: Coffinite

Fuchs and Gebert (1958) have reported an X-ray determined structure for coffinite, $USiO_4$. Energy minimisation calculations have been carried out on this structure using the potentials given in Table 1, and the experimental and calculated structures are given in Table 3. It seen that the structural agreement is comparable to that obtained for zircon, confirming that the potentials are capable of reproducing this structure.

4.2 Structures of zircon-coffinite solid solutions

The calculations reported here have been carried out by starting with the zircon structure, and progressively substituting uranium at the zirconium site. Thus it is possible to predict the lattice parameters for any composition, and also to compare the 100% substituted structures with the experimentally determined structures considered above.

Table 4 gives the calculated lattice parameters for a range of uranium concentrations in zircon, assuming occupancy of the zirconium site (which has been shown to be the preferentially occupied site by Akhtar and Waseem 2001). It is seen that the lattice parameters for 100% occupancy are in agreement with those calculated for coffinite starting with the experimental structure.

4.3 Coffinite structures resulting from radioactive decay of the uranium

The most abundant isotope of uranium, ^{238}U (99.28% abundance), decays radioactively, eventually to ^{206}Pb. Total replacement of the uranium in coffinite by lead would result in $PbSiO_4$, which does not appear to exist as a naturally occurring mineral. An interesting point here is that the lead would be in the 4+ charge state; a change to the 2+ charge state would require charge compensation, and could result in the formation of a different compound. A survey of the literature suggests that the lead content of coffinite is low, around 3.3% (Alexandre and Kyser, 2004). However, it is interesting to predict the effect on the coffinite structure of lead incorporation, using the same procedure as used for looking at uranium in zircon. Table 5 gives the lattice parameters for a range of lead concentrations. This table also contains a prediction of the structure of $PbSiO_4$, which cannot be confirmed by experimental data as it does not appear to be currently available for this material.

Table 5 Coffinite lattice parameters as a function of % lead at the U site.

% Lead	a = b (Å)	c (Å)
0 ($USiO_4$)	6.995	6.262
10	6.836	6.346
20	6.797	6.321
30	6.759	6.295
40	6.720	6.270
50	6.681	6.243
60	6.642	6.216
70	6.603	6.189
80	6.565	6.160
90	6.526	6.131
100 ($PbSiO_4$)	6.489	6.102

5. Conclusions

This chapter has described computer modeling studies of zircon, coffinite and solid solutions of the two. It goes on to look at the structure of coffinite as the uranium decays radioactively to lead. A structure for the compound $PbSiO_4$ is predicted. The procedures described can be applied to other geological materials of interest.

Acknowledgments

The authors thank R. Sneyd, S. Crossley, M. Cook and J. Lawton for useful discussions, and for the provision of a stimulating research environment.

References

Akhtar, M.J., Waseem, S., 2001. Atomistic simulation studies of zircon. Chem. Phys. 274, 109–120.

Alexandre, P., Kyser, T., 2004. Crystal Chemistry of Natural Uraninite: Effects of Substitutions and Recrystallization. American Geophysical Union. Spring Meeting, abstract #V21B-05 (2004). Available online at http://adsabs.harvard.edu/abs/2004AGUSM.V21B..05A.

Amelin, Y., Lee, D.C., Halliday, A.N., Pidgeon, R.T., 1999. Nature of the Earth's earliest crust from hafnium isotopes in single detrital zircons. Nature 399, 252–255.

Baruah, J., Kotoky, P., Sarma, J.N., 1995. Zircons in Jhanji river sediments. Bull. Pure Appl. Sci. 14F, 35–40.

Bea, F., 1996. Residence of REE, Y, Th and U in granites and crustal protoliths; implications for the chemistry of crustal melts. J. Petrol. 37, 521–552.
Chaplot, S.L., Mittal, R., Busetto, E., Lausi, A., 2002. Thermal expansion in zircon and almandine: synchrotron X-ray diffraction and lattice dynamical study. Phys. Rev. B 66, 064302.
Cherniak, D.J., Hanchar, J.M., Watson, E.B., 1997. Diffusion of tetravalent cations in zircon. Contrib. Mineral. Petrol. 127, 383–390.
Deditius, A.P., Utsunomiya, S., Ewing, R.C., 2008. The chemical stability of coffinite, $USiO_4 \cdot nH_2O$; $0 < n < 2$, associated with organic matter: a case study from Grants uranium region, New Mexico, USA. Chem. Geol. 251, 33–49.
Deditius, A.P., Utsunomiya, S., Pointeau, V., Ewing, R.C., 2010. Precipitation and alteration of coffinite ($USiO_4nH_2O$) in the presence of apatite. Eur. J. Mineral. 22, 75–88.
Deditius, A.P., Pointeau, V., Zhang, J.M., Ewing, R.C., 2012. Formation of nanoscale Th-coffinite. Am. Mineral. 97, 681–693.
Dick, B.G., Overhauser, A.W., 1958. Theory of the dielectric constants of alkali halide crystals. Phys. Rev. 112, 90.
Fayek, M., Janeczek, J., Ewing, R.C., 1997. Mineral chemistry and oxygen isotopic analyses of uraninite, pitchblende and uranium alteration minerals from the Cigar Lake deposit, Saskatchewan, Canada. Appl. Geochem. 12, 549–565.
Finch, R., Murakami, T., 1999. Systematics and paragenesis of uranium minerals. Rev. Mineral. Geochem. 38, 91–179.
Fisler, D.K., Gale, J.D., Cygan, R.T., 2000. A shell model for the simulation of rhombohedral carbonate minerals and their point defects. Am. Mineral. 85 (1), 217–224.
Fuchs, L.H., Gebert, E., 1958. X-ray studies of synthetic coffinite, thorite and uranothorites. Am. Mineral. 43, 243–248.
Grimes, C.B., John, B.E., Kelemen, P.B., Mazdab, F.K., Wooden, J.L., Cheadle, M.J., Hanghøj, K., Schwartz, J.J., 2007. Trace element chemistry of zircons from oceanic crust: a method for distinguishing detrital zircon provenance. Geology 35 (7), 643–646.
Harrison, T.M., Blichert-Toft, J., Müller, W., Albarede, F., Holden, P., Mojzsis, S.J., 2005. Heterogeneous Hadean hafnium: evidence of continental crust at 4.4 to 4.5 ga. Science 310, 1947–1950.
Heaman, L.M., Bowins, R., Crocket, J., 1990. The chemical composition of igneous suites: implications for geochemical tracer studies. Geochim. Cosmochim. Acta 54, 1597–1607.
Hoskin, P.W.O., Ireland, T.R., 2000. Rare earth element chemistry of zircon and its use as a provenance indicator. Geology 28, 627–630.
Hower, J.C., Ruppert, L.F., Eble, C.F., 1999. Lanthanide, yttrium, and zirconium anomalies in the Fire Clay coal bed, Eastern Kentucky. Int. J. Coal Geol. 39, 141–153.
Jackson, R.A., Valerio, M.E.G., 2004. A computational study of the structure, lattice and defect properties of F^- and OH^- topaz. J. Phys. Condens. Matter 16, S2771–S2779.
Jackson, R.A., Murray, A.D., Harding, J.H., Catlow, C.R.A., 1986. The calculation of basic defect parameters in UO_2. Philos. Mag. A 53, 27–50.
Janeczek, J., 1999. Mineralogy and geochemistry of natural fission reactors in Gabon. Rev. Mineral. Geochem. 38, 321–392.
Janeczek, J., Ewing, R.C., 1992. Dissolution and alteration of uraninite under reducing conditions. J. Nucl. Mater. 190, 157–173.
Keller, C., 1963. Untersuchungen über die Germanate und Silicate des Typs ABO4 der vierwertigen Elemente Thorium bis Americium. Nukleonik 5, 41–47.
King, H.E., Stimpfl, M., Deymier, P., Drake, M.J., Catlow, C.R.A., Putnis, A., de Leeuw, N.H., 2010. Computer simulations of water interactions with low-coordinated forsterite surface sites: implications for the origin of water in the inner solar system. Earth Planet. Sci. Lett. 300 (1–2), 11–18.

Meis, C., Gale, J.D., 1998. Computational study of tetravalent uranium and plutonium lattice diffusion in zircon. Mater. Sci. Eng. B 57, 52–61.

Mesbah, A., Szenknect, S., Clavier, N., Lozano-Rodriguez, J., Poinssot, C., Den Auwer, C., Ewing, R.C., Dacheux, N., 2015. Coffinite, USiO4, is abundant in nature: so why is it so difficult to synthesize? Inorg. Chem. 54, 6687–6696.

Nardi, L.V.S., Formoso, M.L.L., Müller, I.F., Fontana, E., Jarvis, K., Lamarão, C., 2013. Zircon/rock partition coefficients of REEs, Y, Th, U, Nb, and Ta in granitic rocks: uses for provenance and mineral exploration purposes. Chem. Geol. 335, 1–7.

O'Hara, M.J., Fry, N., Prichard, H.M., 2001. Minor phases as carriers of trace elements in non-modal crystal-liquid separation processes II: illustrations and bearing on behaviour of REE, U, Th and the PGE in igneous processes. J. Petrol. 42, 1887–1910.

Parker, S.C., Price, G.D., 1989. The properties and behaviour of mantle minerals: a computer-simulation approach. Philos. Trans. Royal Soc. A 328 (1599), 391–407.

Pointer, C.M., Ashworth, J.R., Ixer, R.A., 1988. The zircon-thorite mineral group in metasomatized granite, Ririwai, Nigeria 2. Zoning, alteration and exsolution in zircon. Mineral. Petrol. 39, 21–37.

Robinson, K., Gibbs, G.V., Ribbe, P.H., 1971. The structure of zircon: a comparison with garnet. Am. Mineral. 56, 782–789.

Saadoune, I., de Leeuw, N.H., 2009. A computer simulation study of the accommodation and diffusion of He in uranium- and plutonium-doped zircon ($ZrSiO_4$). Geochim. Cosmochim. Acta 73 (13), 3880–3893.

Saadoune, I., Purton, J.A., de Leeuw, N.H., 2009. He incorporation and diffusion pathways in pure and defective zircon $ZrSiO_4$: a density functional theory study. Chem. Geol. 258 (3–4), 182–196.

Salt, D.J., Hornung, G., 1967. Synthesis and X-ray study of hafnium silicate. J. Am. Ceram. Soc. 50, 549–550.

Sawka, W.N., 1988. REE and trace element variations in accessory minerals and hornblende from the strongly zoned McMurry Meadows Pluton, California. Earth Environ. Sci. Trans. R. Soc. Edinb. 79, 157–168.

Saxena, S.K., 1966. Evolution of zircons in sedimentary and metamorphic rocks. Sedimentology 6, 1–33.

Stieff, L.R., Stern, T.W., Sherwood, A.M., 1956. Coffinite, a uranous silicate with hydroxyl substitution: a new mineral. Am. Mineral. 41, 675–688.

Taylor, M., Ewing, R.C., 1978. The crystal structures of the $ThSiO_4$ polymorphs: huttonite and thorite. Acta Crystallogr., Sect. B: Struct. Sci. 34, 1074–1079.

Wilde, S.A., Valley, J.W., Peck, W.H., Graham, C.M., 2001. Evidence from detrital zircons for the existence of continental crust and oceans on the earth 4.4 Gyr ago. Nature 409, 175–178.

Williford, R.E., Begg, B.D., Weber, W.J., Hess, N.J., 2000. Computer simulation of Pu^{3+} and Pu^{4+} substitutions in zircon. J. Nucl. Mater. 278, 207–211.